Lecture Notes in Statistics

Edited by P. Bickel, P. Diggle, S. Fienberg, [illegible]
I. Olkin, N. Wermuth, S. Zeger

Springer
New York
Berlin
Heidelberg
Barcelona
Budapest
Hong Kong
London
Milan
Paris
Santa Clara
Singapore
Tokyo

John E. Kolassa

Series Approximation Methods in Statistics

Second Edition

Springer

John E. Kolassa
Department of Biostatistics
School of Medicine and Dentistry
University of Rochester
Rochester, NY 14642

Kolassa, John Edward.
Series approximation methods in statistics/John E. Kolassa.—
2nd ed.
p. cm.—(Lecture notes in statistics; 88)
Includes bibliographical references and indexes.
ISBN 0-387-98224-8 (pbk: alk. paper)
1. Mathematical statistics—Asymptotic theory. 2. Asymptotic distribution (Probability theory) 3. Edgeworth expansions
I. Title II. Series: Lecture notes in statistics (Springer-Verlag); v. 88.
QA276.K625 1997
519.5—dc21 97-20556

Printed on acid-free paper.

Camera-ready copy provided by the author.
Printed and bound by Braun-Brumfield, Ann Arbor, MI.
Printed in the United States of America.

9 8 7 6 5 4 3 2 1

ISBN 0-387-98224-8 Springer-Verlag New York Berlin Heidelberg SPIN 10573411

Preface

This book was originally compiled for a course I taught at the University of Rochester in the fall of 1991, and is intended to give advanced graduate students in statistics an introduction to Edgeworth and saddlepoint approximations, and related techniques. Many other authors have also written monographs on this subject, and so this work is narrowly focused on two areas not recently discussed in theoretical text books. These areas are, first, a rigorous consideration of Edgeworth and saddlepoint expansion limit theorems, and second, a survey of the more recent developments in the field.

In presenting expansion limit theorems I have drawn heavily on notation of McCullagh (1987) and on the theorems presented by Feller (1971) on Edgeworth expansions. For saddlepoint notation and results I relied most heavily on the many papers of Daniels, and a review paper by Reid (1988). Throughout this book I have tried to maintain consistent notation and to present theorems in such a way as to make a few theoretical results useful in as many contexts as possible. This was not only in order to present as many results with as few proofs as possible, but more importantly to show the interconnections between the various facets of asymptotic theory.

Special attention is paid to regularity conditions. The reasons they are needed and the parts they play in the proofs are both highlighted.

Computational tools are presented to help in the derivation and manipulation of these expansions. The final chapter contains input for calculating many of the results here in Mathematica, a symbolic algebra and calculus program.

This book is organized as follows. First, the notions of asymptotics and distribution approximations in general are discussed, and the present work is placed in this context. Next, characteristic functions, the basis of all of the approximations in this volume, are discussed. Their use in the derivation of the Edgeworth series, both heuristically and rigorously, is presented. Saddlepoint approximations for densities are derived from the associated Edgeworth series, and investigated. Saddlepoint distribution function approximations are presented. Multivariate and conditional counterparts of many of these results are presented, accompanied by a discussion of the extent to which these results parallel univariate results and the points where multivariate results differ. Finally, these results are applied to questions of the distribution of the maximum likelihood estimator, approximate ancillarity, Wald and likelihood ratio testing, Bayesian methods, and resampling methods.

Much of this volume is devoted to the study of lattice distributions, because in representing a departure from regularity conditions they represent an interesting variation on and completion of the mathematical development of the rest of the material, because they arise very frequently in generalized linear models for discrete data and in nonparametric applications, and because many of my research interests lie in this direction. In the interest of not unnecessarily burdening those who wish to avoid the additional complication of lattice distributions, I have tried to place the lattice distribution material as late in each chapter as possible. Those who wish may skip these sections without sacrificing much of the foundation for the rest of the book, but I recommend this material as both useful and inherently interesting.

Prerequisites and Notation

A knowledge of undergraduate real and complex analysis, on the level of Rudin (1976), Chapters 1-9, and Bak and Newman (1982), Chapters 1-12, is presupposed in the text. In particular, an understanding of continuity, differentiation, and integration in the senses of Riemann and Riemann-Stieltjes, is needed, as is an understanding of basic limit theorems for these integrals. An understanding of complex contour integration is also required. Lebesgue integration and integration of differential forms are not required. A background in matrix algebra of comparable depth is also presupposed, but will not be required as frequently.

As far as is possible, statistical parameters will be denoted by Greek characters. In general, upper-case Latin characters will in general refer to random variables, lower-case Latin characters will refer to potential observed values for the corresponding random variables, and bold face will in general denote vector or matrix quantities. Lower case Gothic characters refer to integer constants, and capital Gothic characters refer to sets. For example $\mathfrak{R}$, $\mathfrak{C}$, and $\mathfrak{Z}$ represent the sets of real numbers, complex numbers, and integers, respectively.

I have been unable to follow these conventions entirely consistently; for instance, densities and cumulative distribution functions are generally denoted by f and F, and cumulant generating functions are generally denoted by $\mathcal{K}$. Additionally, estimates of parameter values under various assumptions are denoted by the parameter value with a hat or tilde accent, and random estimators are denoted by upper case counterparts. For instance, ω will represent the signed root of the likelihood ratio, $\hat{\omega}$ will represent its fitted value, and $\hat{\Omega}$ will represent the random variable of which w is an observed value.

Unless stated otherwise, all integrals are Riemann-Stieltjes integrals, and the limiting operation implied by non-absolutely integrable improper integrals are given explicitly. The symbols $\Re$ and $\Im$ are functions returning the real and imaginary parts of complex numbers respectively. The Gamma function is denoted by $\Gamma(x)$, and φ is the di–gamma function $(d/dx)\log(\Gamma(x))$. All logarithms are with respect to the natural base.

Acknowledgments

This edition was completed with support from NIH grant CA63050. I would like to thank Martin Tanner, who suggested and encouraged this project, and whose NIH grant CA35464 supported me for part of the time while I wrote the first edition of this book. I also thank Dr. Robert Griggs, on whose NIH training grant T32NS07338 I began this project. Thanks are also due to David Oakes, Lu Cui, Guang-qin Ma, Shiti Yu, Debajyoti Sinha, and especially Moonseong Heo, who attended the lectures that this book is based on, and who suggested improvements. I am especially indebted to Peter McCullagh and O.E. Barndorff–Nielsen, who read and commented in detail on an early draft of these notes, and others who read and commented more generally on part or all of this work, including Bradley Efron, David Cox, Mitchell Gail, Scott Zeger, Nancy Reid, and Michael Wichura. The efforts of the editor for this series, Martin Gilchrist, and the production staff at Springer–Verlag, are very much appreciated, as are the comments of three anonymous referees. I am very grateful for the support of family and friends while writing this book.

Thanks are also due to those who helped with the second edition, including students who used and commented on preliminary versions of the second edition: Chris Roe, Bo Yang, Peng Huang, Raji Natarajan, Aiyi Liu, Kelly Zou, Leila Chukhina, the editor John Kimmel, and an anonymous reviewer. Special thanks are due to Yodit Seifu and Greg Elmer, whose sharp eye caught many typographical errors.

Contents

List of Illustrations

1
Asymptotics in General

Many authors have examined the use of asymptotic methods in statistics. Serfling (1980) investigates applications of probability limit theorems for distributions of random variables, including theorems concerning convergence almost surely, to many questions in applied statistics. Le Cam (1969) treats asymptotics from a decision-theoretic viewpoint. Barndorff–Nielsen and Cox (1989) present many applications of the density and distribution function approximations to be described below in a heuristic manner. Hall (1992) investigates Edgeworth series with a particular view towards applications to the bootstrap. Field and Ronchetti (1990) treat series expansion techniques in a manner that most closely parallels this work; I have included more detailed proofs and discussion of regularity conditions, and a survey of the use of Barndorff–Nielsen's formula. Their work covers many aspects of robustness and estimating equations not included here. Skovgaard (1990) explores characteristics of models making them amenable to asymptotic techniques, and derives the concept of an analytic statistical model. He also investigates convergence along series indexed by more general measures of information than sample size. Jensen (1995) presents a range of topics similar to that presented here, but with a different flavor.

The question of convergence of various approximations to distribution functions and densities will be considered with as much attention to regularity conditions and rigor as possible. Bhattacharya and Rao (1976) and Bhattacharya and Denker (1990) rigorously treat multivariate Edgeworth series; the present work begins with the univariate case as a pedagogical tool.

Valuable recent review articles include those of Reid (1996) and Skovgaard (1989).

1.1. Probabilistic Notions

Serfling (1980) describes various kinds of probabilistic limits often used in statistics. Probabilists generally model random variables as functions of an unobserved sample point ω lying in some sample space Ω, with a measure $P[\cdot]$ assigning to each set in a certain class $\mathcal{F}$ of subsets of Ω a probability between 0 and 1. Generally sequences of related random variables are modeled as a sequence of functions of ω, and a large part of asymptotic theory is concerned with describing the behavior of such sequences.

The strongest types of convergence are convergence almost surely, and convergence in mean. Random variables X_n are said to converge almost surely, or converge with probability 1, to Y if $\mathrm{P}[X_n(\omega) \to Y(\omega)] = 1$. This notion of convergence crucially involves the sample point ω, and concerns behavior of the functions X_n at ω for all n, or at least for all n sufficiently large, simultaneously. For instance, the strong law of large numbers implies that if X_n is a Binomial random variable representing the number of successes in n independent trials, with each trial resulting in a success with probability π and a failure with probability $1-\pi$, then X_n/n converges almost surely to π. Using this result, however, requires conceptually constructing a sample space Ω on which all of these random variables exist simultaneously. The natural sample space for X_n is $\Omega_n = \{0,1\}^n$ consisting of sequences of zeros and ones of length n. In this case $X_n(\omega) = \sum_i \omega_i$, and probabilities are defined by assigning each $\omega \in \Omega_n$ the probability 2^{-n}. The common sample space Ω must be constructed as a sort of infinite product of the Ω_n; this expanded sample space then bears little relation to the simple Ω_n describing the specific experiment considered.

Random variables X_n are said to converge to Y in r-th mean, for some $r \in (0,\infty)$ if $\mathrm{E}[|X_n(\omega) - Y(\omega)|^r] \to 0$. This type of convergence is less concerned with the simultaneous behavior of the random variables for each ω and more concerned about the overall relationship between $X_n(\omega)$ and $Y(\omega)$ globally on Ω for fixed n. Here still we see that the relative values of X_n and Y for a fixed ω plays a central role.

A weaker form of convergence is convergence in probability. The variables X_n converge to Y in probability, if for every $\epsilon > 0$, then $\lim_{n\to\infty} \mathrm{P}[|X_n - Y| < \epsilon] = 1$. As with convergence in r-th mean, convergence in probability concerns the overall relationship between $X_n(\omega)$ and $Y(\omega)$ globally on Ω for fixed n, but in a weaker sense.

Random variables X_n are said to converge in distribution to Y, if

$$\lim_{n\to\infty} \mathrm{P}[X_n(\omega) \le x] = \mathrm{P}[Y(\omega) \le x]$$

for all x continuity points of $\mathrm{P}[Y(\omega) \le x]$. Of these various convergence notions convergence in distribution is the weakest, in the sense that convergence almost surely and convergence in r-th mean both imply convergence in distribution. It is also weakest in the sense that it relies the least heavily on the classical measure theoretic probability notions. In the binomial example above, then, one can show that $(X_n - n\pi)/\sqrt{n\pi(1-\pi)}$ converges in distribution to a standard normal variable Y, without having to envision, even conceptually, a probability space upon which Y and X_n are simultaneously defined.

If F_n is the cumulative distribution function for X_n and F is the cumulative distribution function for Y then the criterion for convergence in distribution can be written as $F_n(x) \to F(x)$ as $n \to \infty$ for all x at which F is continuous. Often times the limiting distribution F is then used to approximate the distribution F_n in cases when n is considered sufficiently large.

This course will concentrate on variants of the idea of convergence in distribution, and will involve determining when, for a sequence of distributions F_n or equivalently random variables X_n, a limiting distribution G exists. It will then proceed to the question of how to modify G using sample size to produce a more refined approximation G_n to F_n.

At this point it may be useful to introduce order notation. Suppose f and g are two functions of a parameter, and one wishes to describe how much they differ as the parameter approaches some limiting value. One might begin by assessing whether the difference converges to zero or whether the differences are bounded in the limit; a refined analysis might describe the rate at which the difference converges to zero or diverges from zero. The notation $f(n) = g(n)+o(h(n))$ means $(f(n)-g(n))/h(n) \to 0$; the notation $f(n) = g(n) + O(h(n))$ means $(f(n) - g(n))/h(n)$ is bounded as n approaches some limiting value. Usually the implied limit is as $n \to \infty$ if n is a discrete quantity like sample size. If n is a continuous quantity the implied limit is often as $n \to 0$. For example, $1+2t+t^3 = 1+2t+o(t^2)$ as $t \to 0$, and $(n+\log(n)-1)/n^2$ may be described alternatively as $1/n+o(1/n)$ or $1/n+\log(n)/n^2+O(1/n^2)$. As another example, the numbers $a_0, a_1, \ldots, a_l$ are the value and first l derivatives of a function f at zero if and only if $f(t) = \sum_{j=0}^{l} a_l t^l/l! + o(t^l)$ as $t \to 0$.

Probabilistic versions of this order notation also exist. For two sequences of random variables U_n and V_n defined on the same probability space, we say that $V_n = O_p(U_n)$ if for any $\epsilon > 0$ there exist M_ϵ and N_ϵ such that $\mathrm{P}\left[|V_n/U_n| > M_\epsilon\right] < \epsilon$ for $n > N_\epsilon$. We say that $V_n = o_p(U_n)$ if V_n/U_n converges in probability to zero.

1.2. The Nature of Asymptotics

Wallace (1958) discusses the question of approximating a cumulative distribution function F_n depending on an index n by a function G_n also depending on n, in general terms. Often one estimates $F_n(x)$ as the truncation of a nominal infinite series, whose coefficients depend on n. That is,

$$G_{j,n}(x) = \sum_{j=0}^{j} A_j(x) a_{j,n}, \tag{1}$$

where $a_{j,n}$ decreases in n. In the case of sample means often $a_{j,n} = n^{-j/2}$. In many cases, the difference between the target cumulative distribution function and the approximation is of a size comparable to the first term neglected; that is, $|F_n(x) - G_{j,n}(x)| < c(x)a_{j+1,n}$, or in other words,

$$F_n(x) = G_{j,n}(x) + O(a_{j+1,n}). \tag{2}$$

In general the expression

$$F_n(x) = \sum_{j=0}^{\infty} A_j(x) a_{j,n} \tag{3}$$

does not hold. Often such techniques are useful even in cases where the infinite sum does not converge to the target, as noted by Cramér (1925). As an example, consider Stirling's asymptotic expansion for the Gamma function

$$e^x x^{\frac{1}{2}-x}\frac{\Gamma(x)}{\sqrt{2\pi}} = 1+\frac{x^{-1}}{12}+\frac{x^{-2}}{288}-\frac{139x^{-3}}{51840}-\frac{571x^{-4}}{2488320}+\frac{163879x^{-5}}{209018880}+\frac{5246819x^{-6}}{75246796800}-\frac{534703531x^{-7}}{902961561600}-\frac{4483131259x^{-8}}{86684309913600}+\frac{432261921612371x^{-9}}{514904800886784000}+O\left(x^{-10}\right);$$

this is a valid asymptotic expansion as $x \to \infty$, but fixing x and letting the number of terms included increase to infinity eventually degrades performance (Fig. 1).

Error in Stirling's Series as the Order of Approximation Increases

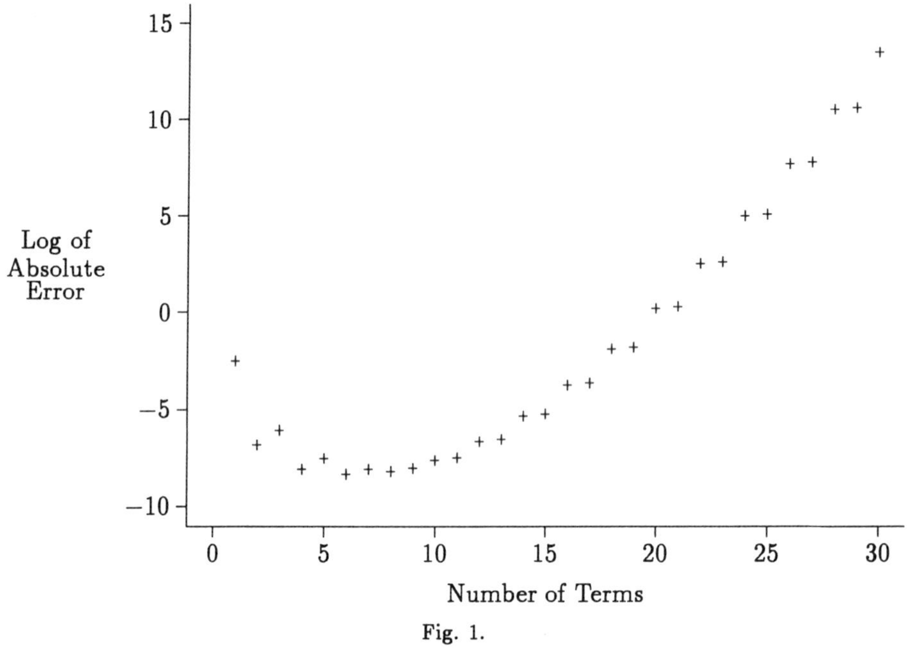

Fig. 1.

Note the distinction between asymptotic expansions and convergence of series like (3). The main concern in this volume is the behavior of the difference between F_n and $G_{j,n}$ as n, rather than j, increases. An asymptotic expansion, then, is a formal series like (3) which when truncated after any number of terms j as in (1) exhibits the behavior described by (2).

2
Characteristic Functions and the Berry-Esseen Theorem

This chapter discusses the role of the characteristic function in describing probability distributions. Theorems allowing the underlying probability function to be reconstructed from the characteristic function are presented. Results are also derived outlining the sense in which inversion of an approximate characteristic function leads to an approximate density or distribution function. These results are applied to derive Berry–Esseen theorems quantifying the error incurred in such an approximation. Finally, the relation between the characteristic function and moments and cumulants are investigated.

2.1. Moments and Cumulants and Their Generating Functions

The characteristic function for a random variable Y taking values in $\Re$ is defined to be

$$\zeta_Y(\beta) = \mathrm{E}\left[\exp(i\beta Y)\right]; \tag{4}$$

$\zeta_Y(\beta)$ is also known as the Fourier transform of the distribution. The characteristic function always exists for $\beta \in \Re$, and if the density for Y is symmetric then $\zeta(\beta) \in \Re$ for all β. It is called "characteristic" because in a sense described below it characterizes the distribution uniquely.

Additionally, the characteristic function has properties convenient when considering transformations of random variables. First, if X_1 and X_2 are independent random variables, a_1, a_2, and b are constants, and $Y = a_1X_1 + a_2X_2 + b$, then

$$\begin{aligned}\zeta_Y(\beta) &= \mathrm{E}\left[\exp(i[a_1X_1 + a_2X_2 + b]\beta)\right] \\ &= \exp(ib\beta)\mathrm{E}\left[\exp(ia_1X_1\beta)\right]\mathrm{E}\left[\exp(ia_2X_2\beta)\right] = \exp(ib\beta)\zeta_{X_1}(a_1\beta)\zeta_{X_2}(a_2\beta)\end{aligned} \tag{5}$$

Hence $\zeta_X(\beta) = \prod_{j=1}^n \zeta_{Y_j}(a_j\beta)$ if $X = \sum_{j=1}^n a_jY_j$ and the variables Y_j are independent, and $\zeta_{\bar{Y}}(\beta) = \zeta_{Y_1}(\beta/n)^n$ if $\bar{Y}$ is the mean of n independent and identically distributed random variables Y_j.

The distribution of the random variable $\sum_{j=1}^n Y_j$ of (b) where $a_j = 1$ for all j is called the convolution of the distributions of the random variables Y_j.

One can recover the moments of a distribution from its characteristic function. By differentiating (4) l times and evaluating the result at zero, one sees that

$$\zeta_Y^{(l)}(0) = \mathrm{E}\left[i^l Y^l \exp(i \times 0 \times Y)\right] = i^l \mathrm{E}\left[Y^l\right], \tag{6}$$

assuming that the orders of integration and differentiation can be interchanged.

The relationship between the characteristic function and moments of the underlying distribution unfortunately involves i. In chapters following this one, we will make use of a more direct generating function, the moment generating function, defined to be $\mathcal{M}_Y(\beta) = \mathrm{E}\left[\exp(\beta Y)\right]$, which is (4) with the i removed. The counterpart of (6) is $\mathcal{M}_Y^{(l)}(0) = \mathrm{E}\left[Y^l\right]$. The function $\mathcal{M}_Y(\beta)$, with β replaced by $-\beta$ is called the Laplace transform for the probability distribution of X. Unlike characteristic functions, these need not be defined for any real $\beta \neq 0$. The range of definition is convex, however, since for any x the function $\exp(x\beta)$ is convex. That is, if $p \in (0,1)$, then

$$\begin{aligned}\mathcal{M}_X(p\gamma + (1-p)\beta) &= \mathrm{E}\left[\exp((p\gamma + (1-p)\beta)X)\right] \\ &\leq \mathrm{E}\left[p\exp(\gamma X) + (1-p)\exp(\beta X)\right] \\ &= p\mathcal{M}_X(\gamma) + (1-p)\mathcal{M}_X(\beta).\end{aligned}$$

Hence if $\mathcal{M}_X(\beta) < \infty$ for any real $\beta \neq 0$, then $\mathcal{M}_X(\beta) < \infty$ for all β in an interval $\mathfrak{Q}$ containing 0, although 0 may lie on the boundary of $\mathfrak{Q}$. The function $\mathcal{M}_X$ may only be defined for values of β lying on one side or the other of the origin. If $\mathcal{M}_X(\beta) < \infty$ on some open interval containing 0, then all moments are finite, and $\mathcal{M}_X$ has a power series expansion about 0 of the form $\mathcal{M}_X(\beta) = \sum_{j=0}^{\infty} \mu_j \beta^j / j!$. The radius of convergence of this series is given by

$$R = \min(\sup(\{\beta : \beta > 0, \mathcal{M}_X(\beta) < \infty\}, \sup(\{-\beta : \beta < 0, \mathcal{M}_X(\beta) < \infty\})). \tag{7}$$

Proof of this claim is left as an exercise. Furthermore, if $\beta \in \mathfrak{Q} \times i\mathfrak{R}$, then $\mathrm{E}\left[|\exp(\beta X)|\right] < \infty$, and hence $\mathcal{M}_X$ exists for these β as well.

The moment generating function will be used for two purposes below. In the remainder of this section it will be used to define a sequence of numbers providing a characterization of a distribution that is more useful than the moments, and in later chapters on saddlepoint approximations, the real part of the moment generating function argument will be used to index an exponential family in which the distribution of interest is embedded, and the immaginary part will be used to parameterize the characteristic function of that distribution.

Normal approximations to densities and distribution functions make use of the expectation, or first moment, and the variance, or second central moment, of the distribution to be approximated. The Berry-Esseen theorem, to be discussed below, which assesses the quality of this approximation involved also the third central moment. Thus far, then, for the purposes of approximating distributions it seems sufficient to describe the distribution in terms of its first few moments.

For reasons that will become clear later, it is desirable to use, rather than moments, an alternate collection of quantities to describe asymptotic properties of distributions. These quantities, which can be calculated from moments, are called cumulants, and can be defined using the power series representation for the logarithm of the characteristic function. Since manipulation of logarithms of complex numbers presents some notational complications, however, we will instead derive cumulants from a real-valued function analogous to the characteristic function, called the moment generating function. Its logarithm will be called the cumulant generating function. Much of the material on this topic can be found in McCullagh (1987), §2.5-2.7, and Kendall, Stuart, and Ord (1987), §3.

Since $\mathcal{M}_X(\beta)$ is real and positive for all β, we can define the cumulant generating function $\mathcal{K}_X(\beta) = \log(\mathcal{M}_X(\beta))$. Let κ_j be its derivative of order j at zero. If derivatives of all orders exist, the formal expansion of $\mathcal{K}_X$ about $\beta = 0$ is $\sum_{j=0}^{\infty} \kappa_j \beta^j / j!$. Since $\mathcal{M}_X(0) = 1$, $\kappa_0 = 0$. The coefficients κ_j for $j > 0$ are called cumulants or semi-invariants.

These terms will be justified below. An important feature of the cumulant generating function is the simple way the cumulant generating function for an affine transformation of independent random variables is related to the underlying cumulant generating functions. Substituting β for $i\beta$ in (5), and taking logs, shows that the cumulant generating function of an affine transformation of one variable is given by

$$\mathcal{K}_{aX+b}(\beta) = \mathcal{K}_X(a\beta) + b\beta,$$

and hence if κ_j and λ_j are the cumulants of X and $aX + b$, $\lambda_j = a^j \kappa_j$ for $j > 1$ and $\lambda_1 = a\kappa_1 + b$. Thus cumulants of order two and higher are invariant under translation, and vary in a regular way with rescaling. This justifies the name semi-invariant.

If X has cumulants $\kappa = (\kappa_1, \kappa_2, \ldots)$, then the cumulant of order j of $(X - \kappa_1)/\sqrt{\kappa_2}$ is $\rho_j = \kappa_j \kappa_2^{-j/2}$ for $j > 1$, and 0 for $j = 1$. Call these quantities the invariants. These do not change under affine transformations of X. Now consider linear combinations of more than one variable. Choose any X and Y independent. Then substituting β for $i\beta$ in (5), and taking logs,

$$\mathcal{K}_{X+Y}(\beta) = \mathcal{K}_X(\beta) + \mathcal{K}_Y(\beta),$$

and hence if κ_j, λ_j, and ν_j are cumulants of X, Y, and $X + Y$ respectively, $\nu_j = \kappa_j + \lambda_j$. Thus the cumulants "cumulate", giving rise to the name.

If Y_j are independent and identically distributed, and $Z = (1/\sqrt{n}) \sum_{j=1}^{n} Y_j$, then

$$\mathcal{K}_Z = n\mathcal{K}_Y(\beta/\sqrt{n}) = \sum_{j=0}^{\infty} \kappa_j n^{(2-j)/2} \beta^j / j!.$$

Using the fact that $\log(1 + z) = \sum_{j=1}^{\infty} (-1)^{j-1} z^j / j$, convergent for $|z| < 1$, one can express cumulants in terms of moments, and using $\exp(z) = \sum_{j=0}^{\infty} z^j / j!$,

Table 1: *Conversions between Moments and Cumulants*

$\kappa_1 = \mu_1,$	$\mu_1 = \kappa_1,$
$\kappa_2 = \mu_2 - \mu_1^2,$	$\mu_2 = \kappa_2 + \kappa_1^2,$
$\kappa_3 = \mu_3 - 3\mu_1\mu_2 + 2\mu_1^3,$	$\mu_3 = \kappa_3 + 3\kappa_1\kappa_2 + \kappa_1^3,$
$\kappa_4 = \mu_4 - 4\mu_1\mu_3 - 3\mu_2^2 + 12\mu_2\mu_1^2 - 6\mu_1^4,$	$\mu_4 = \kappa_4 + 4\kappa_1\kappa_3 + 3\kappa_2^2 + 6\kappa_2\kappa_1^2 + \kappa_1^4.$

convergent for all z, one can express moments in terms of cumulants. These results are tabulated in Table 1.

Conversion in either direction, then, involves forming linear combinations of products of moments and cumulants, with the coefficients derived from the coefficients of the moments or cumulants in the appropriate generating function, and from the number of ways in which a particular term arises in the series exponentiation or logarithm. The constants arising in these conversion relations are more transparent in the multivariate case since the number of times symmetric terms arise in the transformed power series is explicitly recorded. See McCullagh (1987) for a further discussion of these transformations. Kendall, Stuart, and Ord (1987) §3.14 give transformations between these in both directions.

One can define cumulants even when $\mathcal{K}_X$ does not have a positive radius of convergence, using the same conversion rules as above, or using the definition

$$\kappa_j = i^j \left.\frac{d^j \log(\zeta(\beta))}{d\beta^j}\right|_{\beta=0},$$

and some suitable definition for complex logarithms, when this derivative exists. As we will see below, however, existence of derivatives of the characteristic function implies the existence of the moment of the corresponding order only if that order is even; recall the counterexample in which the first derivative of a cumulant generating function exists but the first moment does not. Loosely then we will say that moments to a certain order exist if and only if cumulants of the same order exist.

2.2. Examples of Characteristic Functions

The following are simple examples of calculating characteristic functions.

a. The normal distribution: Recall that the standard normal density has the form

$$f_X(x) = \exp(-x^2/2)/\sqrt{2\pi}.$$

Two options for presenting the corresponding characteristic function present themselves. One might evaluate the integral

$$\begin{aligned}\zeta_X(\beta) &= \int_{-\infty}^{\infty} \exp(ix\beta)\exp(-x^2/2)(2\pi)^{-1/2}\,dx \\ &= \int_{-\infty}^{\infty} (\cos(\beta x) + i\sin(\beta x))\exp(-x^2/2)\,dx/\sqrt{2\pi} \\ &= \int_{-\infty}^{\infty} \cos(\beta x)\exp(-x^2/2)\,dx/\sqrt{2\pi}.\end{aligned}$$

Alternatively one might calculate moments of the random variable and then use the power series expression for the characteristic function to construct $\zeta_X(\beta)$. The moments may be expressed as an integral, which may be evaluated using integration by parts to show:

$$\mu_l = \begin{cases} 0 & l \text{ odd} \\ 2^{-l/2}l!/(l/2)! & l \text{ even.}\end{cases}$$

Since the radius of convergence is infinite,

$$\zeta_X(\beta) = \sum_{l \text{ even}} (i\beta)^l/(2^{l/2}(l/2)!) = \sum_l (-\beta^2/2)^l/l! = \exp(-\beta^2/2).$$

The cumulant generating function is calculated more easily.

$$\begin{aligned}\mathcal{K}_X(\beta) &= \log\left(\int_{-\infty}^{\infty} \exp(\beta x - x^2/2)\,dx/\sqrt{2\pi}\right) \\ &= \log\left(\exp(\beta^2/2)\int_{-\infty}^{\infty} \exp(-(x-\beta)^2/2)\,dx/\sqrt{2\pi}\right) = \beta^2/2,\end{aligned}$$

which exists for $\beta \in \mathfrak{C}$. Hence cumulants of order 3 and above are zero.

b. The uniform distribution on $(-1/2, 1/2)$:

$$\begin{aligned}\zeta_X(\beta) &= \int_{-1/2}^{1/2} (\cos(\beta x) + i\sin(\beta x))\,dx \\ &= [\sin(\beta/2) - \sin(-\beta/2) + i\cos(-\beta/2) - i\cos(\beta/2)]/\beta = 2\sin(\beta/2)/\beta.\end{aligned}$$

Calculation of its cumulant generating function is left as an exercise.

c. The Cauchy distribution: The density $f_X(x) = 1/(\pi(1+x^2))$ has the corresponding characteristic function

$$\zeta_X(\beta) = \exp(-|\beta|), \tag{8}$$

differentiable everywhere *except at* 0. Its derivation is left as an exercise. No moments of order greater than or equal to one exist for this distribution, but expectations of the absolute value of the random variable raised to positive powers less than one do exist. Kendall, Stuart, and Ord (1987) give these as $\mathrm{E}\left[|X|^c\right] = 1/\sin((1+c)\pi/2)$ for $|c| < 1$. For $\beta \in \Re$ such that $\beta \neq 0$, the integral

$$\zeta_X(\beta) = \int_{-\infty}^{\infty} \exp(x\beta)(\pi(1+x^2))\,dx$$

is infinite, and so the cumulant generating function does not exist.

d. The Bernoulli distribution and the binomial distribution: If X is a Bernoulli variable taking the value 1 with probability π and the value 0 otherwise, then its characteristic function is $(1-\pi)+\pi\exp(i\beta)$, and if Y_n has the distribution of the sum of n independent such variables its characteristic function is $((1-\pi)+\pi\exp(i\beta))^n$. The cumulant generating function is $\log((1-\pi)+\pi\exp(\beta))$.

The Bernoulli example illustrates two points. First, this characteristic function has a non-zero imaginary part. In the three preceding examples the distributions were symmetric about zero, eliminating the imaginary part of the integral. This distribution is not symmetric, and so imaginary parts do not cancel out, since, for a fixed value of x, values of the summands or integrands in the expectation calculation are no longer in general the conjugates of the values at $-x$. Second, this characteristic function is periodic. This arises from the fact that possible values for the random variable are restricted to a lattice of equally spaced points. Most of the applications considered in this volume will involve either continuous or lattice distributions. Complications arising from other distributions will be discussed along with the regularity conditions for Edgeworth series. The Bernoulli distribution arises in applications involving testing a binomial proportion, including determining critical regions for the non–parametric sign test.

2.3. Characteristic Functions and Moments

We have seen that under certain regularity conditions, the set of derivatives of ζ determines the moments of Y, which in turn under certain regularity conditions determines the distribution of Y. Billingsley (1986) proves as Theorem 26.2 that two distinct real random variables cannot have the same characteristic function. A portion of this argument is summarized below, in providing an expansion for the characteristic function in terms of moments of the underlying distribution. Two lemmas link the existence of derivatives of the characteristic function at zero to the existence of moments of Y.

Lemma 2.3.1: *If Y has a moment μ_l of order l, then the derivative of ζ of order l exists at zero, with $\zeta^{(l)}(0)=\mu_l i^l$, and $\zeta(\beta)=\sum_{k=0}^{l}\mu_k(i\beta)^k/k!+o(\beta^l)$.*

Proof: We require some inequalities for use in bounding the error when $\exp(y)$ is approximated by partial sums of its Taylor expansion. The equations

$$\exp(iy)=1+i\int_0^x \exp(is)\,ds$$

$$\int_0^x (y-s)^j\exp(is)\,ds=\frac{y^{j+1}}{(j+1)}+\frac{i}{j+1}\int_0^x (y-s)^{j+1}\exp(is)\,ds$$

together imply by induction that

$$\exp(iy)=\sum_{k=0}^{j}(iy)^k/k!+(i^{j+1}/j!)\int_0^x (y-s)^j\exp(is)\,ds.$$

Note that

$$\left|\int_0^x (y-s)^j \exp(is)\, ds\right| \leq |y|^{j+1}. \tag{9}$$

By integration by parts,

$$\begin{aligned}\int_0^x (y-s)^j \exp(is)\, ds &= -iy^j + ij\int_0^x (y-s)^{j-1}\exp(is)\, ds \\ &= ij\int_0^x (y-s)^{j-1}(\exp(is)-1)\, ds\end{aligned}$$

and hence

$$\left|\int_0^x (y-s)^j \exp(is)\, ds\right| \leq 2j\,|y|^j. \tag{10}$$

Relations (9) and (10) will be used in this section and in following sections to bound errors in approximating $\exp(y)$. Using these relations,

$$\left|\zeta(\beta) - \sum_{k=0}^{l} \beta^k i^k \mu_k/k!\right| \leq |\beta|^l \operatorname{E}\left[\min(|\beta|\,|Y|^{l+1}/(l+1)!, 2\,|Y|^l/l!)\right]. \tag{11}$$

Hence if Y has moments $\mu_1, ..., \mu_l$, then $\zeta(\beta) = \sum_{k=0}^{l} \mu_k (i\beta)^k/k! + o(\beta^l)$, and ζ is l-times differentiable at zero with the derivatives claimed.

Q.E.D

Lemma 2.3.2: *If l is an even integer, and if the derivative of ζ of order l exists at zero, then Y has a moment of order l.*

Proof: This proof is essentially that given by Cramér (1946). Let $g_0(\beta, y) = \exp(\beta y)$ and $g_k(\beta, y) = ((\exp(\beta y) - \exp(-\beta y))/(2\beta))^k$ for $k > 0$. Furthermore, let Z be the random variable taking the value of Y if $|Y| < K$ and zero otherwise. One can then show that if the derivative of order k of ζ at 0 exists, then

$$\zeta^{(k)}(0) = \lim_{\beta\to 0} \operatorname{E}\left[g_k(\beta, iY)\right]. \tag{12}$$

This is left as an exercise. Hence

$$\zeta^{(l)}(0) = \operatorname{E}\left[\lim_{\beta\to 0} g_l(\beta, iY)\right] \geq \operatorname{E}\left[\lim_{\beta\to 0} g_l(\beta, iZ)\right] = \operatorname{E}\left[Z^l\right]$$

for all K. By the Monotone Convergence Theorem $\zeta^{(l)}(0) \geq \operatorname{E}\left[Y^l\right]$.

Q.E.D

If l is odd, the Monotone Convergence Theorem does not apply. The claim of the lemma does not hold for odd l.

Left as an exercise is a counterexample in which the first derivative of the characteristic function of a random variable is defined at 0 even though the first absolute moment does not exist. The principal value of the associated sum is 0.

Theorem 2.3.3: *If Y has moments of all orders, and $R = 1/\limsup(\mu_l/l!)^{1/l}$ then ζ has the expansion $\sum_{k=0}^{\infty}(i\beta)^k\mu_k/k!$ valid on $|\beta| < R$. This radius R might be* *zero, in which case the series expansion holds for no $\beta \neq 0$, or R might be ∞, in which case the expansion holds for all $\beta \in \mathfrak{R}$.*

Proof: First, a note about notation. Given a sequence a_n of real numbers, $\limsup a_n$ is the upper bound on the set of limits of subsequences of the sequence a_n, if any convergent subsequences exist, and is infinite otherwise. If the sequence a_n converges, then $\lim_{n\to\infty} a_n = \limsup a_n$.

The theorem follows from (11).

Q.E.D

2.4. Inversion of Characteristic Functions

The following theorem on inverting characteristic functions to recover the underlying cumulative distribution function is found in Billingsley (1986).

Theorem 2.4.1: *If a distribution function F corresponds to a characteristic function ζ and the points b_1 and b_2 have zero probability assigned to them then*

$$F(b_2) - F(b_1) = \lim_{T\to\infty} \frac{1}{2\pi} \int_{-T}^{T} \frac{\exp(-i\beta b_1) - \exp(-i\beta b_2)}{i\beta} \zeta(\beta)\, d\beta. \qquad (13)$$

Proof: Let

$$I_T = \frac{1}{2\pi} \int_{-T}^{T} \frac{\exp(-i\beta b_1) - \exp(-i\beta b_2)}{i\beta} \zeta(\beta)\, d\beta.$$

Then

$$\begin{aligned} I_T &= \frac{1}{2\pi} \int_{-T}^{T} \int_{-\infty}^{\infty} \frac{\exp(-i\beta b_1) - \exp(-i\beta b_2)}{i\beta} \exp(i\beta x)\, dF(x)\, d\beta \\ &= \frac{1}{2\pi} \int_{-T}^{T} \int_{-\infty}^{\infty} \frac{\exp(i\beta(x - b_1)) - \exp(i\beta(x - b_2))}{i\beta}\, dF(x)\, d\beta. \end{aligned}$$

Expanding the complex exponential, one can express I_T as

$$\int_{-T}^{T} \int_{-\infty}^{\infty} \frac{\cos(\beta(x - b_1)) + i\sin(\beta(x - b_1)) - \cos(\beta(x - b_2)) - i\sin(\beta(x - b_2))}{2\pi i\beta}\, dF(x) d\beta.$$

Since the cosine is an even function, the integral of terms involving cosine is zero, leaving

$$I_T = \frac{1}{2\pi} \int_{-T}^{T} \int_{-\infty}^{\infty} \frac{\sin(\beta(x - b_1)) - \sin(\beta(x - b_2))}{\beta}\, d\beta\, dF(x).$$

Letting $S(\theta) = \int_0^\theta \sin(x)/x\, dx$,

$$I_T = \int_{-T}^{T} \frac{1}{2\pi} \lim_{\beta\to\infty} (2S(\beta\,|x - b_1|)\operatorname{sgn}(x - b_1) - 2S(\beta\,|x - b_2|)\operatorname{sgn}(x - b_2))\, dF(x)$$

Since the integrand is bounded, one can pass to the limit:

$$\lim_{T\to\infty} I_T = \frac{1}{2\pi}\int_{-\infty}^{\infty}(2S(\infty)\,\mathrm{sgn}(x-b_1) - 2S(\infty)\,\mathrm{sgn}(x-b_2))\,dF(x)$$
$$= c\int_{-\infty}^{\infty}(\mathrm{sgn}(x-b_1) - \mathrm{sgn}(x-b_2))\,dF(x).$$

The constant $c = S(\infty)/\pi$ is 1, as will be shown below.

Q.E.D

The first factor in the integrand in (13) has a removable singularity at zero; that is, although it is undefined for $\beta = 0$, if the value $b_2 - b_1$ is substituted for $(\exp(-i\beta b_1) - \exp(-i\beta b_2))/i\beta$ when $\beta = 0$ the resulting factor is a differentiable complex function.

A version of this theorem for recovering density functions also exists.

Theorem 2.4.2: *The density of a random variable with the characteristic function ζ satisfying*

$$\int_{-\infty}^{\infty} |\zeta(\beta)|\;d\beta < \infty \tag{14}$$

exists and is given by

$$\frac{1}{2\pi}\int_{-\infty}^{\infty}\exp(-i\beta y)\zeta(\beta)\,d\beta; \tag{15}$$

furthermore, the constant c in the previous theorem has the value 1.

Proof: By (14) one can replace the limit of proper integrals as $T \to \infty$ by the corresponding improper integral over the real line in (13). Furthermore, the bounded convergence theorem assures that the right hand side of (13) is continuous in b_1 and b_2. Then

$$F(b_2) - F(b_1) = \frac{1}{2\pi}c\int_{-\infty}^{\infty}\frac{\exp(-i\beta b_1) - \exp(-i\beta b_2)}{i\beta}\zeta(\beta)\,d\beta$$
$$= \frac{1}{2\pi}c\int_{-\infty}^{\infty}\int_{b_1}^{b_2}\exp(-i\beta y)\,dy\;\zeta(\beta)\,d\beta$$
$$= \int_{b_1}^{b_2}\frac{1}{2\pi}c\int_{-\infty}^{\infty}\exp(-i\beta y)\zeta(\beta)\,d\beta\,dy.$$

Interchange of the order of integration is justified by Fubini's Theorem using the absolute convergence of the integral (14). By comparing the last line with the normal characteristic function, we find that $c = 1$. Hence the density is (15). The bounded convergence theorem implies that the resulting density is continuous.

Q.E.D

A condition like integrability of the first derivative of the density implies the condition on integrability of the characteristic function (14). The presence of probability atoms, or singleton sets of positive probability, implies that condition (14) is violated.

The converse of the preceding theorem is not true. The existence of a density need not imply absolute convergence of the integral of the characteristic function. The question of minimal conditions on a function ζ to make it a characteristic function arises in functional analysis. The interested reader should consult Rudin (1973). We will instead ask the weaker question of what characteristic functions correspond to distributions whose convolutions with itself eventually have densities. For the present purposes this later question is more important, in that most of the density expansion theorems will hold as the number of summands in the quantity whose distribution is approximated is sufficiently large; hence it is necessary to consider when the quantity being approximated actually exists.

Lemma 2.4.3: *The characteristic function ζ of a random variable satisfies*

$$\int_{-\infty}^{\infty} |\zeta(\beta)|^{\mathfrak{r}} \, d\beta < \infty$$

for some $\mathfrak{r} > 1$ if and only if the density of a j-fold convolution of the random variable with itself exists and is bounded, for some integer j.

Proof: The "only if" clause holds by the preceding theorem applied to $j = \lceil \mathfrak{r} \rceil$, where $\lceil \mathfrak{r} \rceil$ is the smallest integer at least as large as $\mathfrak{r}$. The "if" clause holds since for any $T > 0$,

$$\begin{aligned}\int_{-T}^{T} \left|\zeta^{2j}(\beta)\right| \, d\beta &= \int_{-T}^{T} \left[\int_{-\infty}^{\infty} f_j(x) \exp(i\beta x) \, dx\right] \left[\int_{-\infty}^{\infty} f_j(y) \exp(-i\beta y) \, dy\right] \, d\beta \\ &= \int_{-\infty}^{\infty} \int_{-\infty}^{\infty} \int_{-T}^{T} f_j(x) \exp(i\beta x) f_j(y) \exp(-i\beta y) \, d\beta \, dy \, dx;\end{aligned}$$

the interchange in order of integration holds by Fubini's Theorem. Then

$$\begin{aligned}\int_{-T}^{T} \left|\zeta^{2j}(\beta)\right| \, d\beta &= \int_{-\infty}^{\infty} \int_{-\infty}^{\infty} \int_{-T}^{T} f_j(x) f_j(x+z) \exp(-i\beta z) \, d\beta \, dx \, dz \\ &= \int_{-\infty}^{\infty} \int_{-\infty}^{\infty} f_j(x) f_j(x+z) \frac{\exp(-iTz) - \exp(iTz)}{-iz} \, dx \, dz.\end{aligned}$$

Imaginary parts in the previous integral cancel, leaving

$$\begin{aligned}\int_{-T}^{T} \left|\zeta^{2j}(\beta)\right| \, d\beta &= \int_{-\infty}^{\infty} \int_{-\infty}^{\infty} f_j(x) f_j(x+z) \frac{2\sin(Tz)}{z} \, dx \, dz \\ &= \int_{-\infty}^{\infty} \int_{-\infty}^{\infty} f_j(x) f_j(x+v/T) \frac{2\sin(v)}{v} \, dx \, dv \\ &= \int_{-\infty}^{\infty} \int_{-\pi}^{\pi} f_j(x) f_j(x+v/T) \frac{2\sin(v)}{v} \, dv \, dx + \\ &\quad \int_{-\infty}^{\infty} \int_{(-\pi,\pi)^c} f_j(x) f_j(x+v/T) \frac{2\sin(v)}{v} \, dv \, dx.\end{aligned}$$

The first integral above is bounded by the maximum value of f_j times the integral of $2\sin(v)/v$ over $(-\pi, \pi)$; the second integrand is bounded by $f_j(x) f_j(x+v/T) 2\pi^{-1}$, and hence the integrating with respect to v and then with respect to x shows that the second integral can be bounded by $2\pi^{-1}$, which is independent of T.

Q.E.D

The inversion integrand in (15) is the derivative of the inversion integrand in (13), evaluated with $b_1 = b_2$. This will become important in our discussion of saddlepoint series.

2.5. Bounding Distribution Function Differences Using Fourier Transforms

The following material is adapted from Feller (1971). Suppose X_n is a sequence of random variables, with cumulative distribution functions F_n and characteristic functions ζ_n, and suppose X has the cumulative distribution function F and characteristic function ζ, such that $\zeta_n \to \zeta$ pointwise. Then $F_n(x) \to F(x)$ for all x where F is continuous. This in known as weak convergence. A classical elegant probabilistic proof can be found in Billingsley (1986), §26. A classical analytic proof can be found in Feller (1971), §VIII.1. In the present context higher order approximations to $F_n(x)$ as n varies are considered. To generate these approximations additional regularity conditions will be required.

Series expansions will be generated by approximating the appropriate Fourier inversion integral. The first step is to investigate what changes in the cumulative distribution function arise from small changes in a characteristic function. Fundamental is a theorem relating differences in cumulative distribution functions to differences in characteristic functions. Suppose a cumulative distribution function H with mean 0 is to be approximated. Approximations G having properties similar to those of a cumulative distribution function of a continuous random variable will be considered. Specifically, G will be required to have a Fourier transform $\xi(\beta) = \int_{-\infty}^{\infty} \exp(i\beta x)\, dG$ whose derivative will have the correct value of 0 at 0, and the limiting values of $G(x)$ as the argument becomes large will be the same as for a cumulative distribution function. Heuristically the difference between H and G will be bounded by bounding the difference in inversion integrals (13) to obtain

$$H(b_2)-H(b_1)-G(b_2)+G(b_1) = \lim_{T\to\infty} \int_{-T}^{T} \frac{\exp(-i\beta b_1) - \exp(-i\beta b_2)}{2\pi i\beta}(\xi(\beta)-\zeta(\beta))\, d\beta. \tag{16}$$

The coarsest bound for (16), obtained by bounding the numerator of the integrand by 2, is

$$|H(b_2) - H(b_1) - G(b_2) + G(b_1)| \leq \lim_{T\to\infty} \frac{1}{2\pi} \int_{-T}^{T} \frac{2}{|\beta|} |\xi(\beta) - \zeta(\beta)|\, d\beta. \tag{17}$$

The integrand in (17) has a removable singularity at 0 and hence can be integrated over a finite range. The integral over $(-\infty, \infty)$, however, need not converge, and so some device is necessary to allow the limit in (17) to be truncated after some finite T. In fact a result which explicitly bounds the error incurred when the cumulative distribution functions on the left hand side of (16) are replaced by smoothed versions, is proved. The random variable added to smooth these cumulative distribution functions has a distribution which will vary with T. Its characteristic function disappears outside of $(-T,T)$. Since the Fourier transform of a sum of random variables is the product of the individual Fourier transforms, the Fourier transforms of the smoothed variables will then be zero outside of a bounded region, and the limit in (17) can be replaced by the improper integral. The amount of error incurred in the substitution will depend on T. The Edgeworth convergence theorems for cumulative distribution functions later in this book will then be proved by balancing the contribution of this error term with other terms, and so T is here left unspecified. The resulting theorem will be called the Smoothing Theorem.

Lemma 2.5.1: *Suppose H is a cumulative distribution function, and G is a function with the proper limits at infinity, such that G' exists and $|G'| \leq m < \infty$. Let $D = H - G$ and $D^T(\beta) = \int_{-\infty}^{\infty} D(\beta - x)v(xT)T\,dx$ be D smoothed by adding on a little bit of a continuous variate with density $v(x) = \pi^{-1}(1-\cos(x))/x^2$. Let $\eta = \sup|D|$ and $\eta_T = \sup\left|D^T\right|$. Then $\eta_T \geq \eta/2 - 12m/(\pi T)$.*

Proof: Unlike most of the results in this volume, the present lemma provides a lower bound for a maximum difference, rather than an upper bound. The proof will proceed by finding a point x_0 at which the difference should, heuristically, be largest, and use that difference as the lower bound for the maximal difference.

Construct a smoothing variable V with density

$$v(x) = (1-\cos(x))/(\pi x^2) \tag{18}$$

on $(-\infty,\infty)$, and characteristic function

$$\omega(\beta) = \begin{cases} 1-|\beta| & \text{if } |\beta| \leq 1 \\ 0 & \text{if } |\beta| > 1 \end{cases}. \tag{19}$$

Verification that this is indeed the characteristic function is left as an exercise.

The proof is completed by constructing a neighborhood to the left of x_0 where D is bounded below. Since D is right continuous, there exists x_0 such that $|D(x_0)| = \eta$. Without loss of generality assume $D(x_0) = \eta$. Refer to Fig. 2. Note that $D(x_0+s) \geq \eta - ms$ for $s > 0$, since H is non-decreasing and the derivative of G is bounded by m. The supporting line $\eta - ms$ hits the x axis at $x_2 = x_0 + \eta/m$. Let x_1 be midway between these. Call the supporting function just defined

$$D^L = \begin{cases} \eta/2 + (x-x_1)m & \text{if } x \in (x_0,x_2) \\ -\eta & \text{otherwise.} \end{cases}$$

Bound below the convolution D^T at x_1 by the convolution of D^L with V/T, where V has density v. Since on (x_0,x_2) the support function D^L is a constant plus a linear term, and the linear part convolves to 0 by symmetry, the contribution to the convolution here is at least $\eta/2$ times the minimum probability of the interval, $1 - 4/(\pi T(x_2 - x_1))$, and since the probability mass for V/T outside this region is no more than $4/(\pi T(x_2 - x_1))$ the entire convolution is at least $\eta/2 \times [1 - 4/(\pi T(\eta/2m))] - \eta \times (4/\pi T(\eta/2m)) = \eta/2 - 12m/(\pi T)$.

Q.E.D

The Smoothing Theorem is now:

Theorem 2.5.2: *Suppose H is a cumulative distribution function with mean 0 and characteristic function ζ, and G is differentiable, its Fourier transform ξ has a derivative ξ' which takes the value 0 at 0, and G has the limits 0 and 1 as its argument approaches $\pm\infty$ respectively. Then*

$$|H(x) - G(x)| \le \frac{1}{\pi}\int_{-T}^{T} |\xi(\beta) - \zeta(\beta)| \,/\, |\beta| \; d\beta + 24\max(|G'|)/(\pi T). \tag{20}$$

(Feller, 1971, §XVI.4).

Proof: Both H and G will be convolved with a rescaled v to create differentiable functions. The preceding lemma shows that a bound on the smoothed differences yields a bound on the original differences.

Denote by H^T and G^T the convolutions of H and G with the smoothing variable V/T, where V is as in Lemma 2.5.1. Since V/T is a continuous variable, these have derivatives h^T and g^T. Letting $b_1 \to -\infty$ and replacing b_2 by x in (17), and noting that the characteristic functions of H^T and G^T disappear outside of $(-T, T)$,

$$\left|D^T\right| \le \frac{1}{2\pi}\int_{-\infty}^{\infty} |\zeta(\beta) - \xi(\beta)| \, |\omega(\beta/T)| \; d\beta / \, |\beta| \, .$$

Bound $|\omega(\beta/T)|$ by 1. Then $\eta^T \le \frac{1}{\pi}\int_{-T}^{T} |(\zeta(\beta) - \xi(\beta))/\beta| \; d\beta$ and the assertion follows.

Q.E.D

A bounding inequality similar to (17) for densities can be derived from (15). If H and G have Fourier transforms $\xi(\beta)$ and $\zeta(\beta)$ and derivatives h and g, then from (15),

$$|h(x) - g(x)| \le \frac{1}{2\pi}\int_{-\infty}^{\infty} |\xi(\beta) - \zeta(\beta)| \; d\beta. \tag{21}$$

The next step is to measure variation between the true cumulant generating function and the approximation based on its Taylor expansion that will be then used to generate cumulative distribution function approximations. Henceforth this result will be known as the Series Theorem.

Theorem 2.5.3: *For any complex α and v, and any l a non-negative integer,*

$$\left|\exp(\alpha) - \sum_{k=0}^{l} \frac{v^k}{k!}\right| \le \max(\exp(|\alpha|), \exp(|v|)) \left(|\alpha - v| + \left|\frac{v^{l+1}}{(l+1)!}\right|\right).$$

Geometric Construction for Smoothing Lemma; Centered Binomial (4 , 0.5) Example

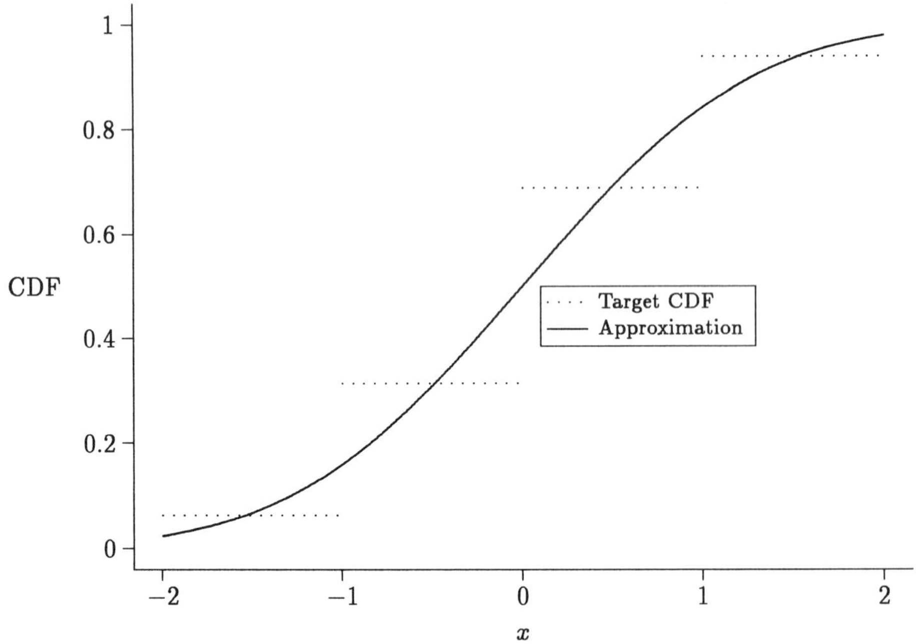

Fig. 2a

Geometric Construction for Smoothing Lemma; Centered Binomial (4 , 0.5) Example

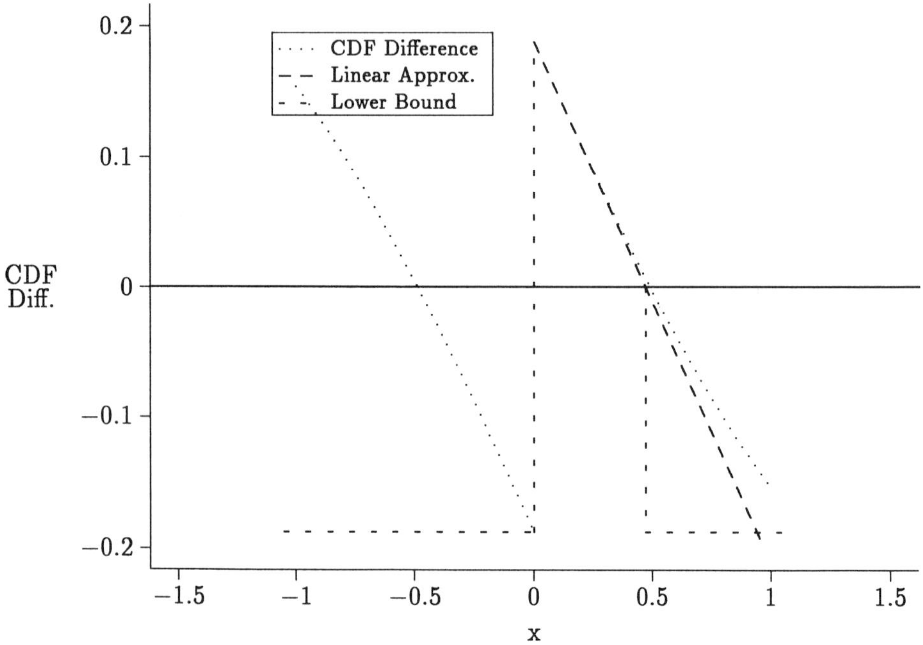

Fig. 2b

Proof: By the triangle inequality,

$$\left|\exp(\alpha) - \sum_{k=0}^{l} \frac{v^k}{k!}\right| \le |\exp(\alpha) - \exp(v)| + \left|\sum_{k=l+1}^{\infty} \frac{v^k}{k!}\right|.$$

The second term is bounded by

$$\begin{aligned}\left|\sum_{k=l+1}^{\infty} \frac{v^k}{k!}\right| &\le \left|\frac{v^{l+1}}{(l+1)!}\right| \sum_{k=0}^{\infty} \left|v^k\right| \frac{(l+1)!}{(k+l+1)!} \\ &\le \left|\frac{v^{l+1}}{(l+1)!}\right| \sum_{k=0}^{\infty} \left|\frac{v^k}{k!}\right| \\ &\le \left|\frac{v^{l+1}}{(l+1)!}\right| \max(\exp(|v|), \exp(|\alpha|)).\end{aligned}$$

Also, by Taylor's Theorem, for some $\theta \in (0,1)$,

$$\begin{aligned}|\exp(\alpha) - \exp(v)| &= |v - \alpha| \exp(\theta\alpha + (1-\theta)v) \\ &\le |v - \alpha| \max(\exp(|\alpha|), \exp(|v|))\end{aligned}$$

Q.E.D

2.6. The Berry-Esseen Theorem

Before proceeding to cumulative distribution function approximations representing refinements of the normal approximation, we assess the error incurred when using the normal approximation. Here of interest is not just the asymptotic order of the error, but actual numeric bounds on the error. Theorems addressing this question are generally known as Berry-Esseen theorems.

Theorem 2.6.1: *Let $\{Y_n\}$ be a sequence of independent and identically distributed random variables with zero mean and unit variance. Let $Z = \sum_{j=1}^{n} Y_j/\sqrt{n}$, and $F_Z(z) = \mathrm{P}[Z \le z]$. Suppose further that $\rho = \mathrm{E}\left[|Y_1|^3\right] < \infty$. Then $|F_Z - \Phi| \le C\rho/\sqrt{n}$, where C is a constant independent of the distributions of the variables Y_n whose value can be taken to be 3.*

Proof: This theorem will be proved by using the Smoothing Theorem. Use the series expansion techniques introduced earlier to bound the characteristic function as $|\zeta(\beta)| \le 1 - \beta^2/2 + |\beta|^3 \rho/6$ if $\beta^2/2 \le 1$. The integrand is bounded over finite regions. The boundary of the finite region, T, is chosen to balance the contributions from the two terms in (20). Choose $T = (4/3)\sqrt{n}/\rho$. Since $\mathrm{E}\left[|Z^2|\right]^{3/2} \le \mathrm{E}\left[|Z|^3\right]$ by Jensen's inequality, $\rho > 1$ and $T \le (4/3)\sqrt{n} < \sqrt{2n}$. Hence the theorem is trivially true for $n < 10$. Using the Smoothing Theorem,

$$\pi |F_Z - \Phi| \le \int_{-T}^{T} \left|\zeta^n(\beta/\sqrt{n}) - \exp(-\beta^2/2)\right| / |\beta| \; d\beta + 24m/T, \tag{22}$$

where $m = 1/\sqrt{2\pi} \leq 2/5$. (This corresponds to a lower bound on π of $25/8 = 3.125$.) By (9) and (10), for l an integer and z real,

$$\left| e^{iz} - \sum_{j=0}^{l} (iz)^j/j! \right| \leq \min(|z|^{l+1}/(l+1)!, 2|z|^l/l!);$$

hence $|\zeta(\beta) - 1 + \beta^2/2| \leq |\beta|^3 \rho/6 \ \forall \beta \in \mathfrak{C}$, yielding, after rescaling, $|\zeta(\beta/\sqrt{n}) - 1 - \beta^2/(2n)| \leq |\beta|^3 \rho n^{-3/2}/6$ and $|\zeta(\beta/\sqrt{n})| \leq 1 - \beta^2/(2n) + |\beta|^3 \rho/(6n^{3/2})$ if $|\beta| \leq T$. Hence

$$\begin{aligned}\left|\zeta(\beta/\sqrt{n})\right| &\leq 1 - \beta^2/(2n) - |\beta|^2 4/(3 \times 6n) \\ &= 1 - 5/(18n)\beta^2 \leq \exp((-5/18n)\beta^2) \text{ if } |\beta| \leq T\end{aligned}$$

and

$$\begin{aligned}\left|\zeta(\beta/\sqrt{n})\right|^{n-1} &= \exp((-5(n-1)/18n)\beta^2) \text{ if } n \geq 10 \\ &\leq \exp(-\beta^2/4).\end{aligned}$$

For α and υ complex, with $|\alpha| \geq |\upsilon|$,

$$|\alpha^n - \upsilon^n| = |\alpha - \upsilon| \left| \sum_{j=0}^{n-1} \alpha^j \upsilon^{n-1-j} \right| \leq |\alpha - \upsilon| \sum_{j=0}^{n-1} |\alpha^j| \left|\upsilon^{n-1-j}\right| = |\alpha - \upsilon| n |\alpha|^{n-1}.$$

Hence the integrand in (22) is bounded by $|\zeta(\beta/\sqrt{n}) - \exp(-\beta^2/2n)| \exp(-\beta^2/4) \times n|\beta|^{-1}$, and

$$\begin{aligned}\left|\zeta(\beta/\sqrt{n}) - \exp(-\beta^2/2n)\right| &\leq \left|\zeta(\beta/\sqrt{n}) - 1 + \beta^2/(2n)\right| + \left|1 - \beta^2/(2n) - \exp(-\beta^2/(2n))\right| \\ &\leq |\beta|^3 \rho/(6\sqrt{n}) + |\beta|^4/(8n)\end{aligned}$$

Hence the integrand is bounded by

$$n\left(\frac{|\beta|^2 \rho}{6\sqrt{n}} + \frac{|\beta|^3}{8n}\right) \exp(-\beta^2/4)(4/3\sqrt{n})/T \leq \left(\frac{2}{9}\beta^2 + \frac{1}{18}|\beta|^3\right) \exp(-\beta^2/4)/T.$$

This can be integrated using integration by parts over $(-T, T)$ to show

$$\pi T |F_Z(z) - \Phi(z)| \leq (8/9)\sqrt{\pi} + 8/9 + 10 < 4\pi$$

Hence the theorem holds with $C = 3$.

Q.E.D

The bound C is constructed rather crudely. It may be replaced by

a. Esseen's original bound 7.59
b. Feller's bound 3
c. Esseen's unpublished bound 2.9
d. Wallace's unpublished bound 2.05.

Esseen (1956) showed by constructing a binomial counterexample that C must be greater than $(3 + \sqrt{10})/(6\sqrt{2\pi}) \approx .40$.

The constant C is required to hold for all distributions F and all sample sizes n. If uniformity is required only over all F as $n \to \infty$ then C may be replaced by a lower value. Beek (1972) shows that $C = 0.7975$ fulfills these less stringent requirements.

2.7. Inversion of Characteristic Functions for Lattice Variables

Now consider inversion integrals for lattice distributions. Suppose that X is a random variable taking values on the lattice $\{a + \Delta\mathfrak{Z}\}$. Assume further, without loss of generality, that $a = 0$ and $\Delta = 1$. The limiting operation (13) can be taken along any sequence of T going to infinity. Taking the limit along T that are integer multiples of 2π plus π will simplify the calculations of this section. Choose $x \in \{a + \Delta\mathfrak{Z}\}$.

$$
\begin{aligned}
\mathrm{P}[X = x] &= \lim_{T\to\infty} \frac{1}{2\pi} \int_{-T}^{T} \frac{\exp(-i\beta(x - \frac{1}{2})) - \exp(-i\beta(x + \frac{1}{2}))}{i\beta} \zeta(\beta)\, d\beta. \\
&= \lim_{T\to\infty} \frac{1}{2\pi} \int_{-T}^{T} \exp(-i\beta x) \frac{\exp(\frac{1}{2}i\beta) - \exp(-\frac{1}{2}i\beta)}{i\beta} \zeta(\beta)\, d\beta. \\
&= \lim_{T\to\infty} \frac{1}{2\pi} \int_{-T}^{T} \frac{\exp(-i\beta x)\sinh(i\beta/2)}{i\beta/2} \zeta(\beta)\, d\beta.
\end{aligned}
$$

The same integral is involved in the inversion here as was involved in (15), with the introduction of $\sinh(iu/2)/(u/2)$. Generation of an asymptotic approximation to this integral will differ, however, because in the lattice case the tails of ζ do not die out. We need to express this as an integral over one period of ζ. The correct expression will be shown to be

$$
\mathrm{P}[X = x] = \frac{1}{2\pi} \int_{-\pi}^{\pi} \exp(-i\beta x)\zeta(\beta)\, d\beta. \tag{23}
$$

This can be derived by working backwards:

$$
\begin{aligned}
\frac{1}{2\pi} \int_{-\pi}^{\pi} \exp(-i\beta x)\zeta(\beta)\, d\beta &= \frac{1}{2\pi} \int_{-\pi}^{\pi} \sum_{j=-\infty}^{\infty} \mathrm{P}[X = x + j] \exp(i\beta(x + j) - i\beta x)\, d\beta \\
&= \frac{1}{2\pi} \int_{-\pi}^{\pi} \left[\mathrm{P}[X = x] + \sum_{j\neq 0} \mathrm{P}[X = x + j] \exp(i\beta j) \right] d\beta \\
&= \frac{1}{2\pi} \int_{\pi}^{\pi} \mathrm{P}[X = x]\ d\beta = \mathrm{P}[X = x]
\end{aligned}
$$

or by proceeding from (13), and noting that characteristic function has period 2π. This periodicity allows the limit of the integral from $-T$ to T to be transformed into an integral over $(-\pi, \pi]$ of an infinite sum. This infinite sum is then shown to equal the integrand in (23).

Of more interest are inversion integrals for tail probabilities. For $x_0,\ x \in \mathfrak{Z}$,

$$\begin{aligned}
\mathrm{P}\left[x_0 > X \geq x\right] &= \sum_{y=x}^{x_0-1} \frac{1}{2\pi}\int_{-\pi}^{\pi} \zeta(\beta)\exp(-i\beta y)\,d\beta \\
&= \frac{1}{2\pi}\int_{-\pi}^{\pi} \zeta(\beta)\sum_{y=x}^{x_0-1} \exp(-i\beta y)\,d\beta \\
&= \frac{1}{2\pi}\int_{-\pi}^{\pi} \zeta(\beta)\frac{\exp(-i\beta x)-\exp(-i\beta x_0)}{1-\exp(-i\beta)}\,d\beta. \qquad (24)
\end{aligned}$$

Letting $x_0 \to \infty$,

$$\begin{aligned}
\mathrm{P}\left[X \geq x\right] &= \frac{1}{2\pi}\int_{-\pi}^{\pi} \zeta(\beta)\frac{\exp(-i\beta(x-\frac{1}{2}))}{2\sinh(i\beta/2)}\,d\beta \\
&= \frac{1}{2\pi}\int_{-\pi}^{\pi} \zeta(\beta)\frac{\exp(-i\beta(x-\frac{1}{2}))}{\beta}\frac{\beta/2}{\sinh(i\beta/2)}\,d\beta. \qquad (25)
\end{aligned}$$

Unfortunately the result (25) is not strictly correct, since it assumes that $\exp(-i\beta x_0)$ converges to zero uniformly as $x_0 \to \infty$. In a later chapter on saddlepoint cumulative distribution function approximations attention will be restricted to cases in which $\zeta(\beta)$ exists for complex β in a neighborhood of 0, and the above manipulations will be justified by deforming the path of integration to make the real part of the argument of the exponential function in $\exp(-i\beta x_0)$ negative. In general (24) is true, and will be used to develop Edgeworth series expansions for lattice random variable cumulative distribution functions.

These equations are easily extended to cases with general positive Δ and general a. If X has characteristic function ζ and is confined to the lattice $\{a+\Delta\mathfrak{Z}\}$ then $X^* = (X-a)/\Delta$ has characteristic function $\zeta(\beta/\Delta)\exp(-i\beta a/\Delta)$ and satisfies the requirements for applying (23), (24), and (25). Hence

$$\begin{aligned}
\mathrm{P}\left[X = x\right] &= \frac{1}{2\pi}\int_{-\pi}^{\pi} \exp(-i\beta(x/\Delta - a/\Delta))\zeta(\beta/\Delta)\exp(-i\beta a/\Delta)\,d\beta \\
&= \frac{1}{2\pi}\int_{-\pi/\Delta}^{\pi/\Delta} \exp(-i\beta x)\zeta(\beta)\Delta\,d\beta \qquad (26)
\end{aligned}$$

$$\begin{aligned}
\mathrm{P}\left[x_0 > X \geq x\right] &= \frac{1}{2\pi}\int_{-\pi}^{\pi} \zeta(\beta/\Delta)\exp(-i\beta a/\Delta)\frac{\exp(-i\beta(x-a)/\Delta)-\exp(-i\beta(x_0-a)/\Delta)}{1-\exp(-i\beta)}\,d\beta \\
&= \frac{1}{2\pi}\int_{-\pi/\Delta}^{\pi/\Delta} \zeta(\beta)\frac{\exp(-i\beta x)-\exp(-i\beta x_0)}{1-\exp(-i\Delta\beta)}\Delta\,d\beta \qquad (27)
\end{aligned}$$

$$\begin{aligned}
\mathrm{P}\left[X \geq x\right] &= \frac{1}{2\pi}\int_{-\pi}^{\pi} \zeta(\beta/\Delta)\exp(-i\beta a/\Delta)\frac{\exp(-i\beta(x/\Delta - a/\Delta - \frac{1}{2}))}{\beta}\frac{\beta/2}{\sinh(i\beta/2)}\,d\beta \\
&= \frac{1}{2\pi}\int_{-\pi/\Delta}^{\pi/\Delta} \zeta(\beta)\frac{\exp(-i\beta(x-\Delta/2))}{\beta}\frac{\Delta\beta/2}{\sinh(i\Delta\beta/2)}\,d\beta. \qquad (28)
\end{aligned}$$

These inversion integrals will be used to generate Edgeworth and saddlepoint series for lattice distributions in later chapters.

2.8. Exercises

1. Justify (12) by showing that if $f_0(\beta)$ is a function with n derivatives on an interval containing 0, and $f_k(\beta)$ is defined recursively by $f_k(\beta) = (f_{k-1}(\beta) + f_{k-1}(-\beta))/(2\beta)$ then the derivative of order n of f_0 at 0 is $\lim_{\beta\to 0} f_n(\beta)$. Consider using L'Hospital's Rule (See Rudin (1976), p. 109 and exercises.)
2. Consider a random variable X taking values on the set of non-zero integers, assigning to the integer j, whether positive or negative, the probability c/j^2.
   a. What is the value of c?
   b. Show that the first derivative of the characteristic function exists at 0 and is 0.
   c. Show that $\mathrm{E}[|X|] = \infty$, and hence that $\mathrm{E}[X]$ does not exist.
   d. In what sense does $\mathrm{E}[X]$ exist? What is its value?
3. Show that the function defined in (19) is actually the characteristic function associated with the density in (18).
4. Show that the density determined in (15) is continuous.
5. Verify that the characteristic function for the Cauchy distribution is as given in (8), possibly making use of the inversion theorem. The right hand side of (8) is the density of the double exponential distribution rescaled and evaluated at β.
6. Prove that the quantity R defined in (7) is actually the radius of convergence of the power series representation of the moment generating function $\mathcal{M}_X(\beta)$.
7. Calculate the characteristic function for the following distributions. Calculate the radius of convergence for the series expansion about the origin, if it exists, and if it doesn't exist explain why.
   a. Exponential.
   b. Logistic, with the cumulative distribution function given by $(1+\exp(-x))^{-1}$. (Hint: You may have to use tabulated definite integral values to do this. The answer is $\log(\pi\beta) - \log(\sin(\pi\beta))$ for $\beta \in (-1, 1)$.)
8. Let U be a random variable uniform on $(-\frac{1}{2}, \frac{1}{2})$.
   a. Show that the cumulant generating function for U is $\log(\sinh(\frac{1}{2}\beta)/(\frac{1}{2}\beta))$, where
   $$\sinh(x) = \tfrac{1}{2}(\exp(x) - \exp(-x)).$$
   b. Show that $\mathcal{K}'(\beta) = -1/\beta + \frac{1}{2} + (\exp(\beta) - 1)^{-1}$.
   c. Let B_k be the coefficients in the power series expansion
   $$\beta/(\exp(\beta) - 1) = \sum_{k=0}^{\infty} B_k\beta^k/k!. \qquad (20)$$

These constants are known as the Bernoulli numbers (Haynsworth and Goldberg (1965), p. 804). Show that the last term in (b) has the expansion

$$\sum_{k=0}^{\infty} B_k\beta^{k-1}/k!, \text{ and that } \mathcal{K}(\beta) = \tfrac{1}{2}\beta + \sum_{k=1}^{\infty} B_k\beta^k/(k\,k!),$$

and hence that the cumulant of order k is B_k/k.

9. Give the moments for the standard lognormal distribution (ie., $Y = \exp(X)$, X standard normal). Comment on convergence properties of the formal power series associated with the characteristic function. Comment on the possibility that there exists another distribution with these same moments.
10. Define the inverse Gaussian distribution to have a density proportional to

$$x^{-3/2}\exp(-\tfrac{1}{2}(\xi/x + \psi x))$$

for positive x. The parameters ξ and ψ are also both positive.
 a. Show that the normalizing constant is $\exp(\sqrt{\psi\,\xi})\sqrt{\xi/(2\pi)}$.
 b. Show that the cumulant generating function for this distribution is

$$\sqrt{\psi\,\xi} - \sqrt{(\psi - 2\beta)\,\xi}.$$

What is the functional inverse of this function? Justify the name inverse Gaussian.

3
Edgeworth Series

Chapter 2 considered the normal approximation to distribution functions. We saw that the normal approximation could be derived by inverting the characteristic function, and that many of its properties could be described in terms of the first few derivatives of the characteristic function at 0. This chapter considers higher-order approximations to cumulative distribution functions and densities. Not surprisingly, these can also be expressed as approximate inversions of the characteristic function. Two heuristic motivations for the Edgeworth series are presented. Their correctness for densities and distribution functions is demonstrated using the methods of the previous chapter. Regularity conditions are investigated and discussed. Examples are given. The standards for assessment of accuracy of these methods are discussed and criticized. The Edgeworth series is inverted to yield the Cornish-Fisher expansion. Extensions of results from the standard application of means of independent and identically distributed continuous random variables to non-identically distributed and lattice cases are presented. Parallels in the lattice case with classical Sheppard's corrections are developed.

Heuristically, if $Z = \sum_{j=1}^{n} Y_j/\sqrt{n}$ is the standardized sum of n independent and identically distributed unit variance zero mean random variables $Y_1, \ldots, Y_n$, then its characteristic function is

$$\exp(-\beta^2/2 + \sum_{j=3}^{\infty} \kappa_j^n (i\beta)^j/j!), \tag{30}$$

where κ_j^n is the cumulant of Z of order j, with $\kappa_j^n = n^{(2-j)/2}\kappa_j$. The Central Limit Theorem in its most basic form, and the Edgeworth series defined below, rely on the heuristic notion of approximating the distribution of Z by truncating the infinite sum in the exponent above, approximating the quantity $\exp(\sum_{j=3}^{j} \kappa_j^n (i\beta)^j/j!)$ as a polynomial $p(\beta)$, and calculating that approximate cumulative distribution function whose Fourier transform is approximately $\exp(-\beta^2/2)p(\beta)$. The result is approximately a cumulative distribution function; it need not be monotonic, nor need it take values in $[0, 1]$. Whether this matters will be discussed below.

After presenting heuristic arguments indicating what the expansion should be, careful attention is paid to quantify both the error incurred in truncating (30) to $\exp(-\beta^2/2)p(\beta)$, and then to the way in which this translates to error in the Fourier inversion. This yields a rigorous proof that the order of error in the resulting Edgeworth series is what the heuristic arguments indicate.

Applications considered in this chapter derive primarily from the Edgeworth series to the cumulative distribution function; these are presented in §§3.13f and in the exercises.

3.1. Heuristic Development

Formally, the Edgeworth series approximation to a density f_X is constructed as a modification of a baseline density f_Y. The following formal construction is due to Davis (1976). Let X and Y be two random variables, such that Y has density f_Y, and X has density f_X. Suppose further that X and Y can be constructed on a common probability space, such that $Z = X - Y$ is independent of Y. Conditionally on $Z = z$, X has density $f_Y(x - z)$, and expanding f_Y as a power series about x, $f_Y(x - z) = \sum_{j=0}^{\infty} f_Y^{(j)}(x)(-z)^j/j!$. Hence the unconditional density of X is $f_X(x) = \sum_{j=0}^{\infty} f_Y^{(j)}(x)(-1)^j\mu_j^*/j!$, where μ_j^* are the moments of Z, again assuming that such a construction is possible. Writing $h_j(x) = (-1)^j f_Y^{(j)}(x)/f_Y(x)$ we observe $f(x) = f_Y(x)\sum_{j=0}^{\infty} h_j(x)\mu_j^*/j!$. The "moments" μ_j^* are the moments of whatever distribution is necessary to add to Y to get X. The cumulant of order j associated with these moments is the cumulant of order j associated with X minus the corresponding cumulant for Y, since $Y + Z = X$ and Z is independent of Y.

The functions h_i are ratios of the derivatives of the baseline density to the density itself. In the case a normal baseline these are polynomials.

The problem with this construction is that in general the random variable Z described above cannot be constructed. For instance, if f_Y is the normal density with the same mean and variance as the density f_X to be approximated, the variance of the constructed variable Z is necessarily zero; hence all cumulants of Z must be zero.

3.2. A Less Untrue but Still Heuristic Development

McCullagh (1987), Chapter 5, presents an alternative heuristic construction of the Edgeworth series. Suppose that f_Y and f_X have moment generating functions $\mathcal{M}_Y(\beta)$ and $\mathcal{M}_X(\beta)$. Express their ratio as a power series in β:

$$\mathcal{M}_X(\beta) = \mathcal{M}_Y(\beta)\sum_{j=0}^{\infty}\beta^j\mu_j^*/j!;$$

this equation defines the coefficients μ_j^*. The inverse Laplace transform will now be applied termwise to this sum to obtain an approximation to f_X.

The equation defining the coefficients μ_j^* can be expressed as

$$\log\left(\sum_{j=0}^{\infty}\beta^j\mu_j^*/j!\right) = \mathcal{K}(\beta) - \mathcal{K}_Y(\beta),$$

where $\mathcal{K}$ and $\mathcal{K}_Y$ are cumulant generating functions associated with f_X and f_Y respectively. Heuristically, μ_j^* are the moments that would be associated with the distribution whose cumulants κ_j^* are the difference between the cumulants of f_X and f_Y; rigorously they are the coefficients of the formal power series obtained when the power series defined by $\exp(z)$ is composed with the power series whose coefficients are the cumulant differences. Since these cumulant differences need not be the cumulants of any distribution, following McCullagh (1987), Chapter 5 we call these pseudo-cumulants and we call the coefficients μ_j^* pseudo-moments. In general these pseudo-moments will not be the moments of any real random variable; μ_2^* will generally be zero, and in some applications it may be negative.

By the definition of $\mathcal{M}_Y(\beta)$ the first term in the inversion is $f_Y(x)$, and

$$\int_{-\infty}^{\infty}\exp(\beta x)\frac{df_Y}{dx}\,dx = \exp(x\beta)f_Y(x)|_{-\infty}^{\infty} - \beta\int_{-\infty}^{\infty}\exp(x\beta)f_Y(x)\,dx = -\beta\mathcal{M}_Y(\beta),$$

where $f_Y^{(j)}(x)$ represents the derivative of order j of $f_Y(x)$ with respect to x. This calculation is predicated on the assumption that $\lim_{x\to\pm\infty}\exp(x\beta)f_Y(x) = 0$ for all β in the domain of $\mathcal{M}_Y$. Weak assumptions, such as uniform continuity of f_Y, will insure this. Similarly

$$\int_{-\infty}^{\infty}\exp(\beta x)f_Y^{(j)}(x)\,dx = \beta^j(-1)^j\mathcal{M}_Y(\beta). \tag{31}$$

Hence the inverse Laplace transform of $\beta^j\mathcal{M}_Y(\beta)$ is $f_Y^{(j)}(x)(-1)^j$. Hence at least formally, $f_X(x)$ can be expressed as $\sum_{j=0}^{\infty} f_Y^{(j)}(x)(-1)^j\mu_j^*/j!$, and

$$f_Y(x)\sum_{j=0}^{\infty}h_j(x)\mu_j^*/j!, \tag{32}$$

where $h_j(x) = f_Y^{(j)}(x)(-1)^j/f_Y$, will be used as a formal expansion for $f_X(x)$

The density approximation (32) gives rise to a cumulative distribution function approximation that is calculated very easily. Suppose F_Y is the cumulative distribution function associated with f_Y. As will be argued later, the coefficient of the leading term in (32), μ_0^*, is 0. Integrating (32) termwise we obtain the formal series approximation to $F(x)$,

$$F_Y(x) + \sum_{j=1}^{\infty} f_Y^{(j-1)}(x)(-1)^j\mu_j^*/j! = F_Y(x) - f_Y(x)\sum_{j=1}^{\infty}h_{j-1}(x)\mu_j^*/j!. \tag{33}$$

Here the first term arises from integrating the the density $f_Y(x)$ by itself. Other terms arise from lowering the order of the derivatives involved by one. Since the order of derivatives changes but the power of (-1) does not, when expressing the integrated approximation in terms of the functions h_j to obtain (33) the sign changes, explaining the subtraction rather than addition of the terms of order one and higher.

3.3. Choice of Baseline

Typically the target density f will in fact depend on some parameter like sample size; here we denote this by n and write f_n for the density. The resulting moments and cumulants for the target density will also depend on n; when necessary denote the dependence by a superscript n. Although (32) is expressed in terms of pseudo-moments, it is often easier to think of it with the pseudo-moments expressed in terms of the pseudo-cumulants, since the pseudo-cumulants are more directly calculated from the target and baseline distributions, and since the dependence of the pseudo-cumulants on n is simpler. As above, in the case of standardized sums of independent and identically distributed random variables, $\kappa_l^n = \kappa_l^1 n^{1-l/2}$. These series expansion methods are also often used in the case of sums of non-independent and identically distributed random variables if $\kappa_l^n = O(n^{1-l/2})$. Note especially that κ_1^n increases rather than decreases as $n \to \infty$. Often the terms in the resulting series can be divided into those that do not decrease as n gets large, and those which get small as n gets large. In the cases when $\kappa_l^n = O(n^{1-l/2})$, an approximating distribution which matches the first and second cumulants of the target distribution makes zero all terms involved in the pseudo-moments, and hence involved in (32), to vanish as $n \to \infty$. When viewing these methods as extensions of the Central Limit Theorem, it is natural to use the approximating normal density as the baseline. This works well as long as $\kappa_j^n = o(1)$ for $j \geq 3$. When f_0 is ϕ, the series (32), truncated after a chosen number of terms, is called the Gram-Charlier series. Expansions based on other than the normal distribution are rare. On occasions expansions based on the χ^2 distribution are used.

3.4. Calculation of Hermite Polynomials and Pseudo-moments

When the approximating normal baseline function is used, the functions h_j of (32) are

$$h_j(x) = (-1)^j \left[\frac{d^j}{dx^j} \exp(-x^2/2) \right] / \exp(-x^2/2).$$

These are called Hermite polynomials. The first few are tabulated in Table 2.

Note the similarity between these equations and the equations for moments in terms of cumulants from Table 1. If in Table 1, x is substituted for κ_1 and -1 is substituted for κ_2, and zero is substituted for all other cumulants, the quantities μ_j are the Hermite polynomials. Recall that the moments are coefficients in the formal power series $\exp(\sum_{j=1}^{\infty} \kappa_j \beta^j / j!)$; McCullagh (1987) exhibits the functions $h_j(x)$ as coefficients in the formal power series $\exp(\beta x - \beta^2/2)$. This relationship extends to the case of a normal baseline with non-zero mean or non-unit variance, and to the multivariate case as well.

Table 2: *The First Six Standard Hermite Polynomials*

$$\begin{aligned} h_1(x) &= x \\ h_2(x) &= x^2 - 1 \\ h_3(x) &= x^3 - 3x \\ h_4(x) &= x^4 - 6x^2 + 3 \\ h_5(x) &= x^5 - 10x^3 + 15x \\ h_6(x) &= x^6 - 15x^4 + 45x^2 - 15 \end{aligned}$$

The pseudo-moments are easily calculated from the underlying baseline cumulants λ_j and target cumulants κ_j. One begins by taking the difference $\kappa_j - \lambda_j$ between respective cumulants. The cumulant to moment conversion formulas in reverse yield expressions for the pseudo-moments μ_j^* as a sum of products of the third and higher order cumulants of the target distribution. That is, the pseudo-moments μ_j^* satisfy the series equation

$$\exp(\sum_{j=1}^{\infty}(\kappa_j - \lambda_j)\beta^j/j!) = \sum_{j=1}^{\infty}\mu_j^*\beta^j/j!. \tag{34}$$

The leading term in both cumulant generating function series expansions are zero; the pseudo-cumulant of order zero is then always zero and the leading term in the exponentiated series, μ_0^*, is always 1. When a normal baseline distribution is used the first two pseudo-cumulants are generally zero and the rest are the unmodified corresponding cumulants of the target distribution. If κ_j is of size $O(n^{(2-j)/2})$, the various terms in the pseudo-moments can be sorted according to their order in n. These are given in Table 3.

3.5. The Expansion Theorem

This section contains a statement of the Edgeworth series theorem for distribution functions and densities of standardized sums. For the present, we may restrict attention to the case in which $\mu = 0$ and $\sigma = 1$. Proofs will follow in the next three sections.

Theorem 3.5.1: *Suppose Y_j are independent and identically distributed random variables, with mean* 0 *and variance* 1, *and that* j *is an integer greater than or equal to two, such that Y_j has a cumulant of order* j. *When* j > 2 *suppose further that*

$$\begin{cases} |\zeta(\beta)| < 1 \ \forall \beta \neq 0 & \textit{if } j = 3, \textit{ or} \\ \limsup_{|\beta|\to\infty} |\zeta(\beta)| < 1 & \textit{if } j > 3. \end{cases} \tag{35}$$

Let $Z = \sum_{j=1}^{n} Y_j/\sqrt{n}$, and let F_Z be the cumulative distribution function for Z. Let $E_j(x, \boldsymbol{\kappa}^n)$ be the result of calculating (33) using only the first j *cumulants, and dropping terms of order $o(n^{1-j/2})$. When $E_j(x, \boldsymbol{\kappa}^n)$ is used to approximate F_Z, the*

Table 3: *Pseudomoments as a Function of Cumulants*

	Order in n: $O(n^{-1/2})$	$O(n^{-1})$	$O(n^{-3/2})$	$O(n^{-2})$	$O(n^{-5/2})$	$O(n^{-3})$	$O(n^{-7/2})$
$\mu_3^* =$	κ_3						
$\mu_4^* =$		κ_4					
$\mu_5^* =$			κ_5				
$\mu_6^* =$		$10\kappa_3^2$		κ_6			
$\mu_7^* =$			$35\kappa_4\kappa_3$		κ_7		
$\mu_8^* =$				$35\kappa_4^2+$ $56\kappa_5\kappa_3$		κ_8	
$\mu_9^* =$			$280\kappa_3^3$		$84\kappa_6\kappa_3+$ $126\kappa_5\kappa_4$		κ_9

absolute error is uniformly of order $o(n^{1-\mathrm{j}/2})$. *If, furthermore,* $\int_{-\infty}^{\infty} |\zeta(\beta)|^{\mathfrak{r}}\, d\beta < \infty$ *for some* $\mathfrak{r} \geq 1$, *then the density* f_n *exists for* Y_n *for* $n \geq \mathfrak{r}$. *Let* $\mathrm{e}_{\mathrm{j}}(x, \boldsymbol{\kappa}^n)$ *be the result of calculating (32) using only the first* j *cumulants, and dropping terms of order* $o(n^{1-\mathrm{j}/2})$. *When* $\mathrm{e}_{\mathrm{j}}(x, \boldsymbol{\kappa}^n)$ *is used to approximate* $f_n(x)$, *the absolute error is uniformly of order* $o(n^{1-\mathrm{j}/2})$.

This notation $E_{\mathrm{j}}(x, \boldsymbol{\kappa}^n)$ and $\mathrm{e}_{\mathrm{j}}(x, \boldsymbol{\kappa}^n)$ is chosen to reflect the fact that the Edgeworth approximation depends on n only through the cumulants of the target distribution.

After substituting in the expressions for pseudo-moments in terms of pseudo-cumulants, and collecting terms according to their power in n, we find

$$E_{\mathrm{j}}(x, \boldsymbol{\kappa}^n) = \Phi(x) - \phi(x)[\kappa_3^n h_2(x)/6 + (\kappa_4^n h_3(x)/24 + 10\kappa_3^{n2} h_5(x)/720) + (\kappa_5^n h_4(x)/120 + 35\kappa_3^n\kappa_4^n h_6(x)/5040) + \cdots]$$

and

$$\mathrm{e}_{\mathrm{j}}(x, \boldsymbol{\kappa}^n) = \phi(x)[1 + \kappa_3^n h_3(x)/6 + (\kappa_4^n h_4(x)/24 + 10\kappa^{n}{}_3^2 h_6(x)/720) + (\kappa_5^n h_5(x)/120 + 35\kappa_3^n\kappa_4^n h_7(x)/5040) + \cdots].$$

Note that (33) was formed from the relationship (31), which holds only when $\mathcal{M}_0(\beta)$ exists for some $\beta \neq 0$. For the sake of completeness, the counterpart of (31) in terms of characteristic functions, which holds generally, is

$$\int_{-\infty}^{\infty} \exp(i\beta x)\frac{d^k f_0}{dx^k}dx = (-i\beta)^k\zeta_0(\beta), \tag{36}$$

where ζ_0 is the characteristic function of f_0.

The above definition of $E_j(x, \boldsymbol{\kappa}^n)$ is appropriate only when $\kappa_1 = 0$ and $\kappa_2 = 1$. For general cumulants $\boldsymbol{\kappa}$, calculate the invariants $\boldsymbol{\rho}$ as in §2.1, and define

$$E_j(x, \boldsymbol{\kappa}^n) = E_j((x - \kappa_1^n)/\sqrt{\kappa_2^n}, \boldsymbol{\rho}^n). \tag{37}$$

When Z is the sum of n independent and identically distributed summands, divided by $\sqrt{n}$, then $\kappa_1^n = \sqrt{n}\kappa_1$ and $\kappa_2^n = \kappa_2$. As we shall see later, however, (37) is applicable in cases when the cumulants of Z depend on n in a more complicated way.

3.6. Approximating the Difference Integrand Near the Origin

Density approximations created by multiplying the normal density times sums of multiples of Hermite polynomials have Fourier transforms of the form the normal characteristic function times a polynomial. In §3.3 a choice for the definition of the Hermite polynomials and their multipliers was motivated, by truncating the power series that resulted from dividing the characteristic function to be inverted by the approximating normal characteristic function after a fixed number of terms, and inverting the Fourier transform termwise. In §3.4 these Hermite polynomials and coefficients were calculated. The error incurred by this series truncation is measured by (21) and (20). Use of these inequalities requires bounding the difference between the true and approximating characteristic functions. In this section the portion of the integrals in (21) and (20) near the origin is bounded using the Series Theorem. In what follows $\alpha(\beta)$ will represent the log of the ratio of the characteristic function to be inverted to the approximating normal characteristic function, and $v(\beta)$ will be the truncated power series for $\alpha(\beta)$.

Near the origin is defined to be for $|\beta| < \delta\sqrt{n}$ for δ determined below. Set $\alpha(\beta) = \log(\zeta^n(\beta/\sqrt{n})) + \beta^2/2$. We need v such that $\alpha - v$ is small. If $v^*(\beta) = \sum_{j=3}^{j} \kappa_j(i\beta)^j/j!$ and $v(\beta) = nv^*(\beta/\sqrt{n}) = \sum_{j=3}^{j} \kappa_j(i\beta)^j n^{1-j/2}/j!$ then $\log(\zeta(\beta)) + \beta^2/2 - v^*(\beta)$ has j continuous derivatives at 0 and all are 0. Hence there exists $\delta > 0$ such that if $|\beta| < \delta$ then

$$\log(\zeta(\beta)) + \beta^2/2 - \sum_{j=3}^{j} \kappa_j(i\beta)^j/j! < \epsilon\,|\beta|^j, \tag{38}$$

and if $|\beta| < \sqrt{n}\delta$ then

$$n\log(\zeta(\beta/\sqrt{n})) + \beta^2/2 - v(\beta) < \epsilon\,|\beta|^j\, n^{1-j/2}. \tag{39}$$

Furthermore we can require that $|\beta| < \delta$ imply that $|\log(\zeta(\beta)) + \beta^2/2| < \beta^2/4$, and hence $|\beta| < \delta\sqrt{n}$ implies that $|\alpha(\beta)| = |n\log(\zeta(\beta/\sqrt{n})) + \beta^2/2| < \beta^2/4$. Also, since $v^*(\beta)$ has a third derivative at 0 there exists C such that $|\beta| < \delta\sqrt{n}$ implies

$$v(\beta) = nv^*(\beta/\sqrt{n}) < C\,|\beta|^3/\sqrt{n}. \tag{40}$$

Inequalities (39) and (40) will be used with the Series Theorem to bound the exponentiated version of the left hand side of (39), and hence to bound the portion of the inversion error integral near zero.

In order to maintain parallelism between derivations of Edgeworth and saddlepoint approximations, observe how (39) and (40) might be strengthened in the presence of additional assumptions on the characteristic function. If it is known that the characteristic function is $\mathrm{j}+1$ times differentiable for $|\beta| < R$, with $\mathrm{j} > 2$, and the derivative of order j is bounded on this region, then δ may be taken to be R and (39) and (40) may be replaced by

$$\log(\zeta(\beta)) + \beta^2/2 - \sum_{j=3}^{\mathrm{j}} \kappa_j (i\beta)^j/j! < \sup\left(\left|\log(\zeta)^{(\mathrm{j})}(\beta)\right|\right) |\beta|^{\mathrm{j}+1}/(\mathrm{j}+1)!$$

and

$$v(\beta) = nv^*(\beta/\sqrt{n}) < \sup\left(\left|\log(\zeta)^{(3)}(\beta)\right|\right) |\beta|^3/(6\sqrt{n}).$$

Furthermore, $|\alpha(\beta)| \leq |\beta| \sup\left(\left|\log(\zeta)^{(3)}(\beta)\right|\right)/6$, and so (40) holds for

$$|\beta| \leq n^{1/2} \min\left(R, 6/\sup\left(\left|\log(\zeta)^{(3)}(\beta)\right|\right)\right).$$

3.7. Rigorous Construction for Cumulative Distribution Functions

The following exposition is based on Chapter XVI of Feller (1971). It makes rigorous the arguments of §3.2, with the characteristic function substituted in place of the moment generating function. An approximation $E_{\mathrm{j}}^*(x, \boldsymbol{\kappa}^{*n})$ will be defined such that $E_{\mathrm{j}}^*(x, \boldsymbol{\kappa}^{*n}) = E_{\mathrm{j}}(x, \boldsymbol{\kappa}^{*n}) + o(n^{1-\mathrm{j}/2})$ and such that its accuracy is easy to verify. Since the cumulative distribution functions can be recovered from their characteristic functions using (13), using the smoothing lemma the difference between F_Z and $E_{\mathrm{j}}^*(x, \boldsymbol{\kappa}^{*n})$ can be bounded by the principal value of the integral

$$\frac{1}{2\pi}\int_{-\infty}^{\infty} \frac{2}{|\beta|} |\zeta_n(\beta) - \xi_n(\beta)|\, d\beta, \tag{41}$$

defined to be

$$\lim_{T\to\infty} \frac{1}{2\pi}\int_{-T}^{T} \frac{2}{|\beta|} |\zeta_n(\beta) - \xi_n(\beta)|\, d\beta,$$

where ξ_n is the characteristic function associated with the approximate distribution function $E_{\mathrm{j}}^*(x, \boldsymbol{\kappa}^{*n})$. Because the integral (41) is derived from (13), which need not converge absolutely, this argument will proceed indirectly.

Three ideas go into this proof, the Smoothing Theorem, the Series Theorem, and a fact about what kind of functions have Fourier transforms looking like polynomials times the normal characteristic function. The Smoothing Theorem is used to convert the possibly non-absolutely convergent integral (41) to an integral over a finite range, to which can be applied a wider range of integral approximation techniques. The

Series Theorem is a really key idea here, since it allows truncation from the infinite series to a finite sum.

The approximation $E_{\mathrm{j}}^*(x, \boldsymbol{\kappa}^{*n})$ is formed as follows. First, form a polynomial approximation to the target characteristic function by deleting terms of size $o(|\beta|^{\mathrm{j}})$ from its power series expansion. Second, separately exponentiate the quadratic term exactly, and using a Taylor series approximation $\exp(x) = 1 + x + \cdots + x^{(\mathrm{j}-2)}/(\mathrm{j}-2)! + o(x^{\mathrm{j}-2})$, involving $\mathrm{j}-2$ terms plus the constant term, to exponentiate the sum of the higher-order terms $\sum_{j=3}^{\mathrm{j}} \kappa_j (i\beta)^j n^{1-j/2}/j!$. The approximation $E_{\mathrm{j}}(x, \boldsymbol{\kappa}^{*n})$ defined earlier is then formed by discarding terms of size $o(n^{1-\mathrm{j}/2})$.

Apply the Smoothing Theorem with the target cumulative distribution function F_Z in place of H, and the cumulative distribution function approximation represented by $E_{\mathrm{j}}^*(x, \boldsymbol{\kappa}^{*n})$ in place of G. The integral (41) heuristically representing error has three ranges. The areas far out in the tails, the area very close to the origin, and the area in between will all be handled separately.

a. Far in the tails ($|\beta| > T$) the contribution of the error in approximating the inversion integral is accounted for by the remainder term in the Smoothing Theorem. A bound on the error in inversion begins by choosing T large enough to make this remainder term small enough. Take T in the Smoothing Theorem to be $2m/(\pi \epsilon n^{1-\mathrm{j}/2})$.
b. Using results (39) and (40) of §3.6, and applying the Series Theorem, we find that the integrand is bounded by

$$\exp(-\beta^2/4)\left[\frac{\epsilon\,|\beta|^{\mathrm{j}}}{n^{\mathrm{j}/2-1}} + \frac{C^{\mathrm{j}-1}\,|\beta|^{3(\mathrm{j}-1)}}{(\mathrm{j}-1)!\,n^{\mathrm{j}/2-1/2}}\right]\frac{1}{|\beta|} \tag{42}$$

for β such that $|\beta| < \delta\sqrt{n}$. Hence the contribution to the integral of the region near the origin is bounded by the integral of (42) over the entire real line, which, since ϵ was arbitrary, shows that this contribution is of order $o(n^{1-\mathrm{j}/2})$.
c. For intermediate values of β ($\beta \in \pm[\delta\sqrt{n}, T]$), regularity conditions (35) on the cumulant generating function are used to bound the integrand. There exists $q < 1$ such that if $|\beta| > \delta\sqrt{n}$ then $|\zeta(\beta/\sqrt{n})| < q$. In the case where $\mathrm{j} = 3$ the weaker condition suffices, since T is proportional to $\sqrt{n}$ and hence the set of $\beta/\sqrt{n}$ such that $|\beta| \in [\delta\sqrt{n}, T]$ is fixed as n varies. The contribution of this part of the range of integration is bounded by the bound on the integrand, times the length of the range of integration, which is bounded. When $\mathrm{j} > 3$ then T increases faster than $\sqrt{n}$ and the stronger condition in (35) is necessary. Since $\sum_{j=1}^{\mathrm{j}-2} [v^*(\beta)]^j / j!$ is a polynomial, it can be bounded over the entire real line by a constant times a power of n, and hence

$$\left|\exp(-\beta^2/2)\sum_{j=1}^{\mathrm{j}-2} [v(\beta)]^j / j!\right| \le C n^l \exp(-\delta^2 n/4) \tag{43}$$

for some C and l independent of n and β. The contribution to the integral (41) in this region is bounded by $2(q^n + \exp(-\delta^2/2)^n C n^l) T\delta/\sqrt{n}$ and approaches 0 geometrically.

3.8. Rigorous Construction for Density Functions

The following proof is almost identical to the last. Since the cumulative distribution functions can be recovered from their characteristic functions using (15), the difference between $f_Z(z)$ and $e_j(z, \boldsymbol{\kappa}^{*n})$ can be bounded by the integral

$$\frac{1}{2\pi}\int_{-\infty}^{\infty} 2\left|\zeta_n(\beta) - \xi_n(\beta)\right| \, d\beta, \tag{44}$$

where ξ_n is the characteristic function associated with the approximate distribution function $e_j(z, \boldsymbol{\kappa}^{*n})$. Because the integral (44) converges absolutely, this argument will not need the Smoothing Theorem. The Series Theorem is still a key idea here.

The integral (44) representing error is the same as (41) of the previous proof, except that the factor of $|\beta|^{-1}$ is gone. The range of integration will now be split in two parts. The area very close to the origin, and the area farther out will be handled separately.

a. Very near the origin ($|\beta| < \delta\sqrt{n}$ for δ determined as before), exactly the same arguments, with the same value of δ, imply that the integrand is bounded by

$$\exp(-\beta^2/4)\left[n^{1-j/2}\epsilon\left|\beta\right|^j + n^{1/2-j/2}C^{j-1}\left|\beta\right|^{3(j-1)}/(j-1)!\right],$$

for β such that $|\beta| < \delta\sqrt{n}$. This is exactly the same bound as found before, except that the factor of $|\beta|^{-1}$ is gone, since it was missing in (44). When integrated over $(-\delta\sqrt{n}, \delta\sqrt{n})$ the result is still of order $o(n^{1-j/2})$.

b. For more extreme values of β ($|\beta| > \delta\sqrt{n}$), the integrability condition on the cumulant generating function is used to bound the integrand. As will be shown in the next section, this condition implies the stronger condition in (35). There exists $q < 1$ such that if $|\beta| > \delta\sqrt{n}$ then $|\zeta(\beta/\sqrt{n})| < q$. Hence the contribution of (44) from this part of the range of integration can be bounded by

$$q^{n-r}\int_{-\infty}^{\infty}\left|\zeta(\beta)\right|^r \, d\beta + \int_{|\beta|>\delta\sqrt{n}}\left|\xi_n(\beta)\right| \, d\beta,$$

and approaches 0 geometrically.

3.9. Regularity Conditions

In these proofs an integral involving the characteristic function was bounded by bounding the integrand near the origin. The regularity conditions were used to ensure that contributions to the integral in the range of integration far from the origin are negligible. The regularity conditions needed for various variants of the approximation theorem are:

1. Distribution functions when $j = 2$: This results in the Central Limit Theorem. No regularity conditions, apart from the existence of the mean and variance,

are needed. In the cumulative distribution function proof, T does not depend on n when $\mathfrak{j} = 2$, and hence one can pick n sufficiently large so that parts (a) and (b) of the range of integration for the error integral cover $\mathfrak{R}$; hence part (c) need not be considered.

2. Distribution functions when $\mathfrak{j} = 3$: Here we require $|\zeta(\beta)| < 1 \ \forall \beta \neq 0$. This suffices for the cumulative distribution function Edgeworth proof, since an upper bound on $|\zeta(\beta)|$ over the whole line is unnecessary. One need bound it only over (δ, T). This requirement is equivalent to requiring the summands have a non-lattice distribution.
3. Distribution functions when $\mathfrak{j} > 3$: Here we require the stronger condition

$$\limsup_{|\beta| \to \infty} |\zeta(\beta)| < 1,$$

since T increases faster than $\sqrt{n}$. This condition is known as Cramér's condition, and rules out perverse examples like cumulative distribution functions that do all of their jumping on a Cantor set, as well as examples where the summands have probability masses on $0, 1, a$ where a is irrational. Examples similar to the latter arise, for instance, in likelihood ratio testing and will be addressed in a later chapter. Both of these examples are of non-lattice singular distributions.
4. Density functions for all orders: We require that $\int_{-\infty}^{\infty} |\zeta(\beta)^{\mathfrak{r}}| \ d\beta < \infty$ for some $\mathfrak{r} \geq 1$. This condition is needed for density proofs, since the Smoothing Theorem is not used to truncate the range of integration for the error integral. This condition implies that the distribution $F_{\mathfrak{r}}$ has a density at every point on $\mathfrak{R}$. Lemma 2.4.3 shows that the finiteness of the integral of the absolute value of the characteristic function over $\mathfrak{R}$ implies the existence of a density, since the characteristic function of a sum of independent and identically distributed random variables is the characteristic function of the summands raised to the power equal to the number of summands. This condition is both necessary and sufficient for a bounded density to exist for some convolution of the random variable with itself.

Condition 4 implies condition 3, since condition 4 implies the existence of a density for Z when $n = \mathfrak{r}$, and the existence of this density implies condition 3 by the Riemann-Lebesgue theorem:

Theorem 3.9.1: *If g is a real function such that $\int_{-\infty}^{\infty} |g(x)| \ dx < \infty$, and if its Fourier transform is given by $\zeta(\beta) = \int_{-\infty}^{\infty} \exp(i\beta x) g(x) \ dx$, then $\lim_{|\beta| \to \infty} \zeta(\beta) = 0$.*

Proof: Choose $\epsilon > 0$. Since $\int_{-\infty}^{\infty} |g(x)| \ dx < \infty$, there exists a step function $g^*(x) = \sum_{j=1}^{J} g_j I_{(a_j, a_{j+1}]}(x)$, where $I_{(a_j, a_{j+1}]}(x)$ is the function taking the value 1 on $(a_j, a_{j+1}]$ and zero otherwise, such that $\int_{-\infty}^{\infty} |g(x) - g^*(x)| \ dx < \epsilon$. Let $\zeta^*(\beta) = \int_{-\infty}^{\infty} \exp(i\beta x) g^*(x) \ dx$. Then

$$|\zeta^*(\beta) - \zeta(\beta)| \leq \int_{-\infty}^{\infty} |g(x) - g^*(x)| \ dx < \epsilon.$$

Since

$$\int_{a_j}^{a_{j+1}} \exp(i\beta x)\,dx = (\exp(i\beta a_{j+1}) - \exp(i\beta a_j))/(i\beta),$$

and $|\exp(i\beta a_{j+1}) - \exp(i\beta a_j)| \leq a_{j+1} - a_j$, then

$$|\zeta^*(\beta)| \leq \sum_j g_j(a_{j+1} - a_j)/\,|\beta| = \int_{-\infty}^{\infty} g^*(x)\,dx/\,|\beta|\,,$$

and hence

$$\zeta(\beta) < \zeta^*(\beta) + \epsilon \leq \int_{-\infty}^{\infty} g^*(x)\,dx/\,|\beta| + \epsilon < \int_{-\infty}^{\infty} g(x)\,dx/\,|\beta| + 2\epsilon$$

for all $\epsilon > 0$, and $\zeta(\beta) \leq \int_{-\infty}^{\infty} g(x)\,dx/\,|\beta|$.

Q.E.D

3.10. Some Examples

The gamma and beta distributions illustrate the behavior of Edgeworth series in approximating distributions that differ from normal in certain fundamental ways.

1. The gamma distribution: A $\Gamma(p,\lambda)$ random variable is one with the density

$$f(x) = \lambda^p x^{p-1}\exp(-\lambda x)/\Gamma(p).$$

Its characteristic function is $\zeta(\beta) = 1/(1 - i\beta/\lambda)^p$, and its cumulant generating function is $\mathcal{K}(\beta) = p\log(\lambda) - p\log(\lambda - \beta)$. Hence the cumulant of order j is $\kappa_j = p\Gamma(j)/\lambda^j$, and the convolution of n $\Gamma(p,\lambda)$ densities is a $\Gamma(np,\lambda)$ density. The standardized mean has a $\Gamma(np,\lambda/\sqrt{n})$ distribution. Note that $\int_{-\infty}^{\infty} |\zeta(\beta)|^\nu\,d\beta < \infty$ if and only if $\nu > 1/p$. This example illustrates the effect of the condition on the integrability of the characteristic function. Part of the claim of the density approximation theorem then is that the density of the standardized sum of n independent replicates of such a random variable exists and is continuous for all $n > 1/p$. In this example, one can explicitly calculate the density of such sums. Those sums with $n < 1/p$ have a density that approaches infinity as $x \to 0$, and sums with $n = 1/p$ have a density discontinuous at zero. Thus, the integrability condition successfully indicates all sums with a continuous density.

Fig. 3a shows the poor performance of the Edgeworth series using two, three, and four cumulants when $p = 1$, yielding the exponential distribution. In this case the target density is discontinuous at zero. The highest order approximation, using four cumulants, behaves poorly in the upper tail. Since it involves a higher order polynomial than the other approximations, oscillations in the tail are more severe. Near the ordinate 3 this approximation is negative. Fig. 3b shows that the performance of the Edgeworth series involving the third and fourth cumulants is much improved for standardized sums of 10 such variables. Distinguishing the $E_4(x,\kappa^{*10})$ approximation from the true density visually is difficult for most of the range plotted. Figs. 3c and 3d show improved behavior when $p = 2$.

2. The beta distribution: A $B(p,q)$ random variable is one with the density

$$f(x) = x^p(1-x)^q/B(p,q)$$

for $x \in (0,1)$, where $B(p,q)$ is the beta function $B(p,q) = \Gamma(p)\Gamma(q)/\Gamma(p+q)$. Its moment of order l is $\Gamma(p+l)\Gamma(p+q)/(\Gamma(p)\Gamma(p+q+l))$. From these moments the cumulants can be calculated. Fig. 4 illustrates the behavior of the Edgeworth series when $p = q = 4$.

Calculations for Beta(4,4);n=1

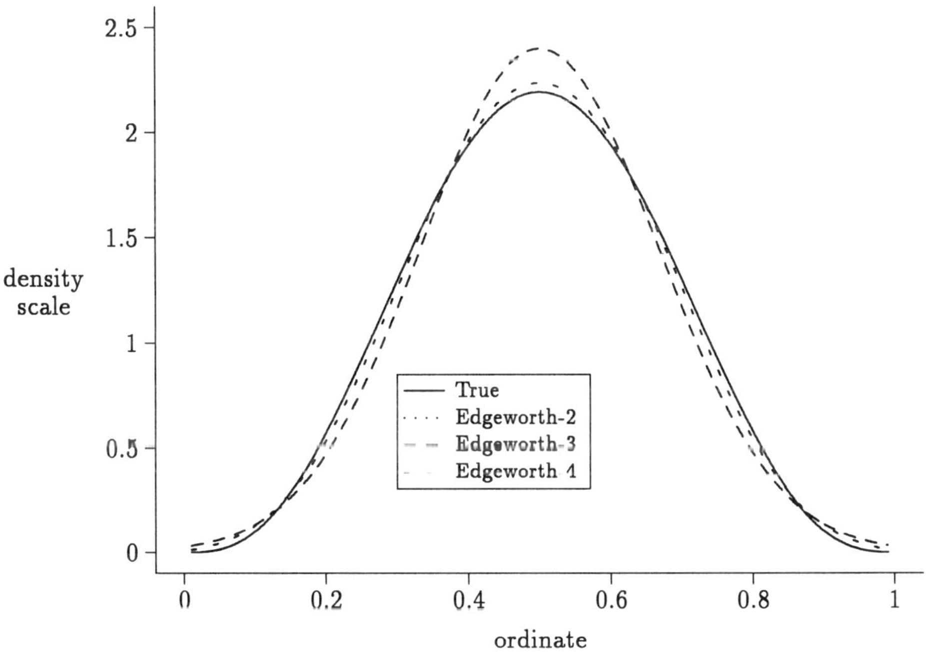

Fig. 4

If $p = q$ the distribution is symmetric about $\frac{1}{2}$. Hence the odd cumulants are zero, and the normal approximation $E_2(x, \kappa^{*n})$ and the next order Edgeworth approximation $E_3(x, \kappa^{*n})$ coincide exactly. The Edgeworth approximation $E_4(x, \kappa^{*n})$ involving kurtosis represents a distinct improvement.

Additional examples occur in later sections on Edgeworth approximation for non-identically distributed and lattice random variables, and in the exercises of this chapter.

Calculations for Gamma(1,1);n=1

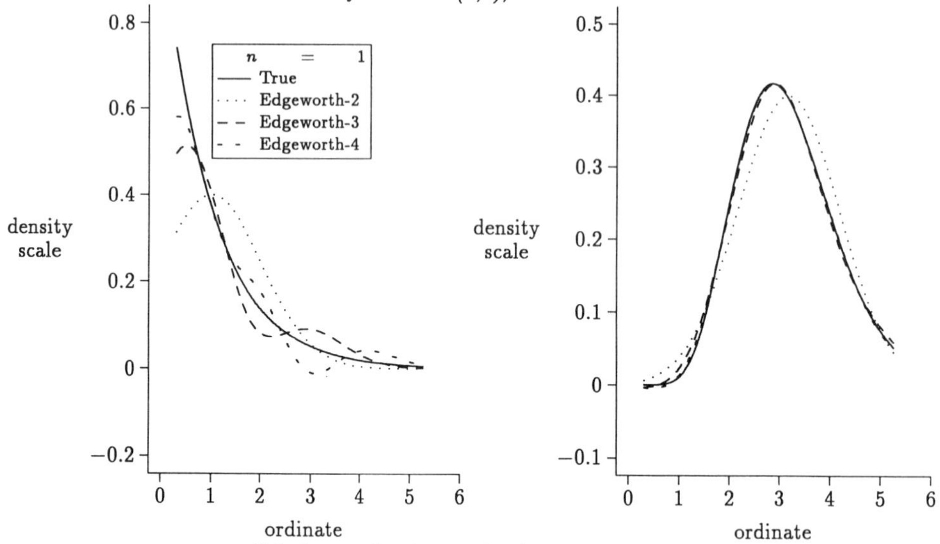

Calculations for Gamma(1,2)

Fig. 3a.

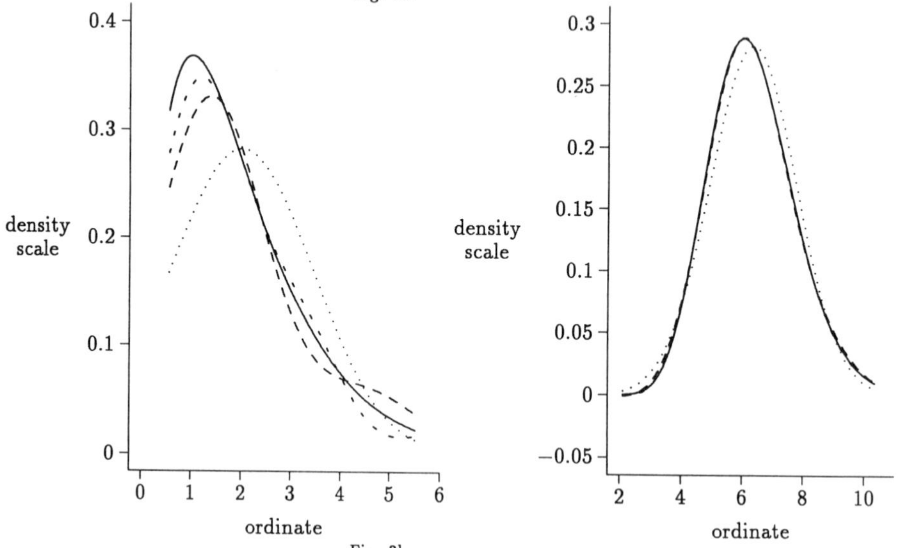

Fig. 3b.

3.11. Relative versus Absolute Error

While absolute error is bounded by our convergence theorem, the relative error, defined as the ratio of the approximation error to what is being approximated, can behave very badly. The first term omitted can be expressed as the normal density approximation times a polynomial, and hence the relative error can usually be expected to be approximately polynomial. In those cases where the approximation becomes negative the relative error exceeds 100%.

3.12. The Non-Identically Distributed Case

Chambers (1967) provides an algorithm for constructing Edgeworth series. Chambers' paper deals primarily with multivariate series, but presents some results applicable here. He gives conditions for Edgeworth approximation to the density to hold other than in the independent and identically distributed standardized mean case using the two conditions:

1. The cumulants are of the same order in n as they are for standardized means;
2. A condition on the size of the integral of the characteristic function in the tails of the distribution.

When leaving the case of independent and identically distributed summands the construction of the Edgeworth series from the cumulants of the resulting variable, rather than from the cumulants of the original summands, becomes important, since these are either no longer the same across summands, or they cumulate in a non-standard manner. Regularity conditions are simpler in the case of independent summands, and are considered below.

As an example, consider the distribution of a least-squares estimate T for the parameter θ in a simple linear regression model $Y_j = \gamma + \theta z_j + E_j$, where the errors E_j are independent and identically distributed with mean zero and j cumulants λ_l. Then $T = \sum_{j=1}^{n}(z_j - \bar{z})Y_j / \sum_{j=1}^{n}(z_j - \bar{z})^2$, and the cumulants κ_l^n are given by $\sum_{j=1}^{n}(z_j - \bar{z})^l \kappa_l / (\sum_{j=1}^{n}(z_j - \bar{z})^2)^l$. In applications where the ordinates z_j do not vary much the estimator should have a distribution as well approximated by an Edgeworth series as is the distribution of means of the errors. A theorem of this section shows that this is true.

First, a lemma adapted from Bhattacharya and Rao (1976):

Lemma 3.12.1: *Suppose Y_j are independent random variables, with characteristic functions $\zeta_j(\beta)$, means zero and variances σ_j^2, and j is an integer greater than or equal to two, such that for all j, Y_j has a cumulant of order j. Let $\varsigma_n^2 = n^{-1}\sum_{j=1}^{n}\sigma_j^2$, and $\bar{\omega}_{n,j} = n^{-1}\sum_{j=1}^{n} \mathrm{E}\left[\left|Y_j^j\right|\right]\varsigma_n^{-j}$. Then there exist K and $\delta > 0$ not depending on the distribution of the summands such that if $|\beta| \le \delta n^{1/2-1/j}\omega_{n,j}$ then*

$$\left|\log(\zeta(\beta/(\sqrt{n}\varsigma_n))) + \beta^2/2 - \eta_j(\beta/(\sqrt{n}\varsigma_n))\right| < K n^{1-j/2}\bar{\omega}_{n,j}\,|\beta|^j. \qquad (45)$$

Proof: Let $v_j(\beta) = \sum_{l=3}^{\mathrm{j}} \kappa_{j,l}(i\beta)^l/l!$, and let $C_j = \mathrm{E}\left[|Y_j|^{\mathrm{j}}\right]^{1/\mathrm{j}}$. By Jensen's Inequality $\sigma_j \leq C_j$. The counterpart of (39) will now be shown to be

$$\left|\sum_{j=1}^{n} \log(\zeta_j(\beta/(\sqrt{n}\varsigma_n))) + \sigma_j^2\beta^2/[2(n\varsigma_j^2)] - v_j(\sigma_j\beta/(\sqrt{n}\varsigma_n))\right| < |\beta|^{\mathrm{j}}$$

for $\beta < \varsigma_n\delta$, for some $\delta > 0$. Let $\beta^* = \beta/(\sqrt{n}\varsigma_n)$. For every non-zero complex a and b, $|\log(a) - \log(b)| \leq |a - b| \max(|1/a|, |1/b|)$. Letting $a = \zeta_j(\beta^*)$ and $b = \exp(-\sigma_j^2\beta^{*2}/2 + v(\beta^*))$, one needs to establish a neighborhood of zero in which a and b are bounded away from zero, and one needs to provide a bound on $|a - b|$. By (11), with $\mathrm{j} = 1$, $|\zeta_j(\beta^*) - 1| < 2|\beta^*|^2 \sigma_j^2 \leq 2|\beta^*|^2 C_j^2$. Hence for $\beta^* < \frac{1}{2}$, $|a| > \frac{1}{2}$. Also, $|b - 1| \leq \exp(p_3(|\beta^*|)) - 1$. Hence there exists $\delta < \frac{1}{2}$ such that $|\beta^*| \leq \delta/C_j$ implies $|b - 1| \leq \frac{1}{2}$ and $a > \frac{1}{2}$. The factor $|a - b|$ may be bounded by

$$\left|\zeta_j(\beta^*) - \sum_{l=2}^{\mathrm{j}} \mu_{j,l}(i\beta^*)^l/l!\right| + \left|\sum_{l=3}^{\mathrm{j}} \mu_{j,l}(i\beta^*)^l/l! - \exp\left(\sum_{l=3}^{\mathrm{j}} \kappa_{j,l}(i\beta^*)^l/l!\right)\right|.$$

The first term above can be bounded by (11), to yield a bound of $2|\beta^*|^{\mathrm{j}} C_j^{\mathrm{j}}/\mathrm{j}!$. The second term has a bound of the form $|\beta^*|^{\mathrm{j}} C_j^{\mathrm{j}} \exp(p_1(C_j|\beta^*|))p_2(|C_j\beta^*|)$, with p_1 and p_2 polynomials not depending on the distributions of summands. Hence there exists K depending only on p_1, p_2, and p_3, such that if $|\beta| \leq \delta\sqrt{n}n^{-1/\mathrm{j}}/[n^{-1}\sum_j(C_j^{\mathrm{j}}/(\varsigma_n^{\mathrm{j}}))]^{1/\mathrm{j}}$ then

$$\left|\log(\zeta(\beta/(\sqrt{n}\varsigma_n))) + \beta^2/2 - v_j(\beta/(\sqrt{n}\varsigma_n))\right| < Kn^{-\mathrm{j}/2}\sum_{j=1}^{n}[C_j^{\mathrm{j}}/\varsigma_n^{\mathrm{j}}]\,|\beta|^{\mathrm{j}}.$$

Q.E.D

This lemma is the key to the following theorem:

Theorem 3.12.2: *Suppose Y_j are independent random variables, with characteristic functions $\zeta_j(\beta)$, means zero and variances σ_j^2, and j is an integer greater than or equal to two, such that for all j, Y_j has a cumulant of order j. Assume that $\bar{\omega}_{n,\mathrm{j}}$ remains bounded as n increases, and that ς_n^2 as defined in Lemma 3.12.1 is bounded away from zero. Further assume that*

$$\prod_{j=1}^{n} \zeta_j(\beta) = o(n^{-a}) \text{ uniformly for } |\beta| > \delta\ \forall\delta, a > 0. \tag{46}$$

Then the approximation $E_{\mathrm{j}}(z, \boldsymbol{\kappa}^{*n})$ formed as in (37) by calculating the associated invariants $\boldsymbol{\rho}_j^n$, and omitting terms of size $O(n^{1-\mathrm{j}/2})$, satisfies $F_Z(z) = E_{\mathrm{j}}(z, \boldsymbol{\kappa}^{*n}) + O(n^{1-\mathrm{j}/2})$, where F_Z is the cumulative distribution function of $Z = \sum_{j=1}^{n} Y_j/\sqrt{n\varsigma_n^2}$.

Proof: This proof will parallel the development of 3.6. Bound (45) replaces (38), and (46) replaces Cramér's condition. The minimum bound on ς_n^2 guarantees that the region near the origin in which the true characteristic function is well–approximated by the Fourier transform of the Edgeworth series expands at the correct rate. The rest of the theorem proceeds as before.

Q.E.D

Returning to the example of regression estimators, T has an Edgeworth approximation valid to $O(n^{1-j/2})$ if

$$\sup_n n^{-1}\sum_{j=1}^{n}(z_j-\bar{z})^{j}\kappa_j/(n^{-1}\sum_{j=1}^{n}(z_j-\bar{z})^2\kappa_2)^{j/2}<\infty,\ \inf_n n^{-1}\sum_{j=1}^{n}(z_j-\bar{z})^2\kappa_2>0.$$

These conditions are trivially true if z_j takes on a finite number of values in a fixed proportion. They are also true with probability one if the regression ordinates are drawn from a population with a finite moment of order j and the regression is considered conditional on the observed collection of ordinates, by the strong law of large numbers.

3.13. Cornish-Fisher Expansions

Related to the problem of approximating tail areas is the inverse problem of approximating quantiles of a distribution. Let Z be the standardized sum of n independent and identically distributed continuous random variables with mean zero and variance one, and a finite fourth moment. Let F_Z be the cumulative distribution function of Z. For each $\alpha \in (0,1)$, there exists at least one solution to $F_Z(y)=\alpha$. Denote this solution by z^n_α. This section provides a series expansion for z^n_α. The result is called the Cornish–Fisher expansion (Cornish and Fisher, 1937, Barndorff–Nielsen and Cox, 1989).

Let $z_\alpha = \Phi^{-1}(\alpha)$. Note that the sequence z^n_α converges to z_α, since otherwise for any $\epsilon > 0$ the distribution F_Z either assigns probability less than $\Phi^{-1}(\alpha)$ to the interval $(-\infty, z_\alpha+\epsilon]$, or assigns probability less than $1-\Phi^{-1}(\alpha)$ to the interval $[z_\alpha-\epsilon,\infty)$, for infinitely many n. This contradicts the central limit theorem.

Applying the Edgeworth cumulative distribution function approximation using the first four cumulants,

$$\begin{aligned}F_Z(z) &= \Phi(z) - \phi(z)\{h_2(z)\kappa_3^n/6 + h_3(z)\kappa_4^n/24 + \kappa_3^{n2}h_5(z)/72\} + C_n(z)/n\\ &= \Phi(z) - \phi(z)\left\{\frac{h_2(z)}{6}\frac{\kappa_3}{\sqrt{n}} + \frac{h_3(z)}{24}\frac{\kappa_4}{n} + \frac{h_5(z)}{36}\frac{\kappa_3^2}{2n}\right\} + C_n(z)/n.\end{aligned}$$

Here $C_{z,n}$ converges to zero uniformly in z. We solve $F_Z(z^n_\alpha)=\alpha$. Expanding in a power series about z_α,

$$\begin{aligned}F_Z(z^n_\alpha) = \alpha - \phi(z_\alpha)\Bigg\{ & -\frac{h_2(z_\alpha)}{6}\frac{\kappa_3}{\sqrt{n}} - \frac{h_3(z_\alpha)}{24}\frac{\kappa_4}{n} - \frac{h_5(z_\alpha)}{36}\frac{\kappa_3^2}{2n} +\\ & \left(1+\frac{h_3(z_\alpha)}{6}\frac{\kappa_3}{\sqrt{n}}+\frac{h_4(z_\alpha)}{24}\frac{\kappa_4}{n}+\frac{h_6(z_\alpha)}{36}\frac{\kappa_3^2}{2n}\right)(z^n_\alpha - z_\alpha) -\\ & \left(h_1(z_\alpha)+\frac{h_4(z_\alpha)}{6}\frac{\kappa_3}{\sqrt{n}}+\frac{h_5(z_\alpha)}{24}\frac{\kappa_4}{n}+\frac{h_7(z_\alpha)}{36}\frac{\kappa_3^2}{2n}\right)\frac{(z^n_\alpha - z_\alpha)^2}{2}\Bigg\} + O((z^n_\alpha - z_\alpha)^3) + \frac{C_n(z^n_\alpha)}{n}.\end{aligned}$$

Setting $F_Z(z_\alpha^n) - \alpha$ equal to 0, and neglecting terms of size $o(n^{-1/2})$, $z_\alpha^n - z_\alpha = (\kappa_3/[6\sqrt{n}])h_2(z_\alpha) + o(1/\sqrt{n}) = O(1/\sqrt{n})$. Setting $F_Z(z_\alpha^n) - \alpha$ to zero and dropping terms of size $o(1/n)$:

$$o\left(\frac{1}{n}\right) = \frac{h_2(z_\alpha)\kappa_3}{6\sqrt{n}} + \frac{h_3(z_\alpha)}{24}\frac{\kappa_4}{n} + \frac{h_5(z_\alpha)}{36}\frac{\kappa_3^2}{2n} - \left(1 + \frac{h_3(z_\alpha)\kappa_3}{6\sqrt{n}}\right)(z_\alpha^n - z_\alpha) + \frac{h_1(z_\alpha)(z_\alpha^n - z_\alpha)^2}{2}.$$

Solving for z_α^n one finds that

$$\begin{aligned} z_\alpha^n &= z_\alpha + \frac{\kappa_3}{\sqrt{n}}\frac{h_2(z_\alpha)}{6} + \frac{\kappa_4}{n}\frac{h_3(z_\alpha)}{24} \\ &\quad + \frac{\kappa_3^2}{2n}\left(\frac{h_5(z_\alpha) - 2h_3(z_\alpha)h_2(z_\alpha) + h_1(z_\alpha)h_2^2(z_\alpha)}{36}\right) + o\left(\frac{1}{n}\right) \\ &= z_\alpha + \frac{\kappa_3}{\sqrt{n}}\frac{h_2(z_\alpha)}{6} + \frac{\kappa_4}{n}\frac{h_3(z_\alpha)}{24} - \frac{\kappa_3^2}{2n}\frac{(2z_\alpha^3 - 5z_\alpha)}{18} + o\left(\frac{1}{n}\right). \end{aligned}$$

For standardized sums of random variables with arbitrary first and second cumulants,

$$z_\alpha^n = \sqrt{n}\kappa_1 + \sqrt{\kappa_2}\left[z_\alpha + \frac{\rho_3}{\sqrt{n}}\frac{h_2(z_\alpha)}{6} + \frac{\rho_4}{n}\frac{h_3(z_\alpha)}{24} - \frac{\rho_3^2}{2n}\frac{(2z_\alpha^3 - 5z_\alpha)}{18}\right] + o(1/n),$$

where the coefficients ρ_j are the invariant cumulants of §2.1. If the summands have a fifth cumulant, $o(1/n)$ may be replaced by $O(n^{-3/2})$, since an additional term in the series might be constructed. Let C_2, C_3, and C_4 represent z_α^n with terms of size $o(1)$, $o(1/\sqrt{n})$, and $o(1/n)$ dropped, respectively.

Fisher and Cornish (1960) provide tables aiding the construction of higher order approximations in the same spirit. Hall (1983) provides an extension to these methods to smooth transformations of random variables with known cumulants; these generalized expansions rely only on the cumulants of the underlying distribution and derivatives of the transformation; limiting properties are demonstrated with full rigor.

Koning and Does (1988) consider the distribution of Spearman's Rank Correlation, used for nonparametric bivariate inference. Spearman's Rank Correlation is an affine transformation of $T = \sum_{j=1}^{n} jR_j/(n+1)$ where R_j are ranks of n independently and identically distributed random variables. This distribution is difficult to calculate exactly, but moments, and hence cumulants and the Cornish-Fisher expansion, are readily available. Koning and Does (1988) give Fortran code implementing this expansion for T. This code is available from STATLIB. In this example the summands are identically distributed but not independent. Also, T has a lattice distribution, while the Cornish-Fisher expansion is continuous. None

Inverse CDF of the Spearman Statistic T for Sample Size 8

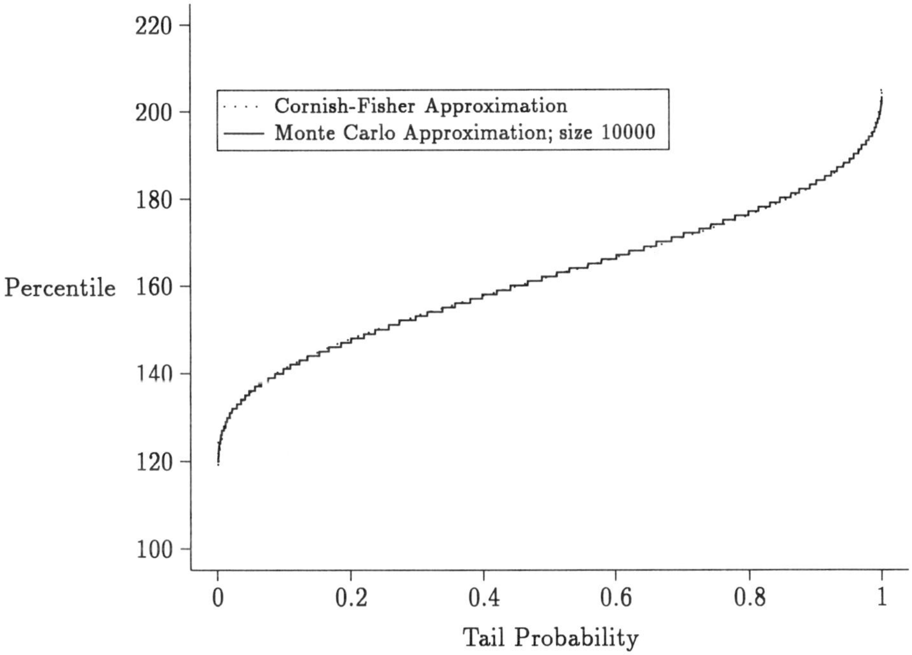

Fig. 5.

the less, the Cornish-Fisher expansion nominally to order $o(1/n)$ performed very well. An example of its application is given in Fig. 5.

As a second example, Barndorff–Nielsen and Cox (1989) approximate quantiles for the χ_n^2 distribution; the cumulants of the square of a standard normal random variable are $\kappa_l = 2^{l-1}(l-1)!$ and the standardized cumulants are $\rho_l = 2^{l/2-1}(l-1)!$; $\rho_3 = 2\sqrt{2}$ and $\rho_4 = 12$. The χ_n^2 distribution is the convolution of n such distributions. True tail probabilities and the Cornish-Fisher expansions are given in Fig. 6.

Konishi, Niki, and Gupta (1988) apply these methods to independently but non-identically distributed sums of random variables. Withers (1988) and Hall (1988) apply these methods to the empirical cumulative distribution function instead of the true cumulative distribution function in order to avoid resampling in bootstrap applications. Such methods are discussed below in the context of saddlepoint approximations.

3.14. The Lattice Case

As before, suppose that $\{Y_j\}$ are independent and identically distributed random variables, where for notational simplicity assume that the summands have zero

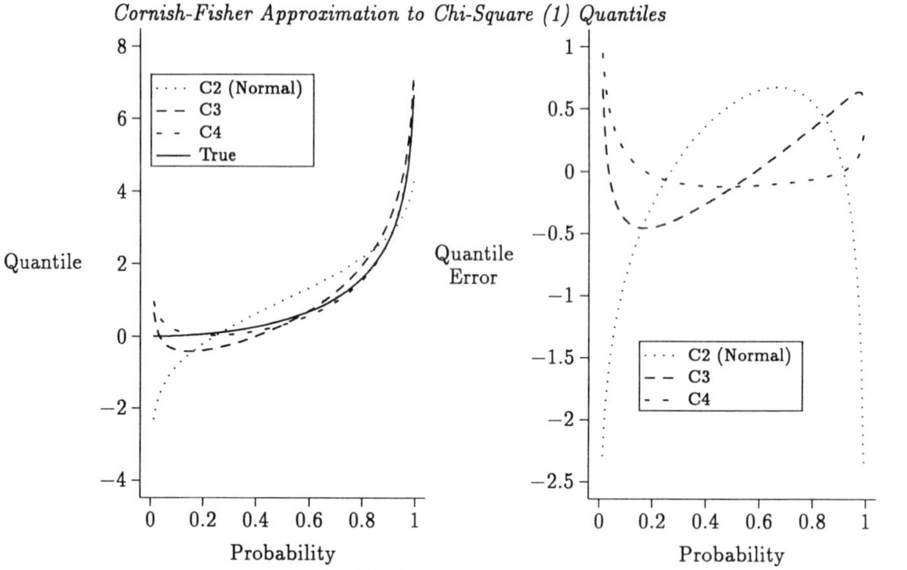
Cornish-Fisher Approximation to Chi-Square (1) Quantiles
C2 (Normal)
C3
C4
True
Quantile
Probability
Quantile Error
C2 (Normal)
C3
C4
Probability

Fig. 6a.

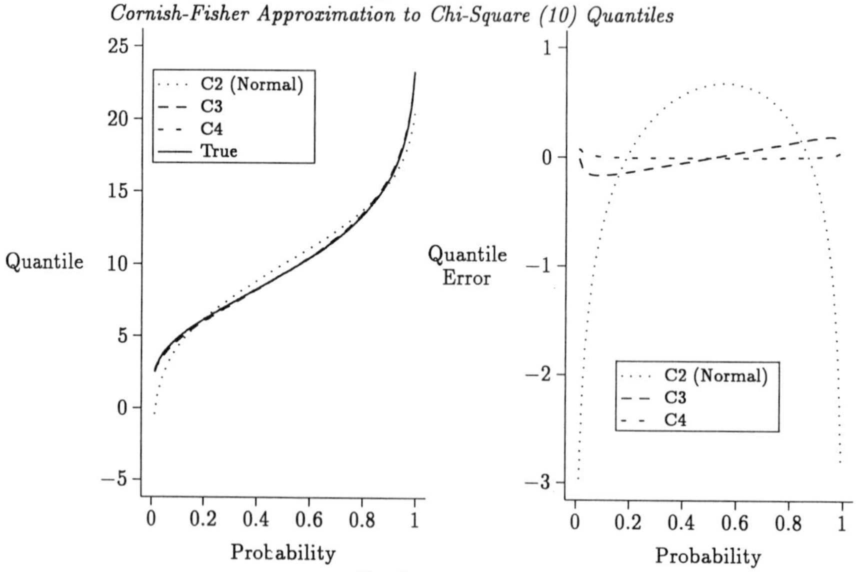
Cornish-Fisher Approximation to Chi-Square (10) Quantiles
C2 (Normal)
C3
C4
True
Quantile
Probability
Quantile Error
C2 (Normal)
C3
C4
Probability

Fig. 6b.

mean and unit variance, and $Z = \sum_{l=1}^{n} Y_l/\sqrt{n}$. Consider again the problem of approximating the cumulative distribution function F_Z of Z. Now, however, suppose that the summands Y_j take values on the lattice $\{a + \Delta\mathfrak{Z}\}$ where $\mathfrak{Z}$ is the set of integers. This violates even the weaker regularity condition of §3.5.

One might be interested in this problem for three reasons. First, many distributions in applied statistics are lattice distributions. These include distributions of sufficient statistics in logistic regression, contingency table applications, and non-parametric inference. Hypothesis testing and confidence interval generation require the approximation of tail probabilities in these cases. Secondly, lattice distributions have applications to fields like queueing theory. Thirdly, the results of this section tie together a number of seemingly unrelated ideas in probability; specifically, they extend the Edgeworth series to the next (and perhaps last) area of regular problems. Asymptotic expansions for lattice distributions exist, and putting them in the same framework as Edgeworth series has theoretical appeal.

A bit of reflection yields reasons why unmodified Edgeworth series will not suffice here. They clearly will not work uniformly, since the cumulative distribution function has jumps of order $O(1/\sqrt{n})$. Feller (1971) shows that the Edgeworth series evaluated only at continuity-corrected points $z^+ = z + \Delta/(2\sqrt{n})$ will yield results accurate to $o(1/\sqrt{n})$.

Esseen (1945) derives an approximation based on the Edgeworth series plus correction terms containing derivatives of the Edgeworth series, involving the functions Q_l given by

$$Q_l(y) = \begin{cases} (l-1)! g_l \sum_{j=1}^{\infty} \cos(2\pi j y)/(2^{l-1}(\pi j)^l) & \text{if } l \text{ is even} \\ (l-1)! g_l \sum_{j=1}^{\infty} \sin(2\pi j y)/(2^{l-1}(\pi j)^l) & \text{if } l \text{ is odd,} \end{cases} \tag{47}$$

with the constants g_l given by

$$g_l = \begin{cases} +1 & \text{if } l = 4j+1 \text{ or } l = 4j+2 \text{ for some integer } j \\ -1 & \text{if } l = 4j+3 \text{ or } l = 4j \text{ for some integer } j. \end{cases}$$

These functions have the following properties:

1. Q_l has period 1,
2. Q_l is piecewise polynomial,
3. Q_l is continuous for $l > 1$ on $\mathfrak{R}$.

Let B_l be the polynomial versions of Q_l on $[0, 1)$. The first few of these polynomials are:

$$B_1(y) = y - \frac{1}{2},\ B_2(y) = y^2 - y + \frac{1}{6},\ B_3(y) = y^3 - \frac{3y^2}{2} + \frac{y}{2}.$$

These polynomials are known as the Bernoulli polynomials and are defined by the relation

$$i\beta \frac{\exp(-i\beta y)}{1 - \exp(-i\beta)} = \sum_{l=0}^{\infty} B_l(y)(-i\beta)^l/l!$$

for $|\beta| < 2\pi$ (Haynsworth and Goldberg, 1965, p. 804).

In the presence of j cumulants, Esseen (1945) constructs an approximation valid uniformly to $o(n^{1-j/2})$, the same order as the original Edgeworth series in the non-lattice case:

Theorem 3.14.1: *If $\{Y_n\}$ are independent and identically distributed random variables with zero mean and unit variance taking values on the lattice $\{a + \Delta\mathfrak{Z}\}$, with finite cumulant of order j, and if $Z = \sum_{l=1}^{n} Y_l/\sqrt{n}$, then Esseen's series*

$$E_j(z;\boldsymbol{\kappa}^n) + D_j(z;\boldsymbol{\kappa}^n) \tag{48}$$

is valid uniformly to $o(n^{1-j/2})$, where $D_j(z;\boldsymbol{\kappa}^n)$ is a discontinuous correction term

$$\sum_{l=1}^{j-2} \frac{1}{l!}(\Delta_n)^l Q_l((z-z_n)/(\Delta_n))(-1)^l \frac{d^l}{dz^l} E_j(z,\boldsymbol{\kappa}^n), \tag{49}$$

$\Delta_n = \Delta/\sqrt{n}$, and z_n is any point on the lattice $\{\sqrt{n}a + \Delta\mathfrak{Z}/\sqrt{n}\}$.

Proof: Let $\zeta_n(\beta)$ be the characteristic function of Z. The inversion integral for lattice tail probabilities with separation $\Delta_n = \Delta/\sqrt{n}$ in (27) is

$$\mathrm{P}\left[z > Z \geq z_0\right] = \frac{1}{2\pi}\int_{-\pi/\Delta_n}^{\pi/\Delta_n} \zeta_n(\beta)\frac{\exp(-i\beta z_0) - \exp(-i\beta z_n)}{1-\exp(-i\Delta_n\beta)}\Delta_n\, d\beta$$

$$= \frac{1}{2\pi}\int_{-\pi/\Delta_n}^{\pi/\Delta_n} \zeta_n(\beta)\left(\frac{\exp(-i\beta z_0)}{1-\exp(-i\Delta_n\beta)} - \exp(-i\beta z)\frac{\exp(-i\beta(z_0 - z))}{1-\exp(-i\Delta_n\beta)}\right)\Delta_n\, d\beta \tag{50}$$

where z_n is the minimal lattice point greater than or equal to z. This proof will be performed in four steps. First, $\zeta_n(\beta)$ in (50) will be replaced by $\xi_n(\beta)$ of §3.7. Second, the remaining factors in (50) will be expanded in a power series and truncated. Third, the inversion integral (50) will be extended over $\mathfrak{R}$. Fourth, the result will be shown to agree with (48).

Since the function $iy/(1-\exp(iy))$ is bounded for $y \in [-\pi, \pi]$, there exists C such that $2\Delta_n/(1-\exp(-i\beta\Delta_n))$ can be bounded by $C/|\beta|$, for $|\beta| \leq \pi/\Delta_n$. The quantity C may be chosen independently of n and Δ. As in §3.7, choose δ such that $|(\zeta_n(\beta) - \xi_n(\beta))/\beta|$ is bounded by (42) for $|\beta| < \delta\sqrt{n}$. Then for $|\beta| < \delta\sqrt{n}$ the difference between the integrand of (50) and this integrand with ζ_n replaced by ξ_n is also of the form (42), and uniformly,

$$\left|\frac{1}{2\pi}\int_{-\delta\sqrt{n}}^{\delta\sqrt{n}} \frac{(\zeta_n(\beta) - \xi_n(\beta))}{\beta}\frac{\beta}{1-\exp(-i\Delta_n\beta)}\Delta_n\, d\beta\right| = o(n^{1-j/2}).$$

Over the region $\delta \leq |\beta| \leq \pi/\Delta$ the function $|\zeta_Y(\beta)|$ is bounded away from 1, and hence over the region $\delta\sqrt{n} \leq |\beta| \leq \pi/\Delta_n$, the function $\zeta_n(\beta\sqrt{n})$ converges uniformly and geometrically to 0. Also, (43) still holds, and $\xi_n(\beta)$ converges uniformly and geometrically to 0. Hence

$$\left|\frac{1}{2\pi}\int_{-\pi/\Delta_n}^{\pi/\Delta_n} (\zeta_n(\beta) - \xi_n(\beta))\frac{1}{1-\exp(-i\Delta_n\beta)}\Delta_n\, d\beta\right| = o(n^{1-j/2}). \tag{51}$$

A polynomial approximation will now be substituted for the terms of form $\exp(-i\beta y)i\beta(1-\exp(-i\beta))^{-1}$ in (50). Choose $\delta > 0$ and $C_1 > 0$ such that

$$\left| i\beta^* \frac{\exp(-i\beta^* y)}{1-\exp(-i\beta^*)} - \sum_{l=0}^{j-2} B_l(y)(-i\beta^*)^l/l! \right| < C_1 |\beta^*|^{j-1}$$

whenever $|\beta^*| < \delta$ and $y \in [0,1)$. Set $\beta^* = \beta\Delta_n$. Choose any real u. Then

$$\left| \frac{1}{2\pi} \int_{-\pi/\Delta_n}^{\pi/\Delta_n} \frac{\xi_n(\beta)}{\beta} \left[\frac{\exp(-i\beta(\Delta_n y + u))\Delta_n \beta}{1-\exp(-i\Delta_n \beta)} - \exp(-iu\beta) \sum_{l=0}^{j-2} \frac{B_l(y)(-i\beta\Delta_n)^l}{l!} \right] d\beta \right|$$
$$\leq \frac{C_1}{2\pi} \int_{-\pi/\Delta_n}^{\pi/\Delta_n} \left| \xi_n(\beta)\beta^{j-2} \right| d\beta \Delta_n^{j-1},$$

where C_1 is independent of n, z, z_n, and δ. The integral on the right-hand side above can be bounded independently of n, z, z_n, and δ, yielding

$$\left| \int_{-\pi/\Delta_n}^{\pi/\Delta_n} \frac{\xi_n(\beta)}{2\pi\beta} \left[\frac{\exp(-i\beta(\Delta_n y + u))\Delta_n \beta}{1-\exp(-i\Delta_n \beta)} - \exp(-iu\beta) \sum_{l=0}^{j} \frac{B_l(y)(-i\beta\Delta_n)^l}{l!} \right] d\beta \right|$$
$$= O(n^{(1-j)/2}). \tag{52}$$

The extension of the above integral from $(-\delta\sqrt{n}, \delta\sqrt{n})$ to $(-\pi/\Delta_n, \pi/\Delta_n)$ is possible by (43).

Applying (52) twice, once with $u = z$ and $y = (z_n - z)/\Delta_n$, and once with $u = z_0$ and $y = 0$, and applying the triangle inequality,

$$\mathrm{P}[z > Z \geq z_0] = \frac{1}{2\pi} \int_{-\pi/\Delta_n}^{\pi/\Delta_n} \frac{\xi_n(\beta)}{-i\beta}$$
$$\times \sum_{l=0}^{j} \left(\exp(-i\beta z) B_l\left(\frac{z_n - z}{\Delta_n}\right) - \exp(-i\beta z_0) B_l(0) \right) \frac{(-i\beta\Delta_n)^l}{l!} d\beta + o(n^{1-j/2}).$$

The third step in the proof is to prove that the above inversion integral is unchanged to $o(n^{1-j/2})$ when the range of integration is extended to the entire real line. Since the above integrand on the compliment of the above range of integration may be bounded by $\exp(-n\pi^2/(4\Delta^2))$ times an integrable function, as in (43), the change occurring in extending the range of integration is of the proper order. By (36),

$$\mathrm{P}[z > Z \geq z_0] = \sum_{l=0}^{j} \frac{\Delta_n^l}{l!} \left(B_l\left(\frac{z_n - z}{\Delta_n}\right) E_j^{*(l)}(z; \boldsymbol{\kappa}^{*n}) - B_l(0) E_j^{*(l)}(z_0; \boldsymbol{\kappa}^{*n}) \right) + o(n^{1-j/2}),$$

where $E_j^{*(l)}(z; \boldsymbol{\kappa}^{*n}) = d^l/(dz^l) E_j^*(z; \boldsymbol{\kappa}^{*n})$. Letting $z_0 \to \infty$, and discarding terms of size $o(n^{1-j/2})$, gives (48).

Q.E.D

Even at the continuity-corrected points the correct asymptotic approximation is not the usual Edgeworth series. Fig. 7 demonstrates the behavior of this approximation.

A lattice analogue for Theorem 3.12.2 is presented by Bhattacharya and Rao (1976):

Theorem 3.14.2: *Suppose Y_j are independent random variables, with characteristic functions $\zeta_j(\beta)$, means zero and variances σ_j^2, and j is an integer greater than or equal to two, such that for all j, Y_j has a cumulant of order j. Assume that $\bar{\omega}_{n,\mathrm{j}}$ as defined in Theorem 3.12.2 remains bounded as n increases, and that ς_n^2 as defined in Lemma 3.12.1 is bounded away from zero. Further assume that each summand takes values in the lattice $\{a+\Delta\mathfrak{Z}\}$, and that this lattice is the minimal one supporting the entire sequence of random variables. Form the Edgeworth series $E_\mathrm{j}(z,\boldsymbol{\kappa}^{*n})$ using (37) as in Theorem 3.12.2. Then (48) approximates $F_Z(z)$ with an accuracy of $O(n^{1-\mathrm{j}/2})$, where F_Z is the cumulative distribution function of $Z=\sum_{j=1}^n Y_j/\sqrt{n\varsigma_n^2}$, (49) is calculated with $\Delta_n=\Delta/\sqrt{n\varsigma_n^2}$, and explicit dependence on sample size in (49) is replaced by a dependence through derivatives of $E_\mathrm{j}(z,\boldsymbol{\kappa}^{*n})$.*

Proof: This theorem parallels the development of §3.6. Bound (45) replaces (38), causing (51) to hold with $O(n^{1-\mathrm{j}/2})$ replacing $o(n^{1-\mathrm{j}/2})$. The rest of the theorem proceeds as before.

Q.E.D

As an example, Albers, Bickel, and van Zwet (1976) explore Edgeworth expansions for cumulative distribution functions of nonparametric one-sample location general score test statistics. For independent and identically distributed random variables $Y_1,\cdots,Y_n$ known to have an absolutely continuous distribution, tests of the hypothesis that their common median is zero may be constructed from the indicators V_i taking the value one if the Y_i is positive and zero otherwise, if the observations are ordered according to the magnitude of their absolute values. Let the general score test statistic be $T_n=\sum_j^n a_j^{(n)}V_j$ for constants $a_j^{(n)}$. In this case the summands are independent but not identically distributed, unless all of the $a_j^{(n)}$ are identical. The first four cumulants are

$$\kappa_1=\sum_{j=1}^n a_j^{(n)}/2,\ \ \kappa_2=\sum_{j=1}^n a_j^{(n)^2}/4,\ \ \kappa_3=0,\ \ \kappa_4=-\sum_{j=1}^n a_j^{(n)^4}/8.$$

A simple example of this test statistic sets $a_j^{(n)}=j$, yielding the Wilcoxon signed rank statistic (Hettmansperger, 1984). In this case,

$$\kappa_1=\tfrac{1}{4}n(n+1),\ \kappa_2=n(n+1)(n+\tfrac{1}{2})/12,\ \kappa_3=0,\ \kappa_4=-n(n+1)(n+\tfrac{1}{2})(n^2+n+\tfrac{1}{3})/120,$$

and generally $\kappa_\mathrm{j}=O(n^{\mathrm{j}+1})$. Then if ρ_j is the invariant of order j, $\rho_\mathrm{j}=O(n^{\mathrm{j}+1-3\mathrm{j}/2})=O(n^{1-\mathrm{j}/2})$, and hence $T_n/\sqrt{\kappa_2}$ has invariants of the same order in n as is usual in Edgeworth series applications. Specifically, Theorem 3.14.2 applies. Furthermore,

Esseen's approximation to the CDF of a Standardized Binomial(1 , 0.5) Variable

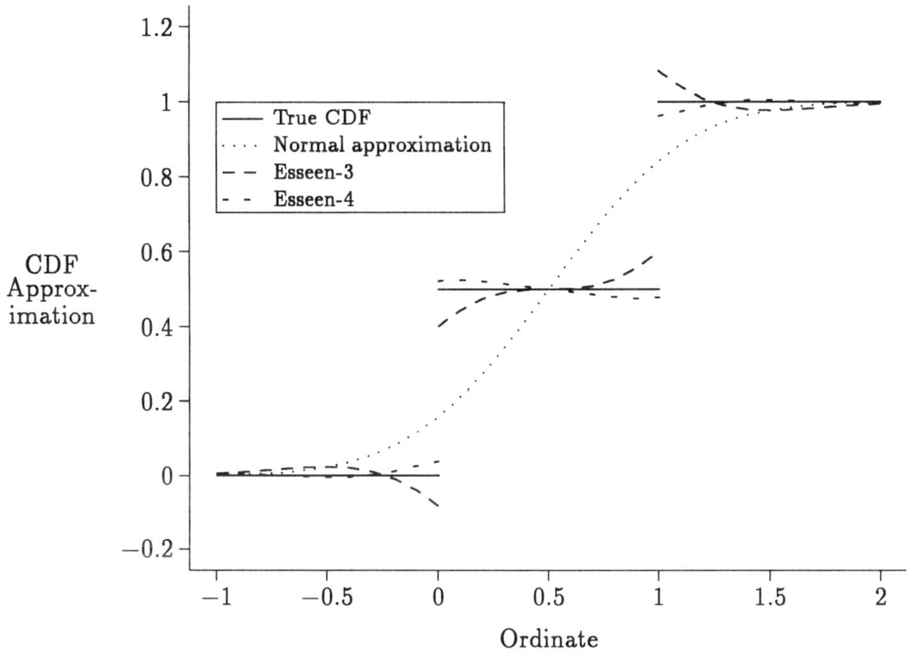

Fig. 7a.

Esseen's approximation to the CDF of a Standardized Binomial(5 , 0.75) Variable

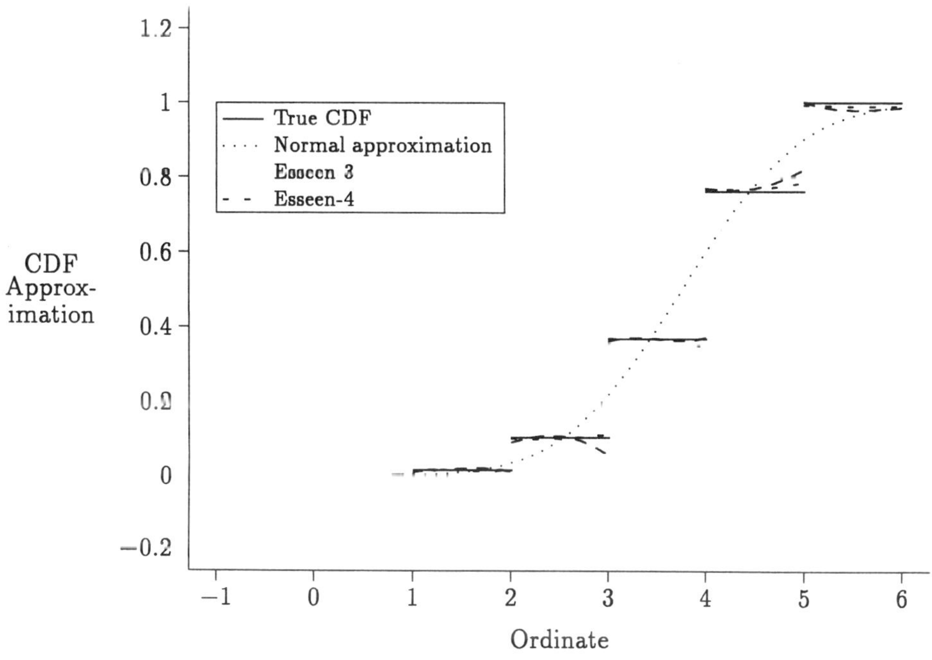

Fig. 7b.

$\Delta_n = O(n^{-3/2})$, indicating that $E_4(z, \boldsymbol{\kappa}^{*n})$ is valid to $O(n^{-3/2})$. Albers, Bickel, and van Zwet (1976) derive the error bound of $O(n^{-5/4})$ for more general scores $a_j^{(n)}$.

As a further example, recall the distribution of T, related to Spearman's Rank Correlation, discussed in §3.13. Its Cornish-Fisher expansion to $o(1/n)$ is dependent on the validity of the standard Edgeworth series to the same order. As the minimal lattice for T has spacings $1/(n+1)$, and $\text{Var}[T] = n^2(n-1)/144$, $T/\sqrt{\text{Var}[T]}$ takes values on a lattice with spacings $O(n^{-5/2})$. In this case errors induced by the lattice nature of the summands at continuity–corrected points are of order $O(n^{-5})$. An Edgeworth series theorem for dependent variables is needed in this case. Prášková–Vizkovaá (1976) addresses this question as well.

Robinson (1982) addresses this problem of calculating tail probabilities for score statistics. Details will follow in a later chapter devoted to resampling and permutation methods. Kong and Levin (1996) address the problem of approximating multivariate distribution functions of sufficient statistics arising in logistic regression, and demonstrate that if the lattice nature of the problem is not too strong, conventional Edgeworth series methodology applies.

3.15. Sheppard's Corrections and Edgeworth Series

If there is an approximation that can be expressed as an Edgeworth series and valid to higher orders the cumulants must be modified. One might try constructing an asymptotically correct series expansion by smoothing the original lattice distribution to a continuous distribution whose cumulative distribution function agrees with the original cumulative distribution function at continuity-corrected points z^+, and using Edgeworth approximation techniques to approximate the smoothed cumulative distribution function. One method for doing this is as follows:

1. Construct a smoothed variable Z_n by choosing U uniform on $[-\frac{1}{2}, \frac{1}{2}]$. Set $Z_n = X_n + (\Delta/\sqrt{n})U$. Then Z_n has a density, and furthermore its cumulative distribution function agrees with F_n at continuity-correction points.
2. The cumulant of order k of $(\Delta/\sqrt{n})U$ is $\epsilon_k^n = (\Delta/\sqrt{n})^k B_k/k$, for $k \geq 2$; the constants B_k are known as the Bernoulli numbers and are derived directly from the series expansion for the characteristic function for U as part of the exercises for Chapter 2.

Since Z_n is not a standardized sum the Edgeworth theorems do not necessarily apply. The density for Z_n has jumps of order $O(1/\sqrt{n})$ at continuity corrected points; approximation by a continuous function must have errors that do not get smaller in n as more cumulants are used.

The previous smoothing attempt failed because the density constructed was not differentiable. One might change the distribution of the added smoother U to produce a result with a density having more derivatives.

A simpler smoothing method is as follows: If Y_n represents a continuous random variable Z_n grouped to a lattice with spacing Δ, then the cumulants κ_k^{n*} of Y_n are given by $\kappa_k^n - \epsilon_k^n$ where κ_k^n are the cumulants of Z_n. These modifications to cumulants of grouped data to recover cumulants of the underlying continuous variate are known as Sheppard's corrections, and are tabulated in Table 4. To obtain Sheppard-corrected cumulants, subtract the last column from the appropriate raw cumulant, after multiplying by the appropriate power of the lattice spacing.

Table 4: *Bernoulli Numbers and Sheppard's Corrections*

k	B_k	B_k/k
any odd integer	0	0
2	1/6	1/12
4	−1/30	−1/120
6	1/42	1/252

Heuristically, then, $\zeta_{Y_n}\zeta_{\Delta U/\sqrt{n}} \approx \zeta_{Z_n}$ and that the discrete variable Y_n and the smoothing $\Delta U/\sqrt{n}$ are approximately independent in the sense measured by multiplicative nature of the characteristic functions. Calculations are performed using calculus of finite differences:

Theorem 3.15.1: *Suppose the random variable C has a continuous density h on $\Re$, and that the first $2l$ derivatives of $h(x)$ exist and converge to zero as $x \to \pm\infty$. This condition is known as high contact. Let $D = \Delta\langle\frac{C}{\Delta}\rangle$ be C rounded to the nearest lattice point, and define $U = (C - D)/\Delta$ to be the standardized difference between C and the nearest lattice point, which takes values on the interval $(-\frac{1}{2}, \frac{1}{2}]$. Here $\langle . \rangle$ denotes the nearest integer, with the convention that integers plus one half are rounded upwards. Let $\zeta_C(\beta_1)$, $\zeta_U(\beta_2)$, and $\zeta_{C,U}(\beta_1, \beta_2)$ be the characteristic functions of C, U, and the pair (C, U) respectively. Then as $\Delta \to 0$,*

$$\zeta_{C,U}(\beta_1, \beta_2) = \zeta_C(\beta_1)\zeta_U(\beta_2) + O(\Delta^{2l}) \text{ and } \zeta_U(\beta_2) = \sinh(\tfrac{1}{2}i\beta_2)/(\tfrac{1}{2}i\beta_2) + O(\Delta^{2l}).$$

Proof: By definition,

$$\begin{aligned}\zeta_{C,U}(\beta_1, \beta_2) &= \int_{-\infty}^{\infty} \exp[i(s\beta_1 + (s/\Delta - \langle s/\Delta\rangle\beta_2))]h(s)\,ds \\ &= \int_{-\infty}^{\infty} \exp[i(s(\beta_1 + \beta_2/\Delta) - \langle s/\Delta\rangle\beta_2)]h(s)\,ds \\ &= \sum_{j\in\mathfrak{Z}} \exp(-ij\beta_2) \int_{(j-1/2)\Delta}^{(j+1/2)\Delta} \exp[is(\beta_1 + \beta_2/\Delta)]h(s)\,ds.\end{aligned}$$

Also,

$$\begin{aligned}\zeta_C(\beta_1)\zeta_U(\beta_2) &= \left[\int_{-\infty}^{\infty} e^{is\beta_1}h(s)\,ds\right]\left[\sum_{j\in\mathfrak{Z}} e^{-ij\beta_2}\int_{(j-1/2)\Delta}^{(j+1/2)\Delta} e^{is\beta_2/\Delta}h(s)\,ds\right] \\ &- \left[\int_{-\infty}^{\infty} \exp(is\beta_1)h(s)\,ds\right]\left[\sum_{j\in\mathfrak{Z}}\int_{\frac{1}{2}}^{\frac{1}{2}} \exp(it\beta_2)h(\Delta(t+j))\Delta\,dt\right].\end{aligned}$$

Hence $\zeta_{C,U}(\beta_1,\beta_2)-\zeta_C(\beta_1)\zeta_U(\beta_2)$ may be expressed as

$$\sum_{j\in\mathfrak{Z}}\int_{-\frac{1}{2}}^{\frac{1}{2}}\left[e^{i(\Delta(j+t)\beta_1+t\beta_2)}-\int_{-\infty}^{\infty}e^{is\beta_1+it\beta_2}h(s)\,ds\right]h(\Delta(j+t))\Delta\,dt$$

$$=\sum_{j\in\mathfrak{Z}}\int_{-\frac{1}{2}}^{\frac{1}{2}}e^{it\beta_2}\left[\int_{-\infty}^{\infty}\left\{e^{i\Delta(j+t)\beta_1}-e^{is\beta_1}\right\}h(s)\,ds\right]h(\Delta(j+t))\Delta\,dt$$

$$=\sum_{j\in\mathfrak{Z}}e^{-ij\beta_2}\int_{(j-1/2)\Delta}^{(j+1/2)\Delta}e^{it\beta_2/\Delta}\left[\int_{-\infty}^{\infty}\left\{e^{it\beta_1}-e^{is\beta_1}\right\}h(s)\,ds\right]h(t)\,dt=\sum_{j\in\mathfrak{Z}}f(j)$$

where

$$f(j)=e^{-ij\beta_2}\int_{(j-1/2)\Delta}^{(j+1/2)\Delta}e^{it\beta_2/\Delta}[\exp(it\beta_1)-\zeta_C(\beta_1)]h(t)\,dt.$$

Note that f depends also on β_1 and β_2.

If $f:\mathfrak{R}\to\mathfrak{R}$ is a function with $2l$ continuous derivatives, and $\lim_{t\to\pm\infty}f^{(k)}(t)$ exists for all $k\le 2l$, then, by the Euler-Maclaurin Summation Formula (Haynsworth and Goldberg, 1965, page 806),

$$\sum_{j\in\mathfrak{Z}}f(j)=\int_{-\infty}^{\infty}f(t)\,dt+\lim_{t\to\infty}\left[\tfrac{1}{2}(f(t)+f(-t))+\sum_{k=1}^{l-1}\frac{B_{2k}}{(2k)!}(f^{(2k-1)}(t)-f^{(2k-1)}(-t))\right]$$
$$+\frac{B_{2l}}{(2l)!}\sum_{j\in\mathfrak{Z}}f^{(2l)}(j+\delta_0)$$

for some $\delta_0\in(0,1)$. The first term on the right hand side above is

$$\int_{-\infty}^{\infty}f(x)\,dx=\int_{-\infty}^{\infty}e^{-ix\beta_2}\int_{(x-1/2)\Delta}^{(x+1/2)\Delta}e^{it\beta_2/\Delta}[\exp(it\beta_1)-\zeta_C(\beta_1)]h(t)\,dt\,dx$$
$$=\int_{-\infty}^{\infty}\int_{t/\Delta-\frac{1}{2}}^{t/\Delta+\frac{1}{2}}e^{-ix\beta_2}e^{it\beta_2/\Delta}[\exp(it\beta_1)-\zeta_C(\beta_1)]h(t)\,dx\,dt$$
$$=\int_{-\infty}^{\infty}(e^{i\beta_2/2}-e^{-i\beta_2/2})/(i\beta_2)[\exp(it\beta_1)-\zeta_C(\beta_1)]h(t)\,dt=0.$$

To evaluate the second term, derivatives of f are needed. Since

$$f(x)=\int_{-\frac{1}{2}}^{\frac{1}{2}}e^{it\beta_2}g(t+x)\,dt,$$

where

$$g(t)=\Delta\int_{-\infty}^{\infty}(e^{it\Delta\beta_1}-e^{is\beta_1})h(s)\,dsh(\Delta t)=\Delta h(\Delta t)[e^{it\Delta\beta_1}-\zeta_C(\beta_1)],$$

then

$$f^{(k)}(x)=\int_{-\frac{1}{2}}^{\frac{1}{2}}e^{it\beta_2}g^{(k)}(t+x)\,dt\to 0$$

as $x \to \pm\infty$ for $k < 2l$, by the high contact condition for h. Also,

$$\sum_{j\in\mathfrak{Z}} f^{(2l)}(j+\delta_0) \leq \Delta^{2l} \sum_{j\in\mathfrak{Z}} \int_{-\frac{1}{2}}^{\frac{1}{2}} \sum_{k=0}^{2l} h^{(k)}(\Delta(t+j))\Delta\, dt$$
$$= \Delta^{2l} \sum_{k=0}^{2l} \int_{-\infty}^{\infty} h^{(k)}(\Delta t)\Delta\, dt = O(\Delta^{2l}).$$

This now proves the first claim of the theorem.

The characteristic function ζ_U of the smoothing variable U must now be calculated.

$$\zeta_U(\beta_2) = \sum_{j\in\mathfrak{Z}} \int_{-\frac{1}{2}}^{\frac{1}{2}} e^{it\beta_2} h(\Delta(t+j))\Delta\, dt = \sum_{j\in\mathfrak{Z}} f(j),$$

where

$$f(j) = \int_{-\frac{1}{2}}^{\frac{1}{2}} e^{it\beta_2} h(\Delta(t+j))\Delta\, dt.$$

The first term in the Euler-Maclaurin Summation Formula is

$$\int_{-\infty}^{\infty} f(x)\, dx = \int_{-\frac{1}{2}}^{\frac{1}{2}} e^{it\beta_2} [\int_{-\infty}^{\infty} h(\Delta(t+x))\Delta\, dx]\, dt = \int_{-\frac{1}{2}}^{\frac{1}{2}} e^{it\beta_2}\, dt = \frac{\sinh(\frac{1}{2}i\beta_2)}{\frac{1}{2}i\beta_2}.$$

Also, $f^{(k)}(x) = \Delta^k \int_{-\frac{1}{2}}^{\frac{1}{2}} e^{it\beta_2} h^{(k)}(\Delta(t+x))\Delta\, dt$. Again the integrability of derivatives of h insures that $\lim_{t\to\pm\infty} f^{(k)}(t) = 0$. Hence

$$\zeta_U(\beta_2) = \frac{\sinh(\frac{1}{2}i\beta_2)}{\frac{1}{2}i\beta_2} + \Delta^{2l} \sum_{j\in\mathfrak{Z}} \int_{-\frac{1}{2}}^{\frac{1}{2}} e^{it\beta_2} h^{(2l)}(\Delta(t+j))\Delta\, dt = \frac{\sinh(\frac{1}{2}i\beta_2)}{\frac{1}{2}i\beta_2} + O(\Delta^{2l}).$$

Q.E.D

3.16. Continuity-Corrected Edgeworth Series

The smoothing argument presented earlier suggests using an Edgeworth series with cumulants adjusted by Sheppard's corrections to approximate the lattice distribution F_Z. Kolassa and McCullagh (1990) prove that by evaluating such an Edgeworth expansion at continuity-corrected points, the resulting errors are as small as are usually obtained in an Edgeworth approximation.

Suppose $\{Y_i\}$ is an independent and identically distributed collection of random variables in $\mathfrak{R}$, with mean 0, variance σ^2, and cumulants $\boldsymbol{\kappa} = (\kappa_j, j = 1, 2, 3, \ldots)$. Suppose further that these variables are confined to the lattice $a + \Delta\mathfrak{Z}$ almost surely. Let $Z = \sum_{i=1}^{n} Y_i/\sqrt{n}$. Then Z has cumulants $\boldsymbol{\kappa}^n = (\kappa_j^n, j = 1, 2, 3, 4, \ldots)$, with $\kappa_j^n = \kappa_j n^{(2-j)/2}$. Let $\lambda_j^n = \kappa_j - \epsilon_j n^{-1}$ be the adjusted cumulants. Let $\boldsymbol{\lambda}^n$ denote the infinite vector $(0, \lambda_2^n, \lambda_3^n, \lambda_4^n, \ldots)$, and let $E_j(z; \boldsymbol{\lambda}^n)$ be the Edgeworth series defined in (33), with pseudo-moments as in (34), but with λ_j^n's in place of the κ_j^n's. Since

$\lambda_2^n - \sigma^2 = -1/12n$, the exponent on the left hand side of (34) now has a quadratic term.

Theorem 3.16.1: *When the cumulative distribution function of a lattice random variable is approximated using an Edgeworth series, with the cumulants adjusted according to Sheppard's Corrections, and the result is evaluated at continuity - corrected lattice points, the result is valid to the same order in n as in standard applications of the Edgeworth series; that is,* $F_Z(z^+) = E_{\rm j}(z^+; \boldsymbol{\lambda}^n) + o(n^{-\frac{{\rm j}-2}{2}})$ *for any lattice point z, with* $z^+ = z + \Delta/(2\sqrt{n})$, *and* $\Delta_n = \Delta/\sqrt{n}$.

Proof: The following is similar to the treatment of series expansions found in Chambers (1967), pages 368-371. The proof is completely algebraic and proceeds by the manipulations of power series generating functions for cumulants.

This proof exploits a relation between power series and Edgeworth-like series. Let $P[\beta]$ be the set of formal power series in the variable β, of form $\sum_{j=0}^{\infty} \alpha_j \beta^j$, where the coefficients α_j are polynomials in $1/\sqrt{n}$. Consider this set to contain all conceivable exponentiated differences of the true cumulant generating function from the approximating normal cumulant generating function, convergent or not; the α_j are the possible cumulant differences. Choose an ordinate z at which to evaluate an Edgeworth series. Let $\psi_{{\rm j},z} : P[\beta] \rightarrow \Re$ be the function that maps $\sum_{j=0}^{\infty} \alpha_j \beta^j$ to the Edgeworth-like series evaluated at z:

$$\alpha_0 \Phi(z, \kappa_2) - \phi(z, \kappa_2) \sum_{j=1}^{\infty} \alpha_j h_{j-1}(z, \kappa_2),$$

with all terms of size $o(n^{-\frac{{\rm j}-2}{2}})$ discarded. This will not be an actually Edgeworth series unless $\alpha_0 = 1$ and some other restrictions among the coefficients hold. Effectively, we apply the Fourier inversion operation termwise, and discard terms that are small enough, to obtain the Edgeworth series. The usual Edgeworth series for the standardized mean of independent and identically distributed random variables, with cumulants $\boldsymbol{\kappa}^n$, applies $\psi_{{\rm j},z}$ to

$$\exp\left[\sum_{j=3}^{k} \frac{1}{j!} \kappa_j^n \beta^j\right] \tag{53}$$

to get

$$E_{\rm j}(z, \boldsymbol{\kappa}^n) = \psi_{{\rm j},z}\left(\exp\left[\sum_{j=3}^{k} \frac{1}{j!} \kappa_j^n \beta^j\right]\right).$$

The function $\psi_{{\rm j},z}$ has the following linearity property:

$$\psi_{{\rm j},z}(\gamma_1 p_1(\beta) + \gamma_2 p_2(\beta)) = \gamma_1 \psi_{{\rm j},z}(p_1(\beta)) + \gamma_2 \psi_{{\rm j},z}(p_2(\beta)).$$

Also,

$$\psi_{{\rm j},z}(\beta^j p(\beta)) = (-1)^j D^j \psi_{{\rm j},z}(p(\beta)). \tag{54}$$

If $\boldsymbol{\kappa}^n$ is replaced with $\boldsymbol{\lambda}^n$, then

$$E_j(z, \boldsymbol{\lambda}^n) = \psi_{j,z}\left(\exp\left[\tfrac{1}{2}(\lambda_2 - \kappa_2)\beta^2 + \sum_{j=3}^{j} \frac{1}{j!}\lambda_k^n \beta^j\right]\right) + o(n^{-\frac{j-2}{2}}).$$

Exact equality does not hold, since some of the terms arising from corrections to the cumulants are of small enough order to be included in the error. For instance, when $\kappa_1 = 0$, and $\kappa_2 = 1$,

$$E_4(z^+, \boldsymbol{\lambda}^n) = \Phi(z^+) - \phi(z^+)\left(\frac{h_2(z^+)\kappa_3}{\sqrt{n}} - \frac{h_1(z^+)\epsilon_2}{n} + \frac{h_3(z^+)}{n}\left(\frac{\epsilon_2^2}{8n} + \frac{\epsilon_4}{24n} + \frac{\kappa_4}{24}\right)\right),$$

but by definition $\psi_{4,z}\left(\exp\left[\frac{1}{2}(\lambda_2 - \kappa_2)\beta^2 + \sum_{j=3}^{4} \frac{1}{j!}\lambda_k^n \beta^j\right]\right)$ omits the terms of size $o(n^{-1})$. The expression $E_4(z^+, \boldsymbol{\lambda}^n)$ has more terms than $E_4(z^+, \boldsymbol{\kappa}^n)$, since second cumulants are no longer matched; instead, following Chambers (1967), pick the baseline variance to be the asymptotic variance.

Esseen's series (48) is a linear combination of derivatives of the Edgeworth series. The periodicity of the functions B_j causes all of the multipliers for these derivatives to be independent of the specific continuity-corrected point under consideration. Using (54) express the derivatives of the Edgeworth series as ψ_{j,z^+} applied to a multiple of a power of β times the power series (53) generating the Edgeworth approximation. Using the linearity of ψ_{j,z^+}, express this as ψ_{j,z^+} applied to the product of a new series times the series (53). Noting that this new series can be expressed as the exponential of the series with Sheppard's corrections completes the proof.

Consider Esseen's series:

$$F_Z(z) = E_j(z, \boldsymbol{\kappa}^n) + \sum_{j=1}^{j-2} \Delta_n^j \frac{1}{j!} B_j\left(\frac{(z - z_n)}{\Delta_n}\right)(-1)^j \frac{d^j}{dx^j} E_j(z, \boldsymbol{\kappa}^n) + o(n^{-\frac{j-2}{2}}).$$

Choose any integer J. Let $z^+ = (na + J\Delta + \Delta/2)/\sqrt{n}$. Then

$$(z^+ - z_n)/\Delta_n = J - \lfloor na/\Delta \rfloor + \tfrac{1}{2},$$

which is an integer plus half. Hence for all positive integers j, $B_j\left((z^+ - z_n)/\Delta_n\right) = B_j(\frac{1}{2})$, by the periodicity of B_j. Straight-forward calculations show that

$$B_j(0) = B_j \text{ and } B_j(\tfrac{1}{2}) = (2^{1-j} - 1)B_j, \tag{55}$$

where again, B_j is the Bernoulli number of order j. Then

$$F_Z(z^+) = \sum_{j=0}^{\mathrm{j}-2} n^{-j/2}\frac{1}{j!}B_j g_j(2^{1-j}-1)D^j E_{\mathrm{j}}(z^+;\boldsymbol{\kappa}^n) + o(n^{-\frac{\mathrm{j}-2}{2}})$$
$$= \sum_{j=0}^{\mathrm{j}-2} n^{-j/2}\frac{1}{j!}B_j(2^{1-j}-1)(-1)^j\psi_{\mathrm{j},z^+}\left(\beta^j\exp\left[\sum_{j=3}^{k}\frac{1}{j!}\kappa_j^n\beta^j\right]\right) + o(n^{-\frac{\mathrm{j}-2}{2}})$$
$$= \psi_{\mathrm{j},z^+}\left(\sum_{j=0}^{\mathrm{j}-2} n^{-j/2}\frac{1}{j!}B_j(2^{1-j}-1)\beta^j\exp\left[\sum_{j=3}^{k}\frac{1}{j!}\kappa_j^n\beta^j\right]\right) + o(n^{-\frac{\mathrm{j}-2}{2}})$$
$$= \psi_{\mathrm{j},z^+}\left(\exp\left[\sum_{j=2}^{\infty}\frac{-B_j}{j\,j!}n^{-\frac{j}{2}}\beta^j\right]\exp\left[\sum_{j=3}^{k}\frac{1}{j!}\kappa_j^n\beta^j\right]\right) + o(n^{-\frac{\mathrm{j}-2}{2}})$$

since

$$\sum_{s=0}^{\infty} n^{-s/2}(2^{1-s}-1)\frac{1}{s!}B_s\beta^s = \frac{s\exp[s/2]}{\exp[s]-1} = \exp\left[\sum_{j=2}^{\infty}\frac{-B_j}{j\,j!}n^{-j/2}\beta^j\right]. \qquad (56)$$

The last equality in (56) follows directly from Exercise 9 of Chapter 2. The first equality will be proved as part of the exercises in this chapter. Powers of (-1) drop out of the second of these equations, because the odd Bernoulli numbers are zero. Sheppard's correction for the cumulant of order j is $-B_j/(jn^{j/2})$. Let $\boldsymbol{\lambda}^n = (\lambda_k^n)$ be the cumulants $\boldsymbol{\kappa}^n$ adjusted by Sheppard's correction.

Hence

$$F_Z(z^+) = \psi_{\mathrm{j},z^+}\left(\exp\left[\sum_{j=3}^{\mathrm{j}}\frac{1}{j!}\lambda_k^n\beta^j\right]\right) + o(n^{-\frac{\mathrm{j}-2}{2}}) = E_{\mathrm{j}}(z^+,\boldsymbol{\lambda}^n) + o(n^{-\frac{\mathrm{j}-2}{2}}).$$

Q.E.D

3.17. Exercises

1. Justify the first equality in (56). Distributing two terms in the factor $(2^{1-s}-1)$ into two separate infinite series should help.
2. Justify the first equality (55) by evaluating the Fourier series expressions (47) for $B_k(0)$, substituting this expression in for B_k on the right hand side of (29), and showing that the result is the left hand side of (29).
3. Justify the second equality(55) by evaluating the Fourier series expressions expression (47) for $B_k(\frac{1}{2})$, and comparing it to the similar expansion for $B_k(0)$.
4. Refer to the example of calculating an Edgeworth expansion for a $\Gamma(p,\lambda)$ random variable, as in §3.10. Prove that $\int_{-\infty}^{\infty}|\zeta(\beta)|^k\,d\beta < \infty$ if and only if $k > 1/p$.
5. Justify relation (36).
6. Justify claim (52).

7. Analytically or numerically calculate the cumulative distribution function for the convolution of 3, 4, 5, and 6 random variables uniform on $[0,1]$. Compare tail probabilities one, two, and three standard deviations above the mean with the Edgeworth estimates.
8. Let F_n be the cumulative distribution function of the sum of n independent random variables each with cumulative distribution function
$$F(x) = \begin{cases} 0 & \text{if } x < 0 \\ \Phi(\sqrt{x}) & \text{otherwise} \end{cases}.$$
Kolassa (1992) constructs a test statistic $S(\boldsymbol{X}, \boldsymbol{\theta})$ whose cumulative distribution function is bounded by F_n if the parameter vector $\boldsymbol{\theta}$ governed a mechanism generating a set of data summarized by $\boldsymbol{X}$ and consisting of n data points. If s satisifes $F_n(s_n) = .95$ then $\{\boldsymbol{\theta} | S(\boldsymbol{X}, \boldsymbol{\theta}) \leq s_n\}$ represents a 95% confidence region for $\boldsymbol{\theta}$. Calculate s_5, s_10, and s_20 numerically, and compare with the Cornish–Fisher expansions to orders $o(1)$, $o(1/\sqrt{n})$, and $o(1/n)$. Also, compare bounds on the coverage rates of the resulting regions when s_n is calculated using the normal approximation, by applying the Edgeworth approximation incorporating the kurtosis term.
9. Compare the accuracies of Cornish-Fisher and normal approximations to the Mann-Whitney-Wilcoxon statistic (Hettmansperger, 1984) percentage points. These approximations are available as FORTRAN subroutines in the statistical archive STATLIB. To receive these subroutines, send two electronic mail messages to the address `statlib@lib.stat.cmu.edu` with the subjects "send 62 from apstat" and "send 234 from apstat", two receive algorithms published in the journal *Applied Statistics.* You will receive algorithm 62 for calculating the distribution of the Mann-Whitney-Wilcoxon statistic, and 234 for calculating the Cornish Fisher inversion of the Mann-Whitney-Wilcoxon statistic. Compare the accuracy of the various Cornish–Fisher approximations for a variety of sample sizes.

4
Saddlepoint Series for Densities

In many statistical applications, approximations to the probability that a random variable exceeds a certain threshold value are important. Such approximations are useful, for example, in constructing tests and confidence intervals, and for calculating p-values. Edgeworth series converge uniformly quickly over the entire possible range of the random variable, when error is measured in an absolute sense. Often times, relative error behavior is more important than absolute error behavior; an error of .005 is of little importance when considering tests of approximate size .05 but is of great importance when considering tests of approximate size .001. Saddlepoint methodology is a method for achieving in many cases uniform bounds on relative error over the range of the distribution. This work was pioneered by Daniels (1954).

This chapter considers the problem of estimating the density of a random variable X, whose distribution depends on n, and whose cumulant generating function is represented by $\mathcal{K}_X$. Throughout this chapter implicitly assume that such a density exists for sufficiently large n. This condition will be strong enough for the introductory material heuristically motivating the saddlepoint density approximation as a rescaled approximation to a member of the natural exponential family containing the true density. When motivating the saddlepoint density as the asymptotic expansion of a complex integral over a properly deformed path, integrability of the characteristic function for sufficiently large n will be assumed, as was done for Edgeworth series density theorems. Recall that the equivalence of these assumptions was given in §2.4. Additional regularity conditions will be added when considering uniform relative error behavior of the saddlepoint approximation.

The saddlepoint approximation is first explored heuristically using an exponential family, and a parallel is developed between the saddlepoint and the maximum likelihood estimator. The question of existence of the saddlepoint is addressed. The saddlepoint approximation is derived using saddlepoint integration techniques applied to the density inversion integral. Advantages of saddlepoint techniques including uniform relative error are discussed. Possibilities of improving the accuracy of the saddlepoint approximation by determining a multiplicative correction to normalize the approximate density to integrate to unity are considered. Finally, numeric hints for applying saddlepoint approximations are given.

The following chapter will concern saddlepoint distribution function approximations. Dividing material on saddlepoint approximations into two chapters, the first for density approximations and the second for distribution function approximations, while Edgeworth density and distribution function approximations are included in one chapter, may appear as a clumsy disruption of a natural parallelism between Edgeworth and saddlepoint approximations. I have chosen to organize the material in this manner precisely because while saddlepoint density approximations may be derived from Edgeworth density approximations, the same does not hold for distribution function approximations. In general Edgeworth distribution function approximations can be expressed as integrals of Edgeworth density approximations, and in general saddlepoint density approximations may not be analytically integrated.

The inversion results presented here will have their primary applications in the following chapter.

4.1. Expression Using Exponential Families

Suppose that an approximation to a density $f_X(x)$ is desired, based on its cumulant generating function $\mathcal{K}_X(\gamma)$. Suppose further that the distribution of X depends on a parameter n, and that one desires to approximate this density far in the tails. Edgeworth series may work poorly in this case. An alternative is to embed the density of interest in an exponential family, and choose a density in this exponential family to estimate. Specifically, let $f_X(x, \gamma) = \exp(x\gamma - \mathcal{K}_X(\gamma))f_X(x)$. Here the density of interest is embedded in an exponential family having the cumulant generating function evaluated at v: $\mathcal{K}_X(v+\gamma) - \mathcal{K}_X(\gamma)$. An approximation for any member of this family easily results in an approximation for the original density, since members of this family differ only by a factor of $\exp(x\gamma - \mathcal{K}_X(\gamma))$. Next one determines which member of this family is easiest to approximate. Recall that Edgeworth series are most accurate for ordinates near the mean, and least accurate for ordinates in the tails. One would expect that the easiest member of this family to approximate is the one whose mean is equal to the ordinate at which we wish to evaluate the approximation. The mean of the generic member of this family whose parameter is γ is $\mathcal{K}'_X(\gamma)$. Let $\hat{\gamma}$ be the parameter associated with the chosen member of the exponential family. Then $\hat{\gamma}$ is given by the solution to $\mathcal{K}'_X(\hat{\gamma}) = x$. One might think of this solution as the maximum likelihood estimator for the exponential family model based on the data x.

Once the element of the exponential family to be approximated is chosen, one much choose an approximation for it. Start with the normal approximation to $f_X(x; \hat{\gamma})$:

$$\exp(-(x - \mathcal{K}'_X(\hat{\gamma}))^2/2)/\sqrt{2\pi\mathcal{K}''_X(\hat{\gamma})} = 1/\sqrt{2\pi\mathcal{K}''_X(\hat{\gamma})},$$

implying the approximation to $f_X(x)$:

$$g_X(x) = \exp(\mathcal{K}_X(\hat{\gamma}) - \hat{\gamma}x)/\sqrt{2\pi\mathcal{K}_X''(\hat{\gamma})}. \tag{57}$$

Although the standard normal approximation to $f_X(x;\hat{\gamma})$ is used, the resulting approximation to $f_X(x)$ uses information from all of the cumulants, because of the way in which $\hat{\gamma}$ is chosen. Unlike the higher-order Edgeworth expansions, this approximation is fortunately always positive. This approximation has a very simple form, but often not calculable explicitly, since an analytic expression for the saddlepoint $\hat{\gamma}$ is not always available.

To improve the accuracy of (57) one might use a higher-order Edgeworth expansion to $f_X(x;\hat{\gamma})$:

$$\begin{aligned} g_X(x) &= \frac{\exp(K_X(\hat{\gamma}) - \hat{\gamma}x)}{\sqrt{K_X''(\hat{\gamma})}}\phi(0)\sum_{j=0}^{\infty} h_j(0)\mu_j^*(\hat{\gamma})/j! \\ &= \frac{\exp(K_X(\hat{\gamma}) - \hat{\gamma}x)}{\sqrt{K_X''(\hat{\gamma})}}\phi(0)\Big[1 + \frac{\rho_3^X h_3(0)}{6} + \left(\frac{\rho_4^X h_4(0)}{24} + \frac{10\rho_3^{X\,2} h_6(0)}{720}\right) \\ &\quad + \left(\frac{\rho_5^X h_5(0)}{120} + \frac{35\rho_3^X\rho_4^X h_7(0)}{5040}\right) + \cdots\Big] \\ &= \frac{\exp(K_X(\hat{\gamma}) - \hat{\gamma}x)}{\sqrt{K_X''(\hat{\gamma})}}\phi(0)[1 + \rho_4^X/8 - 5\rho_3^{X\,2}/24 + \cdots] \end{aligned} \tag{58}$$

using $h_4(0) = 3, h_6(0) = -15$. Here $\rho_j^X = \mathcal{K}_X^{(j)}(\hat{\gamma})/(\mathcal{K}_X^{(2)}(\hat{\gamma}))^{j/2}$.

The above approximation (57) is derived from the cumulant generating function $\mathcal{K}_X$ for the associated random variable. The nature of the dependence of $\mathcal{K}_X$ on n was not important in this heuristic development. When results involving any mathematical rigor are desired, however, assumptions about the form of $\mathcal{K}_X$, or almost equivalently about the distribution of X, are needed. The rigorous development for the Edgeworth series of §§3.7 and 3.8 involved standardized sums of independent and identically distributed random variables. Limit theorems for other standardized sums were also considered in §3. When developing the saddlepoint approximations, it will be useful to take the mean of independent and identically distributed random variables as the prototypical distribution to be approximated. Reasons for this will be given shortly.

When the distribution of X, the mean of independent and identically distributed copies of a random variable Y, is desired, usually an expression in terms of the cumulant generating function of the individual addends is preferred. Since $\mathcal{K}_X(\gamma) = n\mathcal{K}_Y(\gamma/n)$, the saddlepoint $\hat{\gamma}$ satisfies $\mathcal{K}_Y'(\hat{\gamma}/n) = x$, and the approximation (57) becomes $g_X(x) = \sqrt{n}\exp(n[\mathcal{K}_Y(\hat{\gamma}/n) - (\hat{\gamma}/n)x])/\sqrt{2\pi\mathcal{K}_Y''(\hat{\gamma}/n)}$. The leading factor $\sqrt{n}$ arises from $\sqrt{\mathcal{K}''_X(\hat{\gamma}/n)}$. If $\hat{\beta} = \hat{\gamma}/n$, then the usual saddlepoint density approximation is obtained:

$$g_X(x) = \sqrt{n}\exp(n[\mathcal{K}_Y(\hat{\beta}) - \hat{\beta}x])/\sqrt{2\pi\mathcal{K}''_Y(\hat{\beta})}. \tag{59}$$

The saddlepoint $\hat{\beta}$ is given by the solution to

$$\mathcal{K}'_Y(\hat{\beta}) = x. \tag{60}$$

For Edgeworth series the natural scaling for sums of random variables was the standardized sum giving constant variance as n increases. For saddlepoint approximations the natural scaling is the mean, since it gives an expression for $\hat{\beta}$ independent of n, and because the term in the exponent in (59) is n times a factor independent of n. The saddlepoint may be reexpressed in terms of the Legendre transform $\mathcal{K}^*_Y$ of $\mathcal{K}_Y$: For x in the range of $\mathcal{K}'_Y$ define $\mathcal{K}^*_Y(x) = \hat{\beta}x - \mathcal{K}_Y(\hat{\beta})$ for $\hat{\beta}$ satisfying (59). Then $g_X(x) = \sqrt{n}\exp(-n\mathcal{K}^*_Y(x))/\sqrt{2\pi\mathcal{K}''_Y(\hat{\beta})}$. Since $(d/dx)\mathcal{K}^*_Y(x) = \hat{\beta}(x)$ and $d\hat{\beta}/dx = 1/\mathcal{K}''_Y(\hat{\beta})$, then

$$(d^2/dx^2)\mathcal{K}^*_Y(x) = 1/\mathcal{K}''_Y(\hat{\beta}) > 0$$

for all x. Hence $\mathcal{K}^*_Y$ strictly concave as a function of x. These relations will be useful when assessing the error in the renormalized saddlepoint density approximation.

Since $\rho^X_j = n^{1-j}\mathcal{K}^{(j)}_Y(\hat{\beta})/(n^{-1}\mathcal{K}^{(2)}_Y(\hat{\beta}))^{j/2}$, the counterpart of (58) in terms of $\mathcal{K}_Y$ is:

$$f_X(x) = \sqrt{n}\frac{\exp(n[\mathcal{K}_Y(\hat{\beta}) - \hat{\beta}x])}{\sqrt{2\pi\mathcal{K}''_Y(\hat{\beta})}}[1 + (\hat{\rho}_4/8 - 5\hat{\rho}_3^2/24)/n + O(n^{-2})]. \tag{61}$$

Here

$$\hat{\rho}_j = \rho_j(\hat{\beta}) = \mathcal{K}^{(j)}_Y(\hat{\beta})/(\mathcal{K}^{(2)}_Y(\hat{\beta}))^{j/2}. \tag{62}$$

This justifies the following result:

Theorem 4.1.1: *Suppose that f_Y is a density function for a random variable Y, with characteristic function $\mathcal{K}_Y$ defined in an open interval $\boldsymbol{I}$ containing zero, and one desires an approximation to the density f_X of the mean of n independent and identically distributed copies of Y. Suppose that equation (60) has a solution in $\boldsymbol{I}$. Then the expansion (61) represents a valid asymptotic expression for f_X. Furthermore, general higher order terms in (61) are generated using an easy algorithm:*

1. *The saddlepoint $\hat{\beta}$ is determined.*
2. *The cumulant generating function $n\mathcal{K}_Y(\beta/(\sqrt{n}\sqrt{\mathcal{K}''_Y(\hat{\beta})}))$ is expanded about $\hat{\beta}$.*
3. *The quadratic term is removed.*
4. *The resulting series is exponentiated, to form a bivariate series for*

$$\exp\left(n\mathcal{K}_Y\left(\beta/(n\sqrt{\mathcal{K}''_Y(\hat{\beta})})\right) - (\beta - \hat{\beta})^2/2\right)$$

in β about $\hat{\beta}$, and in $1/\sqrt{n}$ about 0.

5. *Powers of $\beta - \hat{\beta}$ are replaced by the corresponding Hermite polynomial evaluated at 0.*

Proof: As this is merely the result of the Edgeworth series theorem for density functions, Theorem 3.5.1, applied to the density proportional to $\exp(\beta x) f_Y(x)$, it suffices to verify the conditions of Theorem 3.5.1. Moments of all orders for this distribution exist, by the arguments of §2.1. Furthermore, since the density f_Y exists, by Lemma 2.4.3, the integrability condition is also fulfilled.

Q.E.D

Here the cumulants of the standardized distribution are varying, and the point at which the Hermite polynomials are evaluated is fixed. Hence uniform asymptotic properties do not follow from the Edgeworth theorem already proved, since the distributions that the Edgeworth series approximation theorems are applied to vary with both x and n. We can say that the series is valid asymptotically as an approximation for the density of X at x as $n \to \infty$ and x changes so as to keep the saddlepoint fixed. In the case of an approximation to the mean of independent and identically distributed random variables, this corresponds to fixing the value of the mean as n varies. Contrast this with the Edgeworth series approximation, in which the value of $n^{-1/2} \sum_j Y_j$ is held fixed.

Later this chapter conditions will be given under which far more can be said about asymptotic bounds on the error for approximations of the form (57), or extensions of the form (61). Heuristically this works because under lax assumptions the standardized cumulants are bounded as $\hat{\beta}$ varies. Hence good relative error behavior rather than merely absolute error behavior can be expected. Recall that for the Edgeworth series, as the ordinate varied over $\Re$, the point at which Hermite polynomials were evaluated also varied over $\Re$, inducing poor relative error properties. Kolassa (1991) shows that an approximate solution to the saddlepoint equation (60), with consequent modifications to (61), provides a compromise between Edgeworth and saddlepoint methodologies, in that good relative error properties are preserved while possibly intractable equations are avoided.

4.2. Examples

The following three examples are cited by Daniels (1954): Consider the saddlepoint density approximation to means of independent and identically distributed random variables, each with a normal distribution with mean μ and unit variance. The cumulant generating function for each summand is $\mathcal{K}_Y(\beta) = \mu\beta + \beta^2/2$, and hence the saddlepoint satisfies $\hat{\beta} + \mu = x$. Then $\hat{\beta} = x - \mu$. Then approximation is $g_X(x) = \exp(-n(x-\mu)^2/2)/n\sqrt{2\pi}$. In this case, the approximation is exactly the correct distribution.

As a further example, consider the gamma distribution. Recall that a $\Gamma(\lambda, p)$ random variable has density

$$f_X(x) = \lambda^p x^{p-1} \exp(-\lambda x)/\Gamma(p)$$

and the cumulant generating function $\mathcal{K}_X(\beta) = p\log(\lambda) - p\log(\lambda - \beta)$. The saddlepoint $\hat{\beta}$ satisfies $x = -p/(\lambda - \hat{\beta})$, and hence $\hat{\beta} = \lambda - p/x$. Then $\mathcal{K}''_X(\hat{\beta}) = p/(\lambda - \hat{\beta})^2 = x^2/p$, and

$$\begin{aligned} g_X(x) &= \sqrt{n}\frac{\exp(n(p\log(\lambda) - p\log(\lambda - \lambda + p/x) - x\lambda + p))}{x\sqrt{2\pi/p}} \\ &= \sqrt{n}\lambda^{np}p^{-np}x^{np-1}\exp(np - x\lambda n)\sqrt{p/2\pi} \\ &= (n\lambda)^{np}x^{np-1}\exp(np - nx\lambda)(np)^{-np+1/2}/\sqrt{2\pi} \end{aligned}$$

From the cumulant generating function one finds that the convolution of n densities of the form $\Gamma(p, \lambda)$ is a $\Gamma(np, \lambda)$ density, and hence the mean has a $\Gamma(np, n\lambda)$ distribution. The saddlepoint approximation agrees with the exact density except that the gamma function is replaced by its Sterling approximation $\Gamma(np) \approx \exp(-np)(np)^{np-1/2}\sqrt{2\pi}$.

As a final example, the moment generating function of the uniform distribution on $(-1/2, 1/2)$ is

$$\mathcal{M}_U(\beta) = (\exp(\beta/2) - \exp(-\beta/2))/(\beta/2) = \sinh(\beta/2)/\beta,$$

and the cumulant generating function $\mathcal{K}_U$ is the logarithm of $\mathcal{M}_U$. Then

$$\mathcal{K}'_U(\beta) = (\exp(\beta/2) + \exp(-\beta/2))/(\exp(\beta/2) - \exp(-\beta/2)) - 1/\beta$$

and $\mathcal{K}''_U(\beta) = 1/\beta^2 - 4/(\exp(\beta/2) - \exp(-\beta/2))^2$. The solution to the saddlepoint equation must either be approximated, or obtained numerically. Here the latter approach is taken. Results show that far in the tails saddlepoint outperforms both the normal and Edgeworth-4 approximations. See Fig. 8.

4.3. Conditions for the Existence of the Saddlepoint

The preceding discussion presupposes the existence of $\hat{\beta}$ solving (60). Daniels (1954) investigates under what circumstances such a $\hat{\beta}$ exists. These conditions will depend on the form of the interval Ω on which the cumulant generating function $\mathcal{K}_Y$ is defined, and also on the form of the interval of support for F_Y. Express Ω as an interval with end points $-c_1 \leq c_2$ where c_1 and c_2 are non-negative. Saddlepoint approximation methods are only considered when these endpoints are not both zero. Express the interval of support for the distribution F_Y as $[-x_1, x_2]$, where values of ∞ for x_1 or x_2 or both are allowed. Then a solution of the saddlepoint equation exists for all values of $x \in (-x_1, x_2)$ if

$$\lim_{\beta \to -c_1} \mathcal{K}'_Y(\beta) = -x_1, \quad \lim_{\beta \to c_2} \mathcal{K}'_Y(\beta) = x_2, \tag{63}$$

since $\mathcal{K}'_Y$ is continuous, and $\mathcal{K}''_Y > 0$, and hence $\mathcal{K}'_Y$ is strictly increasing. Define the conditions $C_i : \lim_{\beta \to (-1)^i c_i} \mathcal{K}'_Y(\beta) = (-1)^i x_i$. Define the conditions C'_i as $c_i = \infty$. Note that C'_i is implied by $x_i < \infty$. As we will see below, the condition $x_i < \infty$

Errors in Approximations to Densities for the Mean of 6 Uniform[-1,1].

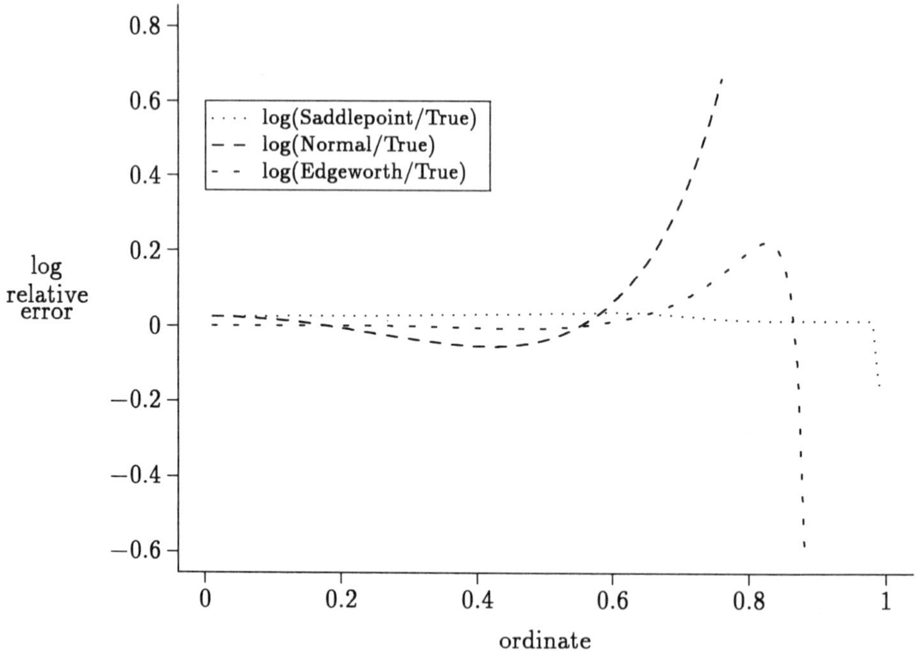

Fig. 8.

forces $c_i = \infty$; however, the following lemma and theorem will be stated in terms of the slightly more immediate conditions. The following lemma holds.

Lemma 4.3.1: *Condition C_i' implies C_i for $i = 1, 2$.*

Proof: This lemma is proved for $i = 2$; the case when $i = 1$ is similar. Consider the two cases.

1. x_2 is finite. Note that

$$x_2 - \mathcal{K}_Y'(\beta) = \int_{-x_1}^{x_2} (x_2 - y)\exp(\beta y)\, dF(y) / \int_{-x_1}^{x_2} \exp(\beta y)\, dF(y) \geq 0.$$

Showing that $\lim_{\beta\to\infty} x_2 - \mathcal{K}_Y'(\beta) \leq 0$ implies C_2 in this case. For any $z < x_2$,

$$\begin{aligned} x_2 - \mathcal{K}_Y'(\beta) &\leq \frac{(x_2 - z)\int_z^{x_2} \exp(\beta(y-z))\, dF(y) + \int_{-x_1}^{z} (x_2 - y)\exp(\beta(y-z))\, dF(y)}{\int_z^{x_2} \exp(\beta(y-z)) dF(y) + \int_{-x_1}^{z} \exp(\beta(y-z)) dF(y)} \\ &= \frac{(x_2 - z) + \int_{-x_1}^{z} (x_2 - y)\exp(\beta(y-z))\, dF(y) / \int_z^{x_2} \exp(\beta(y-z))\, dF(y)}{1 + \int_{-x_1}^{z} \exp(\beta(y-z)) dF(y) / \int_z^{x_2} \exp(\beta(y-z))\, dF(y)} \\ &\to x_2 - z \end{aligned}$$

Hence $\lim_{\beta\to\infty} \mathcal{K}'_Y(\beta) = x_2$.

2. $x_2 = \infty$. Then $\mathcal{K}'_Y(\beta) = \int_{-x_1}^{\infty} y \exp(\beta y)\, dF(y)$. Set $a^* = \int_0^\infty y^2\, dF(y) / \int_0^\infty y\, dF(y)$. Then

$$\mathcal{K}'_Y(\beta) \geq \int_{-x_1}^{0} y \exp(\beta y)\, dF(y) + \int_0^\infty y\, dF(y) \exp(\beta a^*).$$

The limit of the first term is zero as $\beta \to \infty$, and the second term diverges to ∞.

Q.E.D

This lemma leads immediately to the following theorem:

Theorem 4.3.2: *If X is a random variable supported on the possibly infinite interval $[-x_1, x_2]$, whose cumulant generating function is defined on an interval with end points $-c_1$ and c_2, and if for $i = 1$ and 2, $x_i < \infty$ or $c_i = \infty$, then a solution of the saddlepoint equation exists for all values of $x \in (-x_1, x_2)$.*

Proof: Conditions C'_1 and C'_2 implies (63), by Lemma 4.3.1.

Q.E.D

Daniels (1954) presents an example in which these conditions fail.

In this section we have considered only solutions to the saddlepoint approximation for $x \in (-x_1, x_2)$. If $x = -x_1$ or $x = x_2$ one may take the density approximation to be 0, since x sits at the edge of the support of the distribution. When considering cumulative distribution function approximations this choice will be even easier, since by definition the cumulative distribution function evaluated at x will be 0 if $x = -x_1$ and 1 if $x = x_2$. When considering double saddlepoint approximations to conditional densities and distribution functions, other cases with conditioning random variables on the edges of their ranges of definition will be more troublesome. This question will be addressed later.

4.4. Construction of the Steepest Descent Curve

The saddlepoint density approximations (57) and (61) for a random variable X with characteristic function $\zeta_X(\gamma)$ and cumulant generating function $\mathcal{K}_X(\gamma)$ may be derived by using the Fourier inversion formula and approximating the integral involved. Recall from (15) that

$$f_X(x) = \frac{1}{2\pi} \int_{-i\infty}^{+i\infty} \zeta_X(\gamma) \exp(-i\gamma x)\, d\gamma = \frac{1}{2\pi i} \int_{-i\infty}^{+i\infty} \exp(\mathcal{K}_X(\gamma) - \gamma x)\, d\gamma. \quad (64)$$

The variable of integration was changed from β to γ, since later in this section the scale of this variable of integration will be changed, and the variable β will be reserved for this changed scale to conform more closely to standard notation. This expression will be manipulated to yield (61). First, denote the real domain of $\mathcal{K}_X$ by $\boldsymbol{I}_X$. Then within the set $\boldsymbol{I}_X \times i\Re$ the integrand above is differentiable, and by standard complex analytic results the path of integration can be chosen to be any path from $-i\infty$ to $+i\infty$, by the closed curve theorem of complex variables (Bak and Newman (1982), §8.1). That is, the path of integration need not be along the complex axis, but along any path diverting to positive and negative infinity along the complex axis, but having perhaps non-zero real components.

The next step is to show that the path of integration in (64) can be replaced by a path along which the real value of the variable of integration is fixed at any value in the range of definition of the cumulant generating function; that is, that

$$\lim_{T\to\infty}\int_{-iT}^{iT}\exp(\mathcal{K}_X(\gamma)-\gamma x)\,d\gamma=\lim_{T\to\infty}\int_{c-iT}^{c+iT}\exp(\mathcal{K}_X(\gamma)-\gamma x)\,d\gamma \tag{65}$$

for all $c\in \boldsymbol{I}_X$. By the closed curve theorem, all that is required is to show that

$$\lim_{T\to\infty}\int_{iT}^{c+iT}\exp(\mathcal{K}_X(\gamma)-\gamma x)\,d\gamma=0 \text{ and } \lim_{T\to\infty}\int_{-iT}^{c-iT}\exp(\mathcal{K}_X(\gamma)-\gamma x)\,d\gamma=0. \tag{66}$$

At each point γ along the two paths of integration parallel to the real axis, the integrand is the characteristic function of a random variable with density $f_X(y-x)\times\exp(-\mathcal{K}_X(\Re(\gamma))+\Re(\gamma)(y-x))$, evaluated at $\Im(\gamma)$. Here y is a dummy argument for the density, and $\Im(\gamma)=\pm T$. By the Riemann-Lebesgue Theorem of §3.9, this is bounded by $|T|^{-1}$. Hence the integrands of (66) converge uniformly to zero, and so the limits in (66) hold.

Use of steepest descent methods requires that (66) be manipulated so that the exponent of the resulting integrand is approximately quadratic near $\hat{\beta}$. Calculations begin with an expression of the density inversion integral (15) in terms of the cumulant generating function for one summand in the standardized mean of independent and identically distributed components:

$$\begin{aligned} f_X(x)&=\frac{1}{2\pi i}\int_{-i\infty}^{+i\infty}\exp(\mathcal{K}_X(\gamma)-\gamma x)\,d\gamma\\ &=\frac{n}{2\pi i}\exp(n[\mathcal{K}_Y(\hat{\beta})-\hat{\beta}x])\int_{-i\infty}^{+i\infty}\exp(n[\mathcal{K}_Y(\beta)-\beta x-\mathcal{K}_Y(\hat{\beta})+\hat{\beta}x])\,d\beta. \end{aligned} \tag{67}$$

The second equality above follows from the change of variables $\gamma=\beta n$. Relation (65) ensures that (67) can be replaced by

$$f_X(x)=\frac{n\exp(n[\mathcal{K}_Y(\hat{\beta})-\hat{\beta}x])}{2\pi i}\int_{\hat{\beta}-i\infty}^{\hat{\beta}+i\infty}\exp(n[\mathcal{K}_Y(\beta)-\beta x-\mathcal{K}_Y(\hat{\beta})+\hat{\beta}x])\,d\beta. \tag{68}$$

Since the integrand is a characteristic function evaluated at $\Im(\beta)$, when this integration is performed along the path parallel to the complex axis, the result recovered is the Edgeworth series times the leading factor, yielding (61).

The preceding heuristic development of the saddlepoint series implies that a better path of integration runs through the previously defined saddlepoint $\hat{\beta}$. Relation (65) shows that this integration may be performed along the path $\Re(\beta)=\hat{\beta}$. As will be shown later in this section, the integrand in (66) along this path is the characteristic function of the tilted distribution used in the heuristic development of the saddlepoint density approximation in §4.1, and the result will be (61) as before.

This derivation of the saddlepoint series using the shifted vertical path of integration will give the sharpest results, and will be crucial when considering uniformity in the behavior of approximation errors. Never the less, one might ask whether local changes in the path of integration might produce sharper results. To answer this question by showing that the heuristically optimal results are (61), to exhibit the saddlepoint development as introduced by Daniels (1954), and to set the stage for further developments in which these methods will be very useful, this section and the next present the development of saddlepoint series using the method of steepest descent.

This approach can be motivated from a purely analytic point of view, by recourse to saddlepoint integration techniques (Mathews and Walker, 1964). Pick $\hat{\beta}$ along the real axis minimizing $\mathcal{K}(\beta) - \beta x$, and pick the path through $\hat{\beta}$ along which the real part of $\mathcal{K}(\beta) - \beta x$ decreases the fastest as one moves away from the real axis, at least near $\hat{\beta}$. This path is called the path of steepest descent. As will be explained below, along this path $\mathcal{K}(\beta) - \beta x$ is real. See Fig. 9. The steepest descent curve in this figure is denoted by the most finely dotted line. The surface of the real part of $\mathcal{K}$ above the complex plane then has paths through $\hat{\beta}$ along which it rises in either direction, and paths along which it falls in either direction. The surface has the shape of a saddle, giving the name saddlepoint to $\hat{\beta}$. Using the same ideas as found in the proof of the asymptotic validity of Edgeworth series, as n increases, the contribution to the integral of any bit outside of an increasingly small circle about the saddlepoint vanishes exponentially, and the remainder is determined by the derivatives of $\mathcal{K}$ at $\hat{\beta}$.

The existence of a steepest descent curve must be verified. The quantity in the exponent of (68),

$$h(\beta) = n[\mathcal{K}_Y(\beta) - \beta x - \mathcal{K}_Y(\hat{\beta}) + \hat{\beta} x],$$

is an analytic function that is zero and has zero first derivative at $\hat{\beta}$, by the choice of $\hat{\beta}$.

Steepest descent methods involve integration along paths through $\hat{\beta}$ along which the real part decreases fastest near $\hat{\beta}$. The first objective is to determine whether such paths exist. Specifically, a function $\omega : \mathfrak{R} \to \mathfrak{C}$ must be constructed that is continuous and piecewise differentiable, such that $\lim_{a \to -\infty} \omega(a) = \hat{\beta} - i\infty$, $\lim_{a \to +\infty} \omega(a) = \hat{\beta} + i\infty$, and $h(\omega(a))$ falls off as quickly per distance traveled along ω as along any other path through $\hat{\beta}$, in a neighborhood of $\hat{\beta}$. The line integral in (68) is then defined by splitting up this path into its differentiable parts, using the change of variables theorem to express the integral along each part as the sum of two standard one-dimensional integrals, representing the real and complex parts of the transformed integrand

$$\exp(n[\mathcal{K}_Y(\beta(a)) - \beta(a)x - \mathcal{K}_Y(\hat{\beta}) + \hat{\beta} x])(d\beta/da)\, da,$$

performing the separate integrations, and summing the parts. The second objective is to characterize this path.

Path of Integration on β scale, x= 2

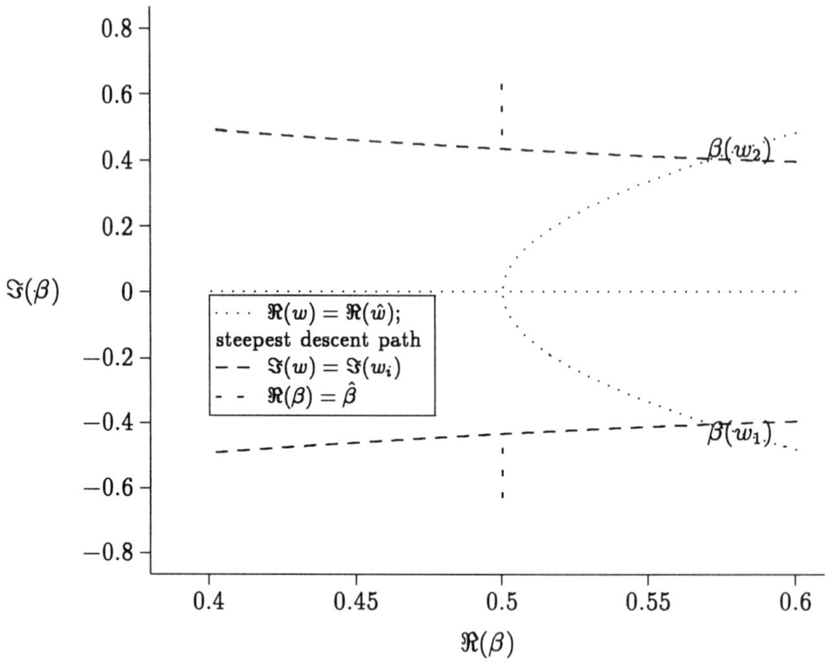

Fig. 9.

Since $h(\beta)$ is analytic, it has a convergent power series representation $h(\beta) = \sum_{k=0}^{\infty} c_k(\beta - \hat{\beta})^k$ for $\left|\beta - \hat{\beta}\right| < \delta$ where $\delta > 0$. By construction $c_0 = c_1 = 0$, and c_2 is real and positive. Then $h(\beta)/(\beta - \hat{\beta})^2$ is analytic for $\left|\beta - \hat{\beta}\right| < \delta$, and its value at $\hat{\beta}$ is c_2. By the composition of functions theorem from complex analysis, $\sqrt{2h(\beta)/(\beta - \hat{\beta})^2}$ is analytic for $\left|\beta - \hat{\beta}\right| < \delta_1$, where $\delta_1 > 0$, and has the value $\sqrt{2c_2} > 0$ at $\hat{\beta}$. An alternate square root function with the opposite sign could also be defined. Let

$$\omega(\beta) - \hat{\omega} = (\beta - \hat{\beta})\sqrt{2h(\beta)/(\beta - \hat{\beta})^2}, \tag{69}$$

where $\hat{\omega} = \sqrt{2(\hat{\beta}x - \mathcal{K}_Y(\hat{\beta}))}$, with the sign the same as that of $\hat{\beta}$. Then $\omega(\beta)$ is an analytic function of β for $\left|\beta - \hat{\beta}\right| < \delta_1$, takes the value $\hat{\omega}$ at $\hat{\beta}$, and has first derivative $\sqrt{2c_2} > 0$ at $\hat{\beta}$. By the inverse function theorem for complex variables, there exist $\delta_2 > 0$ and a function $\beta(\omega)$ analytic for $|\omega - \hat{\omega}| < \delta_2$, such that $\beta(\omega)$ is the inverse of $\omega(\beta)$ in (69).

Through this ball $|\omega - \hat{\omega}| < \delta_2$ runs the path $\Re(\beta(\omega)) = \hat{\beta}$. Integrate instead along the line $\Re(\omega) = \hat{\omega}$, the path of steepest descent. Again, recall that this is the same path along which $h(\beta)$ is real and negative. Within $|\omega - \hat{\omega}| < \delta_2$ will be constructed three curves: The line of steepest descent, along $\Re(\omega) = \hat{\omega}$, and two segments along which the imaginary part of ω is fixed, leading back to the path $\Re(\beta(\omega)) = \hat{\beta}$. Construct these as follows: Let ω_1 and ω_2 be the two places that the path $\Re(\beta(\omega)) = \hat{\beta}$ crosses the circle $|\omega - \hat{\omega}| = \delta_2$, where $\Im(\omega_1) < 0$, $\Im(\omega_2) > 0$. The three paths are then

1. $\Im(\omega) = \Im(\omega_1)$ from $\Re(\omega) = \Re(\omega_1)$ to $\Re(\omega) = \Re(\hat{\omega})$.
2. $\Re(\omega) = \Re(\hat{\omega})$ from $\Im(\omega) = \Im(\omega_1)$ to $\Im(\omega) = \Im(\omega_2)$.
3. $\Im(\omega) = \Im(\omega_2)$ from $\Re(\omega) = \Re(\hat{\omega})$ to $\Re(\omega) = \Re(\omega_2)$.

This reparameterization $\beta(\omega)$ will make finding the path of steepest descent easy. Since $\frac{1}{2}(\omega(\beta) - \hat{\omega})^2 = h(\beta)$, one sees that $\frac{1}{2}(\omega - \hat{\omega})^2 = h(\beta(\omega))$. The path along which $h(\beta)$ decreases most quickly, at least very close to $\hat{\beta}$, is the path along which $\omega - \hat{\omega}$ is pure imaginary. Let $\omega(a) = \beta(\hat{\omega} + ia)$ for $\Im(\omega_1) \leq a \leq \Im(\omega_2)$, representing path 2 above. Along this path ω, when $|a| < \delta_2$, $h(\omega(a))$ is real and negative, since it is the square of a pure imaginary quantity. This gives a second characterization of the path of steepest descent. Complete this path with segments 1 and 3 from above, and segments from $\beta(\omega_2)$ to $\hat{\beta} + i\infty$ and from $\hat{\beta} + i\infty$ to $\beta(\omega_1)$, along which $\Re(\beta) = \hat{\beta}$. See Fig. 9.

For the purposes so far, the choice of $\hat{\omega}$ was irrelevant. Choosing $\hat{\omega} = 0$ would have been as satisfactory. The particular choice of $\hat{\omega}$ will only become critical in later sections when saddlepoint expansions for the cumulative distribution function are constructed. At that point the the fact that the current choice of $\hat{\omega}$ implies not only $\omega(\hat{\beta}) = \hat{\omega}$, but also $\omega(0) = 0$, will become important.

Having determined the path of steepest descent, parameterizing (68) in terms of this path yields

$$f_X(x) = \frac{n}{2\pi i} \exp(n[\mathcal{K}_Y(\hat{\beta}) - \hat{\beta}x]) \int_{\hat{\omega}-i\infty}^{\hat{\omega}+i\infty} \exp(\frac{n}{2}[\omega - \omega]^2) \frac{d\beta}{d\omega} d\omega,$$

plus an error arising from points outside of the circle of radius δ_2; this error converges to zero geometrically quickly.

4.5 Rigorous Derivation Using Steepest Descent

Daniels (1954) uses a special case of Watson's Lemma, a theorem from complex variables (Jeffreys, 1962), in conjunction with the method of steepest descent of the previous section, to show that the order of the Edgeworth series carries over but only pointwise for each value of the mean.

Lemma 4.5.1: *If $\varpi(\omega)$ satisfies $\varpi(\omega) = \sum_{j=0}^{2j} a_j \omega^j / j! + O(\omega^{2j+1})$ as $\omega \to 0$, and A and B are positive numbers, or possibly infinite, then*

$$\left(\frac{n}{2\pi}\right)^{\frac{1}{2}} \int_{-A}^{B} \exp(\frac{n}{2}\omega^2) \varpi(\omega)\, d\omega = \sum_{j=0}^{j} \varpi^{(2j)}(0)/(2n)^j j! + O(n^{-j-1/2}), \tag{70}$$

provided the integrand on the left hand side of (70) converges absolutely for some n. Specifically, there exist constants D_j and N_{j+1} independent of n such that for $n > N_{j+1}$,

$$\left|\left(\frac{n}{2\pi}\right)^{\frac{1}{2}} \int_{-A}^{B} \exp(-\frac{n}{2}\omega^2)\varpi(\omega)\, d\omega - \sum_{j=0}^{j} \varpi^{(2j)}(0)/(2n)^j j!\right| \leq D_j/n^{j+1/2}.$$

Proof: Let $E_{j+1}(\omega) = \varpi(\omega) - \sum_{j=0}^{2j} a_j\omega^j/j!$. Choose $a < A$, $b < B$, and D_j such that on $(-a, b)$, $|E_{j+1}(\omega)| \leq \frac{1}{2}D_j\omega^{2j+1}$. Choose N_{j+1} such that $\frac{1}{2}D_j n^{-j-1/2} \geq \exp(-\max(a,b)n)\varpi^{(j)}(0)$, if $n > N_{j+1}$. Then

$$\left(\frac{n}{2\pi}\right)^{\frac{1}{2}} \int_{-A}^{B} \exp(-\frac{n}{2}\omega^2)\varpi(\omega)\, d\omega = \sum_{j=0}^{2j} \int_{-\infty}^{\infty} \exp(-\frac{n}{2}\omega^2)\frac{\varpi^{(j)}(0)}{j!}\omega^j -$$

$$\sum_{j=0}^{2j} \int_{-\infty}^{-A} \exp(-\frac{n}{2}\omega^2)\frac{\varpi^{(j)}(0)}{j!}\omega^j - \sum_{j=0}^{2j} \int_{B}^{\infty} \exp(-\frac{n}{2}\omega^2)\frac{\varpi^{(j)}(0)}{j!}\omega^j$$

$$+ \int_{-A}^{-a} \exp(-\frac{n}{2}\omega^2)\varpi(\omega)\, d\omega + \int_{b}^{B} \exp(-\frac{n}{2}\omega^2)\varpi(\omega)\, d\omega + \int_{-a}^{b} \exp(-\frac{n}{2}\omega^2)E_{j+1}(\omega)\, d\omega \quad (71)$$

The proof is completed by showing that

1. Even terms in the first summation of (71), multiplied by $(n/(2\pi))^{\frac{1}{2}}$, are the corresponding terms given in the statement of the theorem.
2. Odd terms in the first summation of (71) are zero.
3. The final term in (71) is bounded by $D_j n^{-j-1/2} \int_{-\infty}^{\infty} \exp(-\omega^2/2)\, |\omega|^{2j+1}\, d\omega$.
4. All other terms of (71) converge to zero geometrically, at rate $\exp(-\max(a,b)n)$.

Q.E.D

A rescaled version of this lemma of more immediate use is:

Lemma 4.5.2: *If $\varpi(\omega)$ is analytic in a neighborhood of $\omega = \hat{\omega}$ containing the path $(-Ai + \hat{\omega}, Bi + \hat{\omega})$ with $\infty \geq A > 0$ and $\infty \geq B > 0$, then*

$$i^{-1}\left(\frac{n}{2\pi}\right)^{\frac{1}{2}} \int_{-Ai+\hat{\omega}}^{Bi+\hat{\omega}} \exp(\frac{n}{2}(\omega - \hat{\omega})^2)\varpi(\omega)\, d\omega = \sum_{j=0}^{\infty} \frac{(-1)^j\varpi^{(2j)}(\hat{\omega})}{(2n)^j j!}$$

is an asymptotic expansion in powers of n^{-1}, provided the integral converges absolutely for some n; that is,

$$\left|\left(\frac{n}{2\pi}\right)^{\frac{1}{2}} \int_{-Ai+\hat{\omega}}^{Bi+\hat{\omega}} \exp(\frac{n}{2}(\omega - \hat{\omega})^2)\varpi(\omega)\, d\omega - \sum_{j=0}^{j}(-1)^j\frac{\varpi^{(2j)}(\hat{\omega})}{(2n)^j j!}\right| \leq D_j/n^{j+1}.$$

Proof: This lemma follows immediately from Lemma 4.5.1 and the usual change of variables theorem in integration. Since terms in the expansion of all orders in n exist, the remainder term is of the same order as the last term omitted.

Q.E.D

Temme (1982) extended Watson's Lemma to the case in which ϖ itself involves n.

The rescaled Watson's Lemma will be applied to the center portion of the integral in (68), after changing the path of integration as in §4.4 and parameterizing using (69). This reparameterization introduces the factor $\varpi_x(\omega) = d\beta/d\omega$, which by the arguments of §4.4 is analytic at $\hat{\omega}$. The result is:

Theorem 4.5.3: *Suppose that:*

1. *The random variables $Y_1, \ldots, Y_n$ are independent and identically distributed, each with cumulant generating function $\mathcal{K}_Y$ defined on an open interval containing 0, and each with cumulative distribution function F_Y.*
2. *An ordinate x is chosen such that the equation $\mathcal{K}'_Y(\hat{\beta}) = x$ has a solution.*
3. *There exists $N \in \mathfrak{Z}$ such that the characteristic function for the tilted density*

$$\zeta_N(s) = \exp(N[\mathcal{K}_Y(\beta + is)x - \mathcal{K}_Y(\beta)])$$

is absolutely integrable for each real $\beta \in [0, \hat{\beta}]$: $\int_{-\infty}^{\infty} |\zeta_N(s)| \; ds < \infty$.

Then the mean $X = (Y_1 + \cdots + Y_n)/n$ has a density f_X if $n \geq N$, and the result (61) is a valid asymptotic expansion for $f_X(x)$, in that the error incurred by keeping only the first $\mathfrak{j}$ terms can be bounded by a quantity of the form $D_\mathfrak{j}/n^{\mathfrak{j}+1}$.

Proof: As indicated above, f_X is given by the inversion integral (68). This is integrated along the five-part path described in §4.4.

Recall that the third part is the integral along the path of steepest descent:

$$\sqrt{\frac{n}{2\pi}} \frac{\exp(n[\mathcal{K}_Y(\hat{\beta}) - \hat{\beta}x])}{\sqrt{\mathcal{K}''_Y(\hat{\beta})}} i^{-1} \sqrt{\frac{n}{2\pi}} \int_{\hat{\omega}-i\delta_2}^{\hat{\omega}+i\delta_2} \exp(\frac{n}{2}(\omega - \hat{\omega})) \frac{\sqrt{\mathcal{K}''_Y(\hat{\beta})} d\beta}{d\omega} d\omega.$$

Lemma 4.5.2 is applied to this integral. Verification that $\sqrt{\mathcal{K}''_Y(\hat{\beta})} d\beta/d\omega(\hat{\omega}) = 1$, and that the other coefficients agree with those of (61), will be outlined in the exercises.

The second and fourth parts of the paths of integration contribute a pure imaginary portion. Since the complete resulting integral is real, imaginary contributions must cancel out and can be ignored.

The first and fifth integrals, along rays to $\hat{\beta} \pm i\infty$, are treated exactly as in the Edgeworth series proof for densities, part b, in §3.8. Along the line $\Re(\beta) = \hat{\beta}$ the integrand of (68) is the characteristic function of the sum of n random variables whose probability distribution is given by $dF_Y^*(y) = \exp(\hat{\beta}y) \; dF_Y(y)$, evaluated at $\Im(\beta)$. By condition 3, the limit of the characteristic function for this distribution as the argument diverges to $\pm\infty$ is zero. Hence outside of a fixed interval about the origin the absolute value of the characteristic function is bounded away from one, and the integral over the first and fifth parts of the path is bounded by the least upper bound of the individual characteristic functions over this region, which is less than 1, to the power $n - N$, times a convergent integral. Hence the contribution converges to zero geometrically.

Q.E.D

4.6. Conditions for Uniformly Bounded Relative Error

The preceding steepest descent methods imply that the resulting series are, for a given ordinate, the best approximate cumulant generating function inversion approximations available that require only derivatives of the cumulant generating function at one point. Since the resulting error bounds involve only the invariants ρ_k, one might conjecture that suitable regularity conditions on their behavior imply uniform relative error bounds of the asymptotic expansion. This conjecture is true. Unfortunately steepest descent methods do not provide an easy way to establish this uniformity. Two difficulties arise:

1. The error involved in truncating the power series expansion for $d\beta/d\omega$ near the origin was governed by the behavior of higher-order derivatives of $d\beta/d\omega$. Bounding these coefficients away from the real axis using only behavior along this axis is difficult.
2. The radius of convergence for $d\beta/d\omega$ determined the geometric rate at which parts of the inversion integral away from $\hat{\beta}$ converged to zero. This radius is not constant, and in fact for extreme $\hat{\beta}$ may converge to zero. Consider the logistic distribution. The tilted cumulant generating function $\mathcal{K}(r+is) - (r+is)x$ is, after simplification,

$$\imath \arcsin\left(\frac{\sqrt{2}\,(s\cosh(\pi s)\sin(\pi r) - r\cos(\pi r)\sinh(\pi s))}{\sqrt{(r^2+s^2)\,(\cosh(2\pi s) - \cos(2\pi r))}}\right) - \frac{3}{2}\log(2)$$
$$-\log(\cosh(2\pi s) - \cos(2\pi r)) + \tfrac{1}{2}\log\left(\left(r^2+s^2\right)(\cosh(2\pi s) - \cos(2\pi r))\right) - rx - isx,$$

and at $r = 1$ the imaginary part is

$$-(s\,x) + \arcsin\left(\frac{\sqrt{2}\sinh(\pi s)}{\sqrt{(1+s^2)\,(-1+\cosh(2\pi s))}}\right).$$

The value of s making this imaginary part zero can be plotted as a function of x (Fig. 10). The length of the portion of the path of integration to which arguments involving the derivatives of the integrand apply is shrinking to zero. Fig. 11 illustrates this in the case of the logistic distribution.

A different tactic must be employed demonstrating that the series (61) has desired uniformity properties.

Theorem 4.6.1: *Assume the conditions of Theorem 4.5.3. Suppose further that:*

1. *The standardized absolute moments* $\mathrm{E}\left[|Y_1|^j \exp(Y_1\hat{\beta})\right] / \mathrm{E}\left[|Y_1|^2 \exp(Y_1\hat{\beta})\right]^{j/2}$ *are bounded for each* j *as* $\hat{\beta}$ *varies.*

Height where Steepest Descent Curve for Logistic Density Hits Boundary vs. x

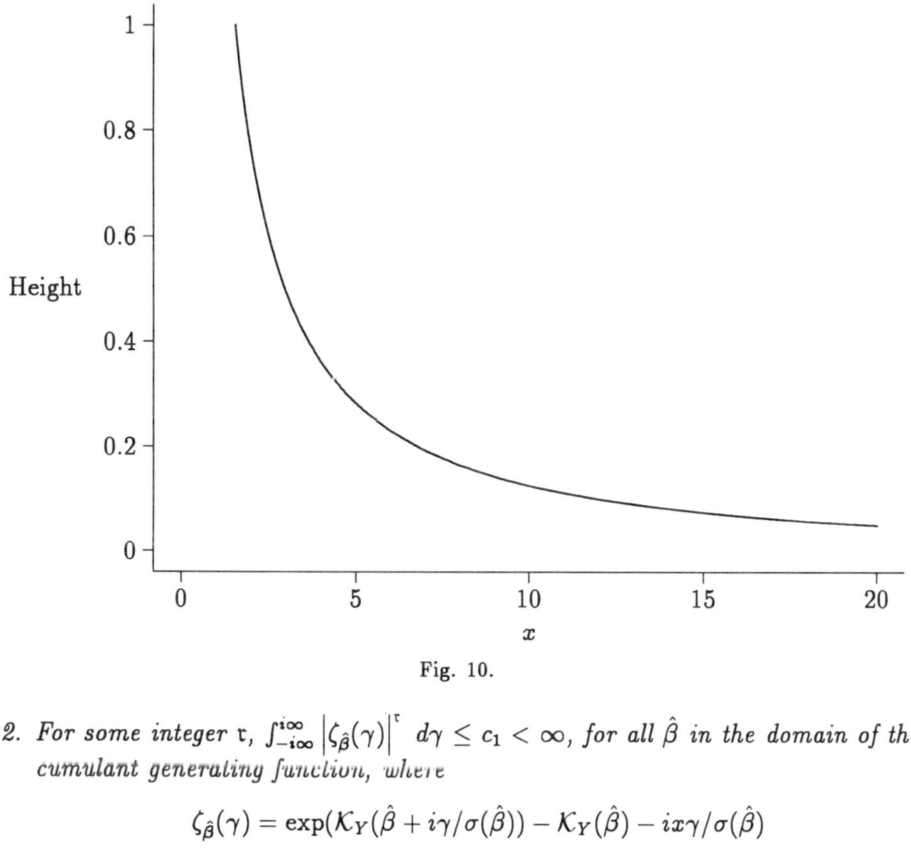

Fig. 10.

2. For some integer $\mathfrak{r}$, $\int_{-i\infty}^{i\infty} \left|\zeta_{\hat\beta}(\gamma)\right|^{\mathfrak{r}} d\gamma \leq c_1 < \infty$, *for all* $\hat\beta$ *in the domain of the cumulant generating function, where*

$$\zeta_{\hat\beta}(\gamma) = \exp(\mathcal{K}_Y(\hat\beta + i\gamma/\sigma(\hat\beta)) - \mathcal{K}_Y(\hat\beta) - ix\gamma/\sigma(\hat\beta))$$

is the characteristic function for each tilted distribution.

3. $\zeta_{\hat\beta}(\gamma)$ *converge uniformly to some function as* $\hat\beta$ *goes to its extreme value,*

Then the bound on the error in the asymptotic series (61) can be chosen independently of x and $\hat\beta$, for all x for which $\hat\beta$ is defined.

Proof: Condition 1 in a sense refines the prior condition 2, in that attention is restricted to ordinates for which $\hat\beta$ exists, and condition 2 extends the prior condition 3. The present condition 1 is slightly stronger than that considered by Daniels (1954), and implies boundedness of the invariants. The result presented is due to Jensen (1988).

Refer to the Edgeworth series density theorem of §3.9. Condition 1, and the comments at the end of §3.6 show that the quantity δ in this proof can be taken independent of x and $\hat\beta$, and that other quantities involved in bounding the portion of the error integral near the origin also can be taken independent of x and $\hat\beta$

Steepest Descent Curve for Logistic Distn, x= 2

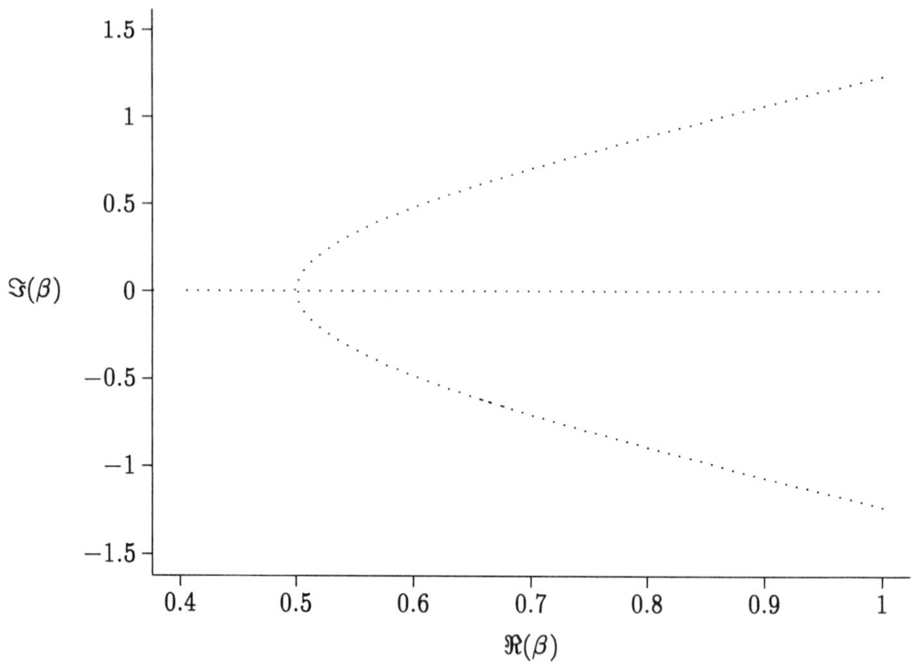

Fig. 11a.

Steepest Descent Curve for Logistic Distn, x= 20

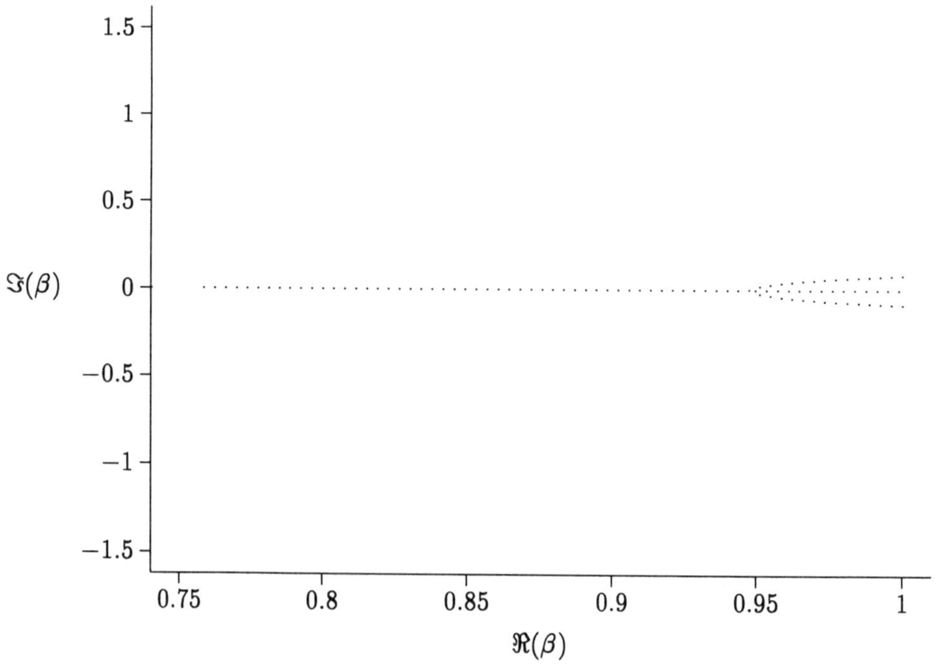

Fig. 11b.

Condition 2 shows that away from the real axis, all of the characteristic functions being approximately inverted converge to 0, and condition 3 shows that this must happen uniformly. Outside of $(-\delta\sqrt{n}, \delta\sqrt{n})$ the characteristic function to be inverted can be bounded by $q^{n-\mathfrak{r}}$ times something integrable, and whose integral is bounded, by condition 2. The constant $q < 1$ is independent of x and $\hat{\beta}$, and the result follows.

Q.E.D

Uniformity in (61) refers to the accuracy of the series multiplying the first-order saddlepoint density approximation. Hence this uniformity refers not to the absolute error of each partial sum in (61), but also to the relative error of the first-order approximation. If $\sqrt{\mathcal{K}_Y''(\beta)}$ remains bounded below, absolute uniformity is recovered, although this theorem demonstrates no improvement over the Edgeworth series result, when considering analytically proven uniform behavior. If $\sqrt{\mathcal{K}_Y''(\beta)}$ goes to infinity then in fact the resulting uniform behavior of the absolute error is better than that for Edgeworth series. Bear in mind that this is all from an analytic point of view; in practice the saddlepoint approximations behave much better than the Edgeworth series.

4.7. Cases with Uniformly Bounded Relative Error

Daniels (1954) lists forms of summand densities such that the relative error of the saddlepoint approximation to the distribution of means of independent and identically distributed copies is uniformly bounded. These are characterized by how the tails behave as the ordinate goes to limiting values. Consider a distribution defined on $(-x_1, x_2)$ and examine behavior as the ordinate approaches x_2. In the first three examples $x_2 = \infty$.

1. Gamma: $f(x) \sim Ax^{\alpha-1}\exp(-cx)$ where $\alpha > 0$, $c > 0$.
2. $f(x) \sim A\exp(\psi x^{\alpha} - cx)$; $\psi > 0$, $c > 0$, $0 < \alpha < 1$. Note Daniels' paper reads $0 < x < 1$.
3. Stable laws: $f(x) \sim A\exp(-\psi x^{\alpha})$; $\psi > 0$, $\alpha > 1$.
4. Beta: $f(x) \sim A(x_2 - x)^{\alpha-1}$ on $(-x_1, x_2)$; $\alpha > 0$.
5. $f(x) \sim A\exp(-\psi(x_2 - x)^{-\gamma})$ on $(-x_1, x_2)$, $\psi > 0$, $\gamma > 0$.

Jensen (1988) presents conditions on density functions that imply the conditions of the theorem of the previous section. These are applied to the cases presented by Daniels.

4.8. The Effect of Renormalization

The quality of the saddlepoint approximation can often be improved by multiplying the density approximation by a constant to make it integrate to 1. Recall from §4.2 that the saddlepoint approximation to a normal density is exact, and the saddlepoint approximation to a gamma density is exact up to a constant of proportionality. There are very few densities with this property, as the following theorem of Blæsild and Jensen (1985) shows.

Theorem 4.8.1: *Suppose that Y_j are independent and identically distributed random variables, and suppose that the saddlepoint approximation to the density of the mean of n such variables is exactly proportional to the true density, for every $n \in \mathfrak{Z}$. Then the distribution of the Y_j is either gamma, normal, or inverse Gaussian.*

Proof: Since §4.2 contains verification that the normal and gamma densities are approximated exactly up to a multiplicative constant, it remains only to prove that the inverse Gaussian also shares this property, and that no other densities share this property. Verifying the exactness of the approximation to the inverse Gaussian is left as an exercise.

From (61), the relative error incurred when using the saddlepoint approximation is

$$\frac{3\hat{\rho}_4 - 5\hat{\rho}_3^2}{24}n^{-1} + \frac{385\hat{\rho}_3^4 - 630\hat{\rho}_3^2\hat{\rho}_4 + 105\hat{\rho}_4^2 + 168\hat{\rho}_3\hat{\rho}_5 - 24\hat{\rho}_6}{1152}n^{-2} + \cdots, \tag{72}$$

where $\hat{\rho}_l$ are given by (62) and depend on $\hat{\beta}$. For this error to be constant, each term in its series (72) must be constant as a function of $\hat{\beta}$. This theorem will be proved by examining the first two terms in (72), setting them equal to a constant, treating them like a system of differential equations for $\mathcal{K}_Y$, and showing that the only solution is the cumulant generating function associated with a normal, gamma, or inverse Gaussian distribution.

If the first term is zero, $\hat{\rho}_4/8 - 5\hat{\rho}_3^2/24 = 0$ and

$$3\mathcal{K}_Y^{(4)}(\hat{\beta}) = 5\mathcal{K}_Y^{(3)}(\hat{\beta})^2/\mathcal{K}_Y^{(2)}(\hat{\beta}). \tag{73}$$

This represents a second-order non-linear differential equation for $\mathcal{K}_Y''(\hat{\beta})$. We seek solutions for which $\mathcal{K}_Y''$ has a power series representation near zero, and for which $\mathcal{K}_Y''(0) > 0$. Expressing $\mathcal{K}_Y''(\hat{\beta})$ as a power series, and substituting into (73) yields a recursion for the power series coefficients of $\mathcal{K}_Y''$ in which higher-order coefficients are completely determined by the first two. The function $\mathcal{K}_Y^{*}{}''(\beta) = (a\beta + b)^{-3/2}$ solves (73), for all $b > 0$. Since $\mathcal{K}_Y^{*}{}''(0) = b^{-3/2}$ and $\mathcal{K}_Y^{*}{}'''(0) = -(3/2)ab^{-5/2}$, the set of possible cumulant generating functions satisfying (73) is contained in the set generated from $\mathcal{K}_Y^{*}$ with a and b varied. Hence solutions of the form $(a\beta + b)^{-3/2}$ for all $b > 0$ exhaust the class of solutions of interest. Showing that these solutions represent cumulant generating functions of normal and inverse Gaussian distributions is left as an exercise.

If the first term is constant but not zero, (73) no longer holds. An alternate equation is derived by examining the conditions of constancy for the coefficients of n^{-1} and n^{-2} in (72), and their derivatives. The following rule for differentiating the $\hat{\rho}_l$ is useful:

$$\begin{aligned} d\hat{\rho}_l/d\hat{\beta} &= \mathcal{K}_Y^{(l)}(\hat{\beta})(-l/2)\mathcal{K}_Y^{(2)}(\hat{\beta})^{-(l+2)/2}\mathcal{K}_Y^{(3)}(\hat{\beta}) + \mathcal{K}_Y^{(l+1)}(\hat{\beta})\mathcal{K}_Y^{(2)}(\hat{\beta})^{-l/2} \\ &= (-l/2)\sqrt{\mathcal{K}_Y''(\hat{\beta})}\hat{\rho}_l\hat{\rho}_3 + \sqrt{\mathcal{K}_Y''(\hat{\beta})}\hat{\rho}_{l+1}. \end{aligned} \tag{74}$$

Setting the first two derivatives of $\hat\rho_4/8-5\hat\rho_3^2/24$ to zero, and setting $(\hat\rho_4/8-5\hat\rho_3^2/24)^2$ to a constant, and $385\hat\rho_3^4-630\hat\rho_3^2\hat\rho_4+105\hat\rho_4^2+168\hat\rho_3\hat\rho_5-24\hat\rho_6/1152$ to a constant, results in the equations

$$15\,\hat\rho_3^3-16\hat\rho_3\hat\rho_4+3\hat\rho_5=0$$

$$\frac{-135\,\hat\rho_3^4}{2}+101\,\hat\rho_3^2\hat\rho_4-16\,\hat\rho_4^2-\frac{47\,\hat\rho_3\,\hat\rho_5}{2}+3\,\hat\rho_6=0$$

$$25\hat\rho_3^4-30\hat\rho_3^2\hat\rho_4+9\hat\rho_4^2=c_2^2$$

$$385\,\hat\rho_3^4-630\,\hat\rho_3^2\,\hat\rho_4+105\,\hat\rho_4^2+168\,\hat\rho_3\,\hat\rho_5-24\,\hat\rho_6=c_1.$$

Multiplying these by $20\,\hat\rho_3/3$, 8, 23/9, and 1, and summing them yields

$$\frac{16\,{\rho_3}^2\,(5\,\hat\rho_3^2-3\hat\rho_4)}{9}=\frac{16\,\hat\rho_3^2\,c_2}{9}$$

on the left hand side, and a constant on the right. Hence in this case, $\hat\rho_3$ is constant, and $\mathcal{K}_Y''$ is determined by the first order differential equation $\mathcal{K}_Y^{(3)}(\hat\beta)=c\mathcal{K}_Y''(\hat\beta)^{3/2}$ for some constant c. A solution to this equation is $\mathcal{K}_Y^{*}{}''(\hat\beta)=(-\frac{1}{2}c\hat\beta+b)^{-2}$ for $b>0$. For the same reasons as in the first case solutions of this form include all solutions of interest. Showing that these solutions represent cumulant generating functions of normal and gamma distributions is left as an exercise.

Q.E.D

Near exactness up to a multiplicative constant is fairly common for saddlepoint density approximations. Such near exactness holds if the coefficients in (61) vary slowly enough. Recovering the constant multiplier and normalizing has potential for improving the behavior of the saddlepoint approximation. This constant may be derived by numerically integrating the approximation, and Durbin (1980) shows that in some cases the order of the error is divided by $\sqrt{n}$.

Theorem 4.8.2: *Suppose that the conditions of Theorem 4.6.1 hold. Then the saddlepoint density approximation (59) may be rescaled so that its relative error is* $O(n^{-3/2})$ *as the value of the sum standardized to constant variance is held fixed. This rescaling may be calculated by integrating (59) and dividing by the result.*

Proof: Without loss of generality assume that $\mathrm{E}[Y]=0$. From (61),

$$f_n(x)=\sqrt{n}\frac{\exp(n[\mathcal{K}_Y(\hat\beta)-\hat\beta x])}{\sqrt{2\pi\mathcal{K}_Y''(\hat\beta)}}[1+b(\hat\beta)/(2n)+O(n^{-2})].$$

Here c is a sum of products of invariants; hence its derivative is the sum of products of invariants and $\sqrt{\mathcal{K}_Y''(\hat\beta)}$, by (74). The derivative of $b(\hat\beta)$ with respect to x is $db/dx=b'(\hat\beta)/\sqrt{\mathcal{K}_Y''(\hat\beta)}$, which is bounded on a set of the form $(-\delta,\delta)$ for some

$\delta > 0$. By the strict concavity of the Legendre transform, $\mathcal{K}_Y(\hat{\beta}) - \hat{\beta}x$ is bounded away from zero for x outside of $(-\delta, \delta)$. Using Taylor's theorem, for $x^\dagger \in [0, x]$,

$$f_n(x) = \sqrt{n}\frac{\exp(n[\mathcal{K}_Y(\hat{\beta}) - \hat{\beta}x])}{\sqrt{2\pi\mathcal{K}_Y''(\hat{\beta})}}[1 + b(0)/(2n) + y(db/dx)(x^\dagger)/(2n\sqrt{n}) + O(n^{-2})] \tag{75}$$

where $y = x\sqrt{n}$. For any fixed y, then $x \in (-\delta, \delta)$ for sufficiently large n. Hence multiplying (59) by $1 + b(0)/(2n)$ gives the standardization resulting in a relative error of size $O(n^{-3/2})$, non-uniformly. Furthermore, since $\sqrt{n}\bar{X}$ has a limit in distribution, the limit of

$$\int_{-\infty}^{\infty} \sqrt{n}\frac{\exp(n[\mathcal{K}_Y(\hat{\beta}) - \hat{\beta}x])}{\sqrt{2\pi\mathcal{K}_Y''(\hat{\beta})}}(\sqrt{n}x)(db/dx)(x^\dagger)\, dx$$

is finite. Integrating both sides of (75),

$$1 = (1 + b(0)/(2n))\int_{-\infty}^{\infty} \sqrt{n}\frac{\exp(n[\mathcal{K}_Y(\hat{\beta}) - \hat{\beta}x])}{\sqrt{2\pi\mathcal{K}_Y''(\hat{\beta})}}\, dx(1 + O(n^{-3/2})).$$

Hence $\int_{-\infty}^{\infty} \sqrt{n}\exp(n[\mathcal{K}_Y(\hat{\beta}) - \hat{\beta}x])/\sqrt{2\pi\mathcal{K}_Y''(\hat{\beta})}\, dx = (1 + b(0)/(2n))(1 + O(n^{-3/2}))$.
Q.E.D

4.9. Numeric Hints for Solving Saddlepoint Equations

In many applications the saddlepoint equation (60) cannot be solved analytically, even when the solution $\hat{\beta}$ exists. Usually saddlepoint methods can still be applied by solving (60) numerically.

Often one uses Newton-Raphson derivative-based methods to calculate the saddlepoint. These methods in general will work well, since the function $\mathcal{K}_Y(\beta) - \beta x$ to be minimized is convex. The second derivative exists and is always positive. An iterative procedure is then used to calculate the saddlepoint. An initial approximation β_0 is chosen. In general this will not exactly solve $\mathcal{K}_Y'(\hat{\beta}) = x$, and so the initial approximation must be modified. Approximating the saddlepoint equation linearly in the neighborhood of β_0, the saddlepoint $\hat{\beta}$ satisfies $\mathcal{K}_Y'(\beta_0) + \mathcal{K}_Y''(\beta_0)(\hat{\beta} - \beta_0) \approx x$. The solution to this equation is

$$\hat{\beta} = (x - \mathcal{K}_Y'(\beta_0))/\mathcal{K}_Y''(\beta_0) + \beta_0. \tag{76}$$

The solution to this equation is then used as an updated approximate solution, and (76) is iterated until the approximation is considered close enough to the true solution to (60). This convergence may be assessed by referring to the difference $|x - \mathcal{K}_Y'(\beta_0)|$, or by referring to changes in subsequent values of β_0. The former method is preferred, because in cases where $\mathcal{K}_Y$ is defined only on a finite range, and diverges to infinity as its argument approaches the end of this range, $\mathcal{K}_Y(\hat{\beta})$ and $\mathcal{K}_Y''(\hat{\beta})$ are far more sensitive to small errors in $\hat{\beta}$ near the ends of the range than in the middle. These quantities $\mathcal{K}_Y(\hat{\beta})$ and $\mathcal{K}''_Y(\hat{\beta})$ are needed to calculate the saddlepoint density approximation. Hence a policy of terminating iterations of (76) when changes in $\hat{\beta}$ become small on an absolute scale is problematic, since it is likely to be unnecessarily fine for moderate values of $\hat{\beta}$ and too coarse for extreme values of $\hat{\beta}$.

The Newton-Raphson method is not guaranteed to yield a solution $\hat{\beta}$. The method depends on the cumulant generating function being approximately quadratic, at least in a region containing the true and approximate saddlepoint. If the cumulant generating function is exactly quadratic one iteration of (76) will result in an exact solution to the saddlepoint equation. If the second derivative of the cumulant generating function decreases as one moves from β_0 to $\hat{\beta}$, iterations of (76) will tend to result in steps toward $\hat{\beta}$ that are too small. In this case the iterative saddlepoint approximations typically move monotonically toward $\hat{\beta}$. If the second derivative of the cumulant generating function increases as one moves from β_0 to $\hat{\beta}$, iterations of (76) will tend to result in steps toward $\hat{\beta}$ that overshoot $\hat{\beta}$. In this case the iterative saddlepoint approximations typically first cross to the other side of $\hat{\beta}$, and then move monotonically toward $\hat{\beta}$. If the second derivative of $\mathcal{K}_Y$ is lower on either side of the saddlepoint than it is at the saddlepoint, successive saddlepoint approximations may move from one side of $\hat{\beta}$ to the other while never converging.

Since the Newton-Raphson method performs best when the cumulant generating function is approximately quadratic, rescaling the cumulant generating function to make it closer to quadratic speeds convergence. Especially to be avoided are cases in which the cumulant generating function exists only on a finite interval, and in which Newton's method may give iterations falling outside this domain of definition. In this case rescaling the cumulant generating function argument often avoids such problems.

For example, consider the logistic distribution whose cumulant generating function and derivatives are

$$\mathcal{K}_Y(\beta) = \log(\pi\beta/\sin(\pi\beta))$$
$$\mathcal{K}'_Y(\beta) = 1/\beta - \pi\cot(\pi\beta)$$
$$\mathcal{K}''_Y(\beta) = -1/\beta^2 + \pi^2\csc^2(\pi\beta)$$

for $\beta \in (-1, 1)$. Solving the saddlepoint equation in this case by iterating (76) gives rise to two problems. First, note that the values at zero are $\mathcal{K}'_Y(0) = 0$ and $\mathcal{K}''_Y(0) = \pi^2/3$. Hence for $x > \pi^2/3$, when starting from $\beta_0 = 0$, one iteration of (76) gives a second saddlepoint approximation outside the domain of $\mathcal{K}_Y$. Second, supposing that for a large x a more reasonable starting value β_0 is found. Since $\mathcal{K}_Y$ is far from quadratic, the Newton-Raphson method can be expected to converge very slowly.

Solving instead the equation $\mathcal{K}'_Y(2\tan^{-1}(\hat{s})/\pi) = x$ using the Newton-Raphson method, and setting $\hat{\beta} = 2\tan^{-1}(\hat{s})/\pi$, avoids both of these problems. The new quantity to be solved for can take any value on the real line, so no problem of provisional approximations lying outside the domain of the cumulant generating function arises. Furthermore, no poles near the end points slow the convergence. See Fig. 12.

Alternatively, one might use the secant method to solve for the saddlepoint. Begin with two initial values, β_0 and β_1, are determined such that $\mathcal{K}'_Y(\beta_0) < x < \mathcal{K}'_Y(\beta_1)$. If $\mathcal{K}'(\beta)$ were linear, the saddlepoint would then be given by solving $(\beta_2 - \beta_0)/(x - \mathcal{K}'_Y(\beta_0)) = (\beta_2 - \beta_1)/(x - \mathcal{K}'_Y(\beta_1))$ for β_2, to yield $\beta_2 = (x[\beta_1 - \beta_0] + \beta_0\mathcal{K}'_Y(\beta_1) - \beta_1\mathcal{K}'_Y(\beta_0))/(\mathcal{K}'_Y(\beta_1) - \mathcal{K}'_Y(\beta_0))$. Graphically, one determines $\hat{\beta}$ by observing where the secant joining $(\beta_0, \mathcal{K}'_Y(\beta_0))$ to $(\beta_1, \mathcal{K}'_Y(\beta_1))$ crosses the horizontal line through x. One then evaluates $\mathcal{K}'_Y(\beta_2)$ and determines whether $\mathcal{K}'_Y(\beta_2) > x$, implying β_2 replaces β_1, or whether $\mathcal{K}'_Y(\beta_2) < x$, implying β_2 replaces β_0. This process continues until β_0 and β_1 are close enough to consider the common value the saddlepoint.

This secant method generally only produces the correct answer in one iteration if $\mathcal{K}'_Y$ is linear, as is the case with the Newton–Raphson method. Like the Newton–Raphson method, if $\mathcal{K}''_Y$ varies quickly enough, convergence can be very very slow, although divergence is not possible for the secant method.

4.10. Exercises

1. Complete the proof of Watson's Lemma by following the steps:
    1. Prove the first and second assertions in the proof of Watson's Lemma in §4.5, by using integration by parts, or otherwise.
    2. Prove the third assertion in the proof of Watson's Lemma in §4.5.
    3. Prove the fourth assertion in the proof of Watson's Lemma in §4.5.
2. Show that the expansion resulting from applying the modified version of Watson's Lemma to $d\beta/d\omega$ gives the results (61) by following the steps:
    1. Show that if a function $f(\beta)$ has a series expansion $1 + \sum_{j=1}^{\infty} a_j\beta^j$, then $\sqrt{f(\beta)}$ has the expansion
    $$1 + \tfrac{1}{2}a_1\beta^1 + (\tfrac{1}{2}a_2 - \tfrac{1}{8}a_1^2)\beta^2 + (\tfrac{1}{2}a_3 - \tfrac{1}{4}a_1a_2 + \frac{1}{16}a_1^3)\beta^3 + \cdots.$$
    2. Use the preceding result to show that if a function $f(\beta)$ has a series expansion $\sum_{j=2}^{\infty} a_j\beta^j$, then $\sqrt{f(\beta)}$ has the expansion
    $$\sqrt{a_2}(\beta + \tfrac{1}{2}a_3\beta^2 + (\tfrac{1}{2}a_4 - \tfrac{1}{8}a_3^2)\beta^3 + (\tfrac{1}{2}a_5 - \tfrac{1}{4}a_3a_4 + \frac{1}{16}a_3^3)\beta^4 + \cdots. \quad (77)$$
    3. Show that if a function $f(\beta)$ has a series expansion $\sum_{j=1}^{\infty} a_j\beta^j$, and $f(\beta) = s$, then
    $$\beta = a_1^{-1}s - a_1^{-3}a_2s^2 + a_1^{-5}(2a_2^2 - a_1a_3)s^3 + \cdots. \quad (78)$$

Derivative of Logistic Cumulant Generating Function

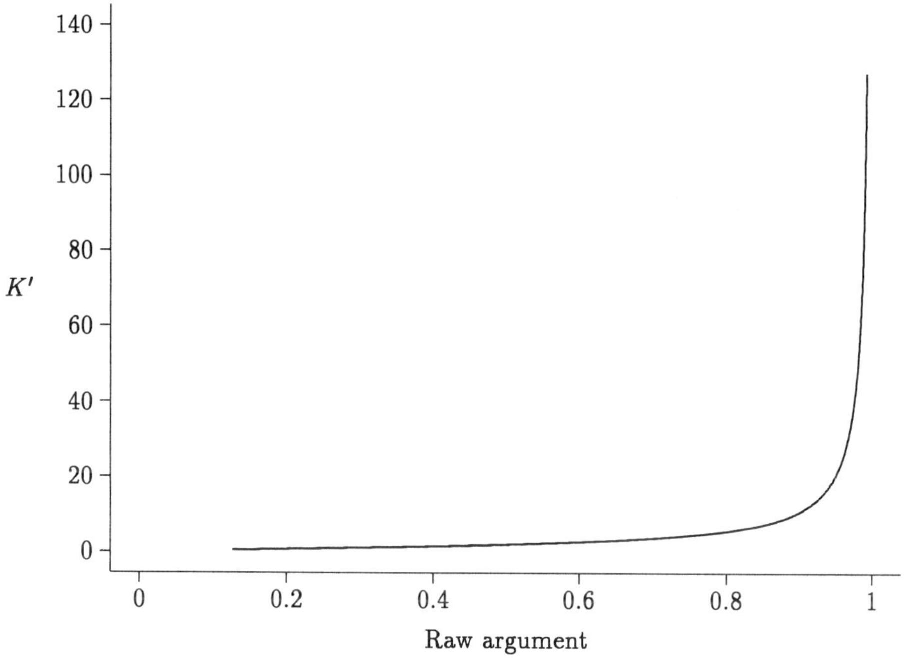

Fig. 12a.

Derivative of Logistic Cumulant Generating Function

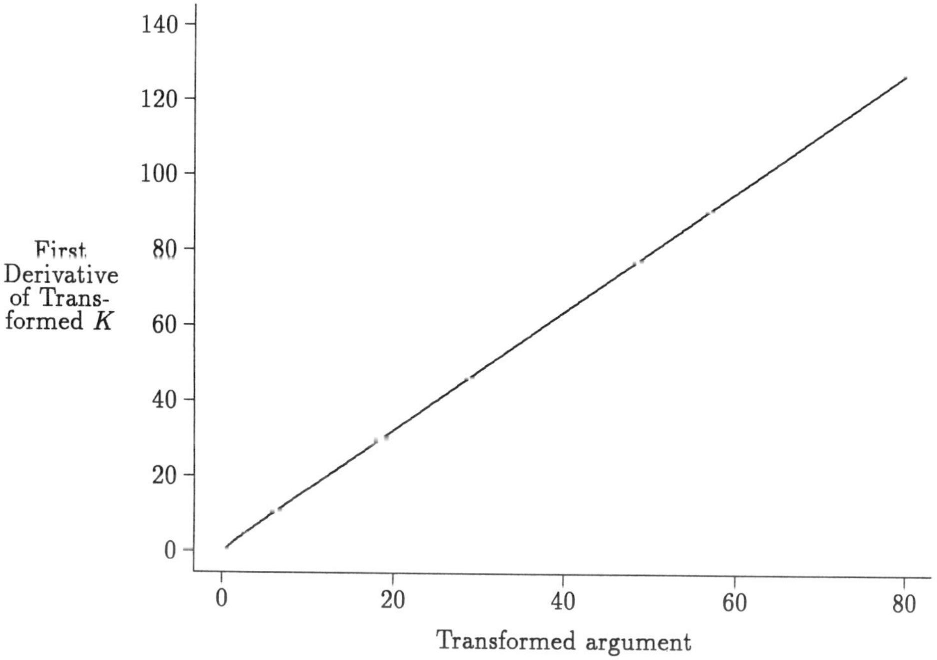

Fig. 12b.

4. Use the result (77) to get a series expansion for ω in terms of β and derivatives of $\mathcal{K}_Y(\hat{\beta})$, and use (78) to give an expansion for β in terms of ω. Take a derivative of this to get the expansion for $d\beta/d\omega$.

3. Complete the proof of the theorem of §4.8 using the following steps.
   1. Show that normal and inverse Gaussian distributions have cumulant generating functions satisfying $\mathcal{K}''(\beta) = (a\beta + b)^{-3/2}$ for $b > 0$.
   2. Show that normal and gamma distributions have cumulant generating functions satisfying $\mathcal{K}''(\beta) = (-\frac{1}{2}c\hat{\beta} + b)^{-2}$ for $b > 0$.
   3. Show that the saddlepoint approximation to the inverse Gaussian distribution is exact.

5
Saddlepoint Series for Distribution Functions

Recall from §3 that calculating distribution function approximations from Edgeworth density approximations was a simple matter. The Edgeworth series for the density is a linear combination of derivatives of the normal distribution function, and hence is easily integrated to give a corresponding cumulative distribution function approximation. This cumulative distribution function approximation inherits many good properties from the density approximation.

Throughout this chapter, assume that the random variable X results as the mean of n independent and identically distributed random variables, each of which has a cumulant generating function $\mathcal{K}_Y(\beta)$ defined in an open interval containing zero. Hence X also has a cumulant generating function $\mathcal{K}_X(\beta)$. Until noted to the contrary, assume that X has a density f_X.

Integrals involving saddlepoint density approximations are in general not tractable analytically. One might approximate the cumulative distribution function by integrating the saddlepoint density (59) from $-\infty$ to x numerically. Many techniques for numeric integration exist (Thisted, 1988). Most of these techniques involve calculating the integrand at a large number of points $x_0, \ldots, x_j$ between $-\infty$ and x, evaluating the integrand at each of these points to obtain $y_0, \ldots, y_j$, and estimating the integral as a linear combination of $y_0, \ldots, y_j$. Each evaluation of the saddlepoint density may require an iterative solution to (60). Daniels (1987) notes that this numeric integration of the saddlepoint density can be simplified by changing variables from x to $\hat{\beta}$:

$$
\begin{aligned}
Q_X(x) &= \mathrm{P}\left[X \geq x\right] \\
&= \int_x^{\max(X)} f_X(z)\, dz && (79) \\
&= \int_x^{\max(X)} \exp(\mathcal{K}_X(\hat{\beta}(z)) - z\hat{\beta}(z))/\sqrt{2\pi\mathcal{K}_X''(\hat{\beta}(z))}\, dz + O(n^{-1}) && (80) \\
&= \int_{\hat{\beta}(x)}^{\max(\beta)} \exp(\mathcal{K}_X(\beta) - \beta\mathcal{K}_X'(\beta))\sqrt{\mathcal{K}_X''(\beta)/2\pi}\, d\beta + O(n^{-1})
\end{aligned}
$$

to avoid having to calculate $\hat{\beta}$ for each value x_i, as would be required to numerically evaluate (80). Although quite good results can be obtained in this way, this method has the inelegant drawback of embodying both numeric and analytic approximation. The rest of this chapter will be concerned with alternative analytic methods for deriving saddlepoint cumulative distribution function approximations. No strong heuristic parallels exist to guide us here. Principle results are derived directly from the appropriate inversion integrals. This task is complicated by a pole at 0 in the integrand. Approximations are derived ignoring and accounting for this pole, and are compared. Finally, extensions of these techniques to lattice distributions are presented. Reid (1996) provides a recent thorough review of these questions.

Rather than approximating the cumulative distribution function of a random variable directly, we begin by calculating tail areas for the random variable. This is done because some of the saddlepoint cumulative distribution function approximations degenerate for ordinates near the mean, as will be shown in the next sections.

Many analytic approximations to tail areas begin with integration of the exact expression for the density (15). This results in a double integral, with respect to the dummy ordinate representing potential values of the random variable, and with respect to the cumulant generating function argument:

$$\begin{aligned}
Q_X(x) &= \int_x^\infty \left[\frac{1}{2\pi i}\int_C \exp(\mathcal{K}_X(\beta) - \beta y)\, d\beta\right] dy \\
&= \frac{1}{2\pi i}\int_C \left[\int_x^\infty \exp(\mathcal{K}_X(\beta) - \beta y)\, dy\right] d\beta && (81) \\
&= \lim_{b\to\infty} \frac{1}{2\pi i}\int_C \exp(\mathcal{K}_X(\beta))(\exp(-\beta x) - \exp(-\beta b))\, d\beta/\beta && (82) \\
&= \frac{1}{2\pi i}\int_C \exp(\mathcal{K}_X(\beta) - \beta x)\, d\beta/\beta \\
&= \frac{1}{2\pi i}\int_C \exp(n\,[\mathcal{K}_Y(\beta) - \beta x])\, d\beta/\beta && (83)
\end{aligned}$$

The integration interchange in (81) can be justified using Fubini's Theorem by selecting a path of integration C for which all points have a positive real part bounded away from zero. The result (83) of the limiting operation in (82) also requires such a path. This inversion integral (83) lacks the factor n appearing in the density inversion integral (67) from the linear transformation creating $d\beta$, since it is eliminated by an identical factor in β.

5.1. A Large-Deviations Result

Bahadur and Ranga Rao (1960) approximate tail areas in a manner similar to that used to derive (80). Instead of forming the Edgeworth series after tilting the distribution separately for each y in the range of integration, one tilted Edgeworth series is calculated; this is chosen so that the mean is at x. Recall that when it can be constructed, the Edgeworth series e_j has error uniformly of order $o(n^{1-j/2})$ in general, but in this case, since Edgeworth series of all orders can be constructed, e_j

has error uniformly of order $O(n^{1/2-j/2})$. Let $\hat{\omega} = \sqrt{2(\hat{\beta}x - \mathcal{K}_Y(\hat{\beta}))}$, with the sign the same as that of $\hat{\beta}$. From (79),

$$\begin{aligned}
Q_X(x) &= \exp(\mathcal{K}_X(\hat{\beta}) - \hat{\beta}x) \int_x^\infty \exp(-(\hat{\beta}y - \hat{\beta}x)) \exp((\hat{\beta}y - \mathcal{K}_X(\hat{\beta}))) f_X(y)\, dy \\
&= \exp(-\frac{n}{2}\hat{\omega}^2) \int_x^\infty \exp(-(\hat{\beta}y - \hat{\beta}x)) f_X(y; \hat{\beta})\, dy \\
&= \exp(-\frac{n}{2}\hat{\omega}^2) \int_x^\infty \frac{\exp\left(-(\hat{\beta}y - \hat{\beta}x)\right) \left[e_j((y-x)[\mathcal{K}_X''(\hat{\beta})]^{-1/2}; \boldsymbol{\rho}(\hat{\beta})) + O(n^{1/2-j/2})\right]}{\sqrt{\mathcal{K}_X''(\hat{\beta})}}\, dy \\
&= \exp(-\frac{n}{2}\hat{\omega}^2) \int_0^\infty \exp\left(-\hat{\beta}y\sqrt{n\mathcal{K}_Y''(\hat{\beta})}\right) \left[e_j(y; \boldsymbol{\rho}(\hat{\beta})) + O(n^{\frac{1}{2}-j/2})\right]\, dy \qquad (84)
\end{aligned}$$

Here $\boldsymbol{\rho}$ are the invariants of §2.1, and implicitly involve n. Bahadur and Ranga Rao (1960) expand the integral in (84) as an asymptotic series in n, effectively using Watson's Lemma, which is justified for any $\hat{\beta} > 0$.

This calculation may also be derived from (83). The same change of variables as in (69) is used:

$$\omega(\beta) - \hat{\omega} = (\beta - \hat{\beta})\sqrt{2h(\beta)/(\beta - \hat{\beta})^2}$$

for ω near $\hat{\omega}$. Then

$$Q_X(x) = \frac{1}{2\pi i} \exp(n[\mathcal{K}_Y(\hat{\beta}) - \hat{\beta}x]) \int_{\hat{\omega}-i\infty}^{\hat{\omega}+i\infty} \exp(\frac{n}{2}[\omega - \hat{\omega}]^2)\beta^{-1}\frac{d\beta}{d\omega}\, d\omega. \qquad (85)$$

Since the factor of β^{-1} does not have an exponent of n this term is included with $d\beta/d\omega$ in the function $\varpi(\omega) = (1/\beta)(d\beta/d\omega)$ in Lemma 4.5.1. The function ϖ has a pole at 0. Furthermore, if $x = \mathrm{E}[Y]$, and hence $\hat{\beta} = 0$, this pole coincides with $\hat{\omega}$. Otherwise $\varpi(\omega)$ is analytic at $\hat{\omega}$, and Watson's Lemma can be used. As before the value of the integral is determined by the derivatives of ϖ at $\hat{\omega}$. These derivatives will be calculated by expanding $\varpi(\omega) = (1/\beta)(d\beta/d\omega)$ about $\hat{\omega}$. Question 2 of §4.10 described how the series expansion for $d\beta/d\omega$ could be found, by:

1. determining the relation between the series expansion for a function $f(\beta)$ and the expansion for its square root $\sqrt{f(\beta)}$.
2. determining the relation between a series expansion and the expansion of the functional inverse.

Let $u = \log(\beta)$. To expand $du/d\omega = \beta^{-1}d\beta/d\omega$, an additional step is needed. Step 1 is used to calculate ω as a function of β. This expansion is composed with the expansion for β in terms of u to express ω in terms of u. The resulting series is

$$\hat{z}\,(u - \hat{u}) + \frac{\hat{z}\,(3 + \hat{z}\,\rho_3)\,(u - \hat{u})^2}{6} + \left(\frac{\hat{z}}{6} + \frac{\hat{z}^2\,\rho_3}{6} - \frac{\hat{z}^3\,\rho_3^2}{72} + \frac{\hat{z}^3\,\rho_4}{24}\right)(u - \hat{u})^3 + \mathrm{O}(u - \hat{u})^4$$

where $\hat{u} = u(\hat{\beta})$, $\hat{z} = \hat{\beta}\sqrt{\mathcal{K}_Y''(\hat{\beta})}$. This series is then reversed as in step 2 to express u in terms of ω, and the derivative is taken to get the desired series:

$$\left(1 + \frac{(\omega - \hat{\omega})^2\,(24\,\hat{z}^{-2} + 12\,\hat{z}^{-1}\,\rho_3 + 5\,\rho_3^2 - 3\,\rho_4)}{24}\right)/\hat{z}$$

The above series has odd order terms in $\omega - \hat{\omega}$ removed, since these do not affect the value of the integral as approximated by Watson's Lemma.

Alternatively, rather than expanding the composition of the logarithm function with $\beta(\omega)$, one might expand the series $d\beta/d\omega$ and $\beta(\omega)$, and calculate the resulting quotient.

This gives rise to the tail probability approximation

$$\frac{\exp(\mathcal{K}_X(\hat{\beta}) - \hat{\beta}x)}{\hat{\beta}\sqrt{\mathcal{K}_X''(\hat{\beta})}}\left\{1 + (1/n)\left[\frac{1}{8}\rho_4 - \frac{5}{24}\rho_3^2 - \frac{\rho_3}{2\hat{z}} - \frac{1}{\hat{z}^2}\right] + O(n^{-2})\right\}. \tag{86}$$

This is valid only for $x > 0$, since the calculations resulting in (83) only hold for integration paths C consisting of points with positive real parts. One expects the expansion to become inaccurate as x moves close to 0, since as $x \to 0$ the radius of convergence for ϖ shrinks, and the error incurred in the outer parts of the inversion integral becomes more important.

Jensen (1988) includes uniform relative error bounds for (86) for ordinates with $\hat{\beta}$ bounded away from 0 as a consequence of the similar result for densities presented in §4.6, under the same conditions. Since (86) diverges to infinity as $\hat{\beta} \to 0$, it is inappropriate for tail areas corresponding to ordinates near the mean.

5.2. Direct Saddlepoint Series for Distribution Functions

Robinson (1982) presents a much different approach. He expresses the tail probability $Q_X(x)$ as (84), but unlike Bahadur and Ranga Rao (1960), explicitly accounts for the end of the range of integration at zero, by expressing (84) as the sum of integrals of the form $I_j = \int_0^\infty \exp(-ay)h_j(y)\phi(y)\,dy$, in essence deriving the Laplace transform of the Edgeworth series. Using integration by parts, and noting that $(d/dy)h_j(y)\phi(y) = -h_{j+1}(y)\phi(y)$, one finds that $I_j = (2\pi)^{-1/2}h_{j-1}(0) - aI_{j-1}$ and $I_0 = \exp(a^2/2)(1 - \Phi(a))$. Then

$$\begin{aligned} I_1 &= (2\pi)^{-1/2} - a\,\exp(a^2/2)\,(1 - \Phi(a)), \\ I_2 &= -a\left((2\pi)^{-1/2} + a\exp(a^2/2)(1 - \Phi(a))\right), \\ I_3 &= a^2\left((2\pi)^{-1/2} - a\exp(a^2/2)(1 - \Phi(a))\right) - (2\pi)^{-1/2}. \end{aligned}$$

When j in (84) is 3:

$$Q_X(x) = \exp\left(n(\hat{z}^2 - \hat{\omega}^2)/2\right) \times \left[[1 - \Phi(\sqrt{n}\hat{z})]\left(1 - n\rho_3\hat{z}^3/6\right) + \phi(\sqrt{n}\hat{z})\rho_3(n\hat{z}^2 - 1)/(6\sqrt{n}) + O(n^{-1})\right]. \tag{87}$$

This is valid to same order as the Edgeworth series it uses, since the error rate of $e_j(y; \boldsymbol{\kappa}(\hat{\beta}))$ is uniform. The method is adaptable to situations in which the Edgeworth series is valid, and particularly has applications outside the realm of sums of independent and identically distributed random variables.

Both approximations (86) and (87) are derived from the integral (84). The uniform error behavior of the Edgeworth series and the bounded convergence theorem might lead one to believe that the integral represented by (84) will converge uniformly at least on intervals bounded above zero. This is not necessarily true, as the coefficients of the Edgeworth series depend on the ordinate. One expects the accuracy of (87) to deteriorate as $\hat{\beta} \to 0$, since the quantity multiplying the error term increases as $\hat{\beta}$ decreases. Unlike the previous large-deviations result (86), however, (87) is still defined when $\hat{\beta} = 0$.

Daniels (1987) notes that (84) may be formally derived by expanding $(\omega/\beta) \times (d\beta/d\omega)$ about $\hat{\omega}$, and (85) may be expressed as an infinite linear combination of integrals of the form

$$J_k = \int_{\hat{\omega}-i\infty}^{\hat{\omega}+i\infty} \frac{\exp(n[\frac{1}{2}\omega^2 - \omega\hat{\omega}])(\omega - \hat{\omega})^k}{2\pi i \omega}\, d\omega,$$

which can be calculated analytically, by noting that

$$\begin{aligned} J_k &= \int_{\hat{\omega}-i\infty}^{\hat{\omega}+i\infty} \exp(n[\tfrac{1}{2}\omega^2 - \omega\hat{\omega}])(\omega-\hat{\omega})^{k-1}\, d\omega - \hat{\omega} J_{k-1}, \\ J_0 &= \frac{\exp(-\frac{n}{2}\hat{\omega}^2)}{2\pi i} \int_{\hat{\omega}-i\infty}^{\hat{\omega}+i\infty} \exp(\frac{n}{2}[\omega - \hat{\omega}]^2)\, d\omega/\omega = \int_{-i\infty}^{+i\infty} \frac{\exp(\frac{1}{2}v^2 - \sqrt{n}v\hat{\omega})}{2\pi i}\, dv/v \\ &= \int_{-i\infty}^{+i\infty} \int_{\sqrt{n}\hat{\omega}}^{\infty} \frac{\exp(\frac{1}{2}v^2 - vu)}{2\pi i}\, du\, dv \\ &= \int_{\sqrt{n}\hat{\omega}}^{\infty} \int_{-i\infty}^{+i\infty} \frac{\exp(v^2 - vu)}{2\pi i}\, dv\, du \\ &= \int_{\sqrt{n}\hat{\omega}}^{\infty} \phi(u)\, du = 1 - \Phi(\sqrt{n}\hat{\omega}). \end{aligned} \tag{88}$$

As $\hat{\omega} \to 0$, these integrals become larger, and one expects the accuracy of approximating the infinite series by a truncated series to deteriorate. Furthermore, if Q^* represents the right hand side of (87), and $Q^\dagger$ represents the same quantity calculated for the distribution of $-X$, then $Q^*(x) \neq 1 - Q^\dagger(x)$, illustrating that (87) is not the appropriate expansion for lower tails.

Robinson (1982) notes that his approach and that of Bahadur and Ranga Rao (1960) differ in that his series contains terms involving the normal distribution function and the equivalent terms in the Bahadur and Ranga Rao (1960) series have this replaced by a first order approximation. The results of Robinson (1982) correspond to inversion integrals with β in the denominator of the integrand, while Bahadur and Ranga Rao (1960) expand β^{-1} about $\hat{\beta}$, removing β from the denominator.

Heuristically, in the work of Bahadur and Ranga Rao (1960), factors like $\Phi(x)$ are replaced by approximations.

5.3. Less Direct Saddlepoint Series for Distribution Functions

Lugannani and Rice (1980) provide an alternate expansion again using the methods of steepest descent. The form of this expansion is very simple, and it is valid over the entire range of possible ordinates. Again the same steepest descent path (69) is used, and the transformed distribution function inversion integral is still (85): An expansion is now developed for this integral that will differ from (86) in that it will explicitly account for the singularity in $\varpi(\omega) = (1/\beta)(d\beta/d\omega)$ at zero. This will illustrate why care has been taken in the selection of $\hat{\omega}$. Until now the only value of ω of interest has been that value corresponding to $\beta = \hat{\beta}$. In the development of this section the value of ω corresponding to $\beta = 0$ is also of interest, since explicit allowance for the singularity here is desired.

Perform the integration in two steps, first with $\varpi^*(\omega) = (1/\beta)(d\beta/d\omega) - (1/\omega)$ and second with $\varpi^\dagger(\omega) = 1/\omega$. In the first instance the function is differentiable for ω in a region about $\hat{\omega}$ which does not shrink as $\hat{\omega}$ approaches zero. To see this, note that $\omega\varpi^*(\omega)$ is analytic and hence $\varpi^*(\omega) = \sum_{j=-1}^{\infty} c_j \omega^j$ for some coefficients $\{c_j\}$. Also, $\lim_{\omega\to 0} \omega\varpi^*(\omega) = \lim_{\omega\to 0}(\omega/\beta)(d\beta/d\omega) - 1 = 0$ by the definition of first derivative. This implies that $c_{-1} = 0$ and $\varpi^*(\omega)$ is analytic at 0. Hence the use of Watson's Lemma does not have the problems associated with it as does the process yielding (86).

As noted in §4.5, $(d\beta/d\omega)(\hat{\omega}) = 1/\sqrt{\mathcal{K}_Y''(\hat{\beta})}$, and hence $\varpi^*(\hat{\omega}) = 1/\hat{z} - 1/\hat{\omega}$. Hence Watson's Lemma shows that

$$\frac{\exp(-\frac{n}{2}\hat{\omega}^2)}{\sqrt{2\pi}} \int_{\hat{\omega}-iA}^{\hat{\omega}+iB} \exp\left(\frac{n}{2}[\omega - \hat{\omega}]^2\right)\left(\frac{1}{\beta}\frac{d\beta}{d\omega} - \frac{1}{\omega}\right) d\omega = \frac{\exp(-\frac{n}{2}\hat{\omega}^2)}{\sqrt{2n\pi}}\left(\frac{1}{\hat{z}} - \frac{1}{\hat{\omega}}\right) + O(n^{-1})$$

$$= \phi(\sqrt{n}\hat{\omega})\left(\frac{1}{\hat{z}} - \frac{1}{\hat{\omega}}\right)/\sqrt{n} + O(n^{-1}).$$

Integration of $\varpi^\dagger(\hat{\omega})$ can be done exactly; (88) implies that

$$\frac{\exp(-\frac{n}{2}\hat{\omega}^2)}{2\pi} \int_{\hat{\omega}-i\infty}^{\hat{\omega}+i\infty} \exp(\frac{n}{2}[\omega - \hat{\omega}]^2) d\omega/\omega = 1 - \Phi(\sqrt{n}\hat{\omega}).$$

Hence the Lugannani and Rice approximation to $Q_X(x)$ is

$$Q_n^*(x) = 1 - \Phi(\sqrt{n}\hat{\omega}) + \phi(\sqrt{n}\hat{\omega})(1/\hat{z} - 1/\hat{\omega})/\sqrt{n}(1 + O(n^{-1})), \qquad (89)$$

where again $\hat{\omega}$ and $\hat{z}$ are given in §5.1. For example, when this approximation is applied to calculate tail areas for logistic variables, even for a mean of one variable the Lugannani and Rice approximation performs very well. See Fig. 13.

If $Q^\dagger$ represents Q^* calculated for the distribution of $-X$, then $Q^*(x) = 1 - Q^\dagger(-x)$. Hence the Lugannani and Rice approximation to upper tails applies equally well to lower tails, unlike the previous tail probability approximations.

Saddlepoint Approximations to the CDF of Mean of 10 Logistics.

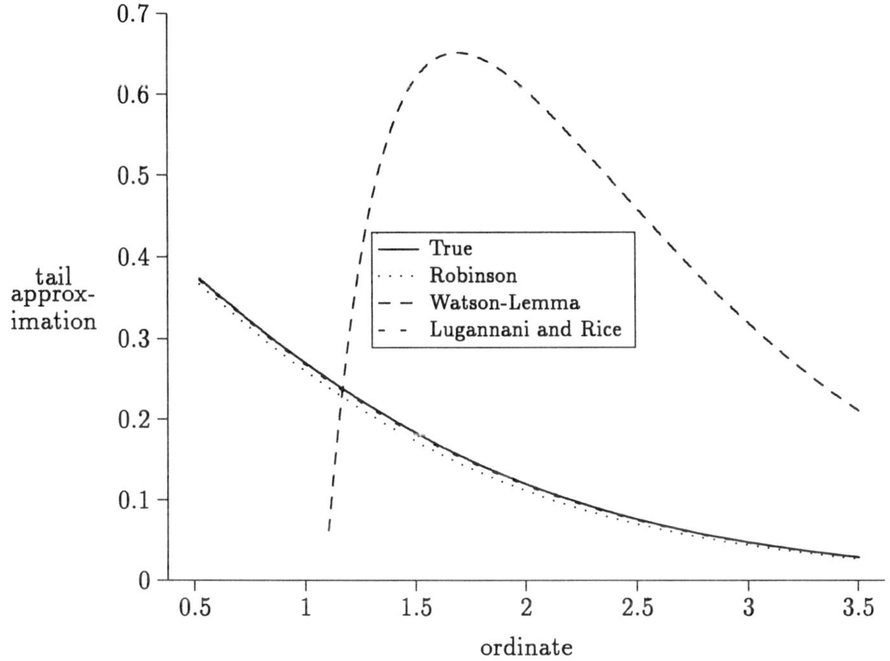

Fig. 13a.

Saddlepoint Approximations to the CDF of Mean of 10 Logistics.

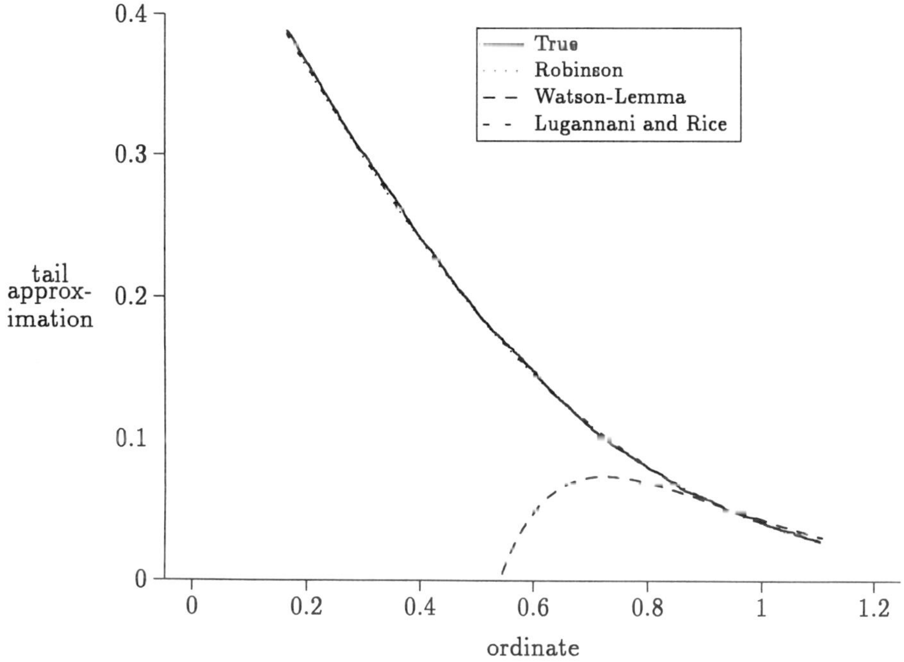

Fig. 13b.

5.4. An Adjusted Approximation

Note that $\hat{\omega}$ is the value of $\hat{\Omega}$, the signed root of the log likelihood ratio statistic, corresponding to the potential data value x. Since $\hat{\omega}$ is an increasing function of x, the cumulative distribution function of $\hat{\Omega}$ satisfies

$$F_{\hat{\Omega}}(\hat{\omega}) = \Phi(\sqrt{n}\hat{\omega}) + \phi(\sqrt{n}\hat{\omega})(1/\hat{\omega} - 1/\hat{z})/\sqrt{n}(1 + O(n^{-1})),$$

providing an additive correction to the cumulative distribution function approximation for $\hat{\Omega}$. Barndorff–Nielsen (1986, 1990a) provides corrections additive on the ordinate scale instead, in order to facilitate conditional inference. Jensen (1992) demonstrates that these corrections hold in great generality, by demonstrating the following result:

Lemma 5.4.1: *Let g_1 and g_2 be smooth increasing functions from $\Re$ to $\Re$, such that $g_1(0) = g_2(0) = 0$ and $g_1'(0) = g_2'(0) > 0$. Let $g_3(s) = g_1(s)^{-1} \log(g_2(s)/g_1(s))$, and let $g_4(s) = g_1(s) - g_3(s)/n$. Then for $s_0 > 0$,*

$$\Phi\left(\sqrt{n}g_4(s)\right) = \Phi\left(\sqrt{n}g_1(s)\right) - \frac{\phi(\sqrt{n}g_1(s))}{\sqrt{n}}\left\{\frac{1}{g_2(s)} - \frac{1}{g_1(s)} + \frac{e_{1,n(s)}}{g_1(s)n}\right\},$$

with

$$\sup\{e_{1,n}(s)|0 \le s \le s_0, n > 0\} < \infty, \ \sup\{\sqrt{n}e_{1,n}(s)|0 \le s \le s_0/\sqrt{n}, n > 0\} < \infty. \tag{90}$$

Proof: For $s < s_0/\sqrt{n}$, the quantities $g_1(s)\sqrt{n}$, $g_2(s)\sqrt{n}$, $|g_1(s)/g_2(s) - 1|\, s^{-1}$, and $g_1(s)^{-1}\,|g_1(s)/g_2(s) - 1|$ are bounded. The power series expansion for $\log(z)$ is $(z-1) - \frac{1}{2}(z-1)^2 + \frac{1}{3}z^{*-3}(z-1)^3$ for z^* between 1 and z, so there exist bounded functions $e_{2,n}(s)$ and $e_{3,n}(s)$ such that

$$g_4(s) - g_1(s) = \frac{n^{-1}}{g_1(s)}\left\{\left(\frac{g_1(s)}{g_2(s)} - 1\right) - \frac{1}{2}\left(\frac{g_1(s)}{g_2(s)} - 1\right)^2 + e_{2,n}(s)\left(\frac{g_1(s)}{g_2(s)} - 1\right)^3\right\},$$

and

$$\Phi(\sqrt{n}g_4(s)) - \Phi(\sqrt{n}g_1(s)) = \sqrt{n}(g_4(s) - g_1(s))\times$$
$$\phi(\sqrt{n}g_1(s))\left[1 - \frac{n}{2}(g_4(s) - g_1(s))g_1(s)\right] + e_{3,n}(s)n^{3/2}(g_4(s) - g_1(s))^3.$$

Substituting,

$$\Phi(\sqrt{n}g_4(s)) - \Phi(\sqrt{n}g_1(s)) = \frac{\phi(\sqrt{n}g_1(s))}{\sqrt{n}g_1(s)}\left\{\left(\frac{g_1(s)}{g_2(s)} - 1\right) - \frac{1}{2}\left(\frac{g_1(s)}{g_2(s)} - 1\right)^2\right.$$
$$\left.+e_{2,n}(s)\left(\frac{g_1(s)}{g_2(s)} - 1\right)^3\right\}\left[1 - \frac{1}{2}\left(\frac{g_1(s)}{g_2(s)} - 1\right) + \frac{1}{4}\left(\frac{g_1(s)}{g_2(s)} - 1\right)^2 - \frac{e_{2,n}(s)}{2}\left(\frac{g_1(s)}{g_2(s)} - 1\right)^3\right]$$
$$+e_{3,n}(s)n^{3/2}(g_4(s) - g_1(s))^3.$$

Thus the second relation in (90) holds.

For $s > s_0/\sqrt{n}$, replace the difference in the normal cumulative distribution function values by an integral of the density over the interval between the ordinates, replace the density using the first relationship of Table 2, use integration by parts, to find

$$\Phi\left(\sqrt{n}g_4(s)\right) - \Phi\left(\sqrt{n}g_1(s)\right) = \int_{\sqrt{n}g_1(s)}^{\sqrt{n}g_4(s)} \phi(x)\,dx = -\int_{\sqrt{n}g_1(s)}^{\sqrt{n}g_4(s)} \phi'(x)x^{-1}\,dx$$

$$= \frac{\phi(\sqrt{n}g_1(s))}{\sqrt{n}g_1(s)} - \frac{\phi(\sqrt{n}g_4(s))}{\sqrt{n}g_4(s)} - \int_{\sqrt{n}g_1(s)}^{\sqrt{n}g_4(s)} \phi(x)x^{-2}\,dx$$

$$= \frac{\phi(\sqrt{n}g_1(s))}{\sqrt{n}g_1(s)} - \frac{\phi(\sqrt{n}g_4(s))}{\sqrt{n}g_4(s)} + g_3(s)\frac{\phi(\sqrt{n}g_1(s))}{\sqrt{n}g_1(s)^2} - \int_{\sqrt{n}g_1(s)}^{\sqrt{n}g_4(s)} \left[\frac{\phi(x)}{x^2} - \frac{\phi(\sqrt{n}g_1(s))}{ng_1(s)^2}\right] dx$$

$$= \frac{\phi(\sqrt{n}g_1(s))}{\sqrt{n}g_1(s)} - \frac{\phi(\sqrt{n}g_4(s))}{\sqrt{n}g_4(s)} + g_3(s)\frac{\phi(\sqrt{n}g_1(s))}{\sqrt{n}g_1(s)^2}$$

$$- \phi(\sqrt{n}g_1(s)) \int_{\sqrt{n}g_1(s)}^{\sqrt{n}g_4(s)} \int_{\sqrt{n}g_1(s)}^{x} \frac{\exp(ng_1^2(s)/2 - u^2/2)(2+u^2)}{u^3}\,du\,dx.$$

The final integral may be expressed as

$$\frac{\phi(\sqrt{n}g_1(s))}{\sqrt{n}g_1(s)} \int_{\sqrt{n}g_1(s)}^{\sqrt{n}g_4(s)} \int_{\sqrt{n}g_1(s)}^{x} \frac{\exp(\frac{1}{2}(ng_1(s) - \sqrt{n}u)(g_1(s) + u/\sqrt{n}))\sqrt{n}g_1(s)(2+u^2)}{u^3}\,du\,dx.$$

The integrand above is bounded, and hence the integral is bounded by a constant times $n\,|g_4(s) - g_1(s)|^2$, which in turn is bounded by a constant times $1/n$. Note that

$$\frac{\phi(\sqrt{n}g_4(s))}{\sqrt{n}g_4(s)} = \frac{\phi(\sqrt{n}g_1(s))}{\sqrt{n}g_1(s)} \exp(-\tfrac{1}{2}g_3(s)/\sqrt{n})/(1 + n^{-1}g_3(s)/g_1(s));$$

this allows expansion of the remaining terms, demonstrating the result.

Q.E.D

Lemma 5.4.1 justifies the approximation

$$F_X(x) = F_{\hat{\Omega}}(\hat{\omega}) = \Phi(\sqrt{n}\hat{\omega}^*) + \phi(\sqrt{n}\hat{\omega})O(n^{-1}), \tag{91}$$

for $\hat{\omega}^* = \hat{\omega} + (n\hat{\omega})^{-1}\log(\hat{z}/\hat{\omega})$; (91) is known as the r^* approximation, since the symbol r is often used in place of $\hat{\omega}$.

5.5. Secant Approximations

Consider a random variable X_n with density $f_n(x) = c_n\sqrt{n}\exp(-np(x))q(x)$, with $p(x) \to \infty$ as $x \to a$ or $x \to b$, and with its unique minimum at $\hat{x}$, and with q positive. Tail probabilities might be calculated for X under conditions weaker than those assumed so far. Change variables to obtain $y(x) = (x - \hat{x})\sqrt{2(p(x) - p(\hat{x}))/(x - \hat{x})^2}$. This is similar to the change of variables in (69). Let $Y_n = y(X_n)$. Then $dy/dx =$

$p'(x)/y$, the density of Y_n is given by $\exp(-np(\hat{x})-\frac{n}{2}y^2)g(y)$, with $g(y)=q(x(y))\times y/p'(x(y))$, and the integral to be approximated is $\int_{y_0}^{\infty}\exp(-\frac{n}{2}y^2)g(y)\,dy$. The function $g(y)$ could be expressed as a power series in y, and the resulting integrand could be treated as in (84). Such a treatment would have the disadvantages outlined at the end of §5.3. Skates (1993) presents an alternative to (89) for such cases, and refers to it as the secant approximation. No cumulant generating function is assumed to exist in this example, and hence usual saddlepoint inversion techniques do not apply. The hard analytical work for these results is contained in the following lemma.

Lemma 5.5.1: *If g is a function of bounded variation defined on $\Re$, having the expansion $g(y)=\sum_{j=1}^{2\mathrm{j}} a_j y^j/j! + O(y^{2\mathrm{j}+1})$ as $y\to 0$, and j continuous derivatives in a neighborhood of y_0, such that*

$$I_{n,0}(y_0)=\sqrt{n}\int_{y_0}^{\infty}\phi(\sqrt{n}y)g(y)\,dy \tag{92}$$

is finite for some integer n, and if $I_{0,j}(-\infty)$ is as in (92) with the range of integration replaced by the whole real line, then

$$\frac{I_{n,0}(y_0)}{I_{n,0}(-\infty)}=1-\Phi(\sqrt{n}y_0)+\left(\sum_{j=0}^{\mathrm{j}} a_l n^{-l}+O(n^{-\mathrm{j}-\frac{1}{2}})\right)\phi(\sqrt{n}y_0)/\sqrt{n}, \tag{93}$$

for quantities a_l defined in terms of y_0 below. Furthermore, if the functions m_l as defined below, and their derivatives, exist and are bounded on $\Re$, then the error term given above is uniform.

Proof: By symmetry about 0, it suffices to prove this for $y_0>0$. Let

$$I_{n,j}(y_0)=\sqrt{n}\int_{y_0}^{\infty}\phi(\sqrt{n}y)g_j(y)\,dy.$$

Following Bleistein (1966), Skates (1993) notes that when the function $g(y)$ is approximated by the affine function of y taking the correct value both at $y(\hat{x})$ and at zero, the integral of the remainder is of a lower order in n. Specifically, let $g_0(y)=g(y)$, and recursively, let $m_j(y)=(g_j(y)-g_j(0))/y$ and $g_{j+1}(y)=m_j'(y)$. Since $g_j(y)=g_j(0)+ym_j(y_0)+y\,[m_j(y)-m_j(y_0)]$, then for $j\le \mathrm{j}$,

$$I_{n,j}(y_0)=\sqrt{n}\int_{y_0}^{\infty}\phi(\sqrt{n}y)\{g_j(0)+ym_j(y_0)+y\,[m_j(y)-m_j(y_0)]\}\,dy$$

$$=g_j(0)(1-\Phi(\sqrt{n}y_0))+\frac{m_j(y_0)}{\sqrt{n}}\phi(\sqrt{n}y_0)+$$

$$\sqrt{n}\int_{y_0}^{\infty}\phi(\sqrt{n}y)y[m_j(y)-m_j(y_0)]\,dy.$$

Integrating by parts, using the identity $\int_a^b u\ dv = uv|_a^b - \int_a^b v\ du$ with $v = \phi(\sqrt{n}y)$, and $dv = ny\phi(\sqrt{n}y)$ yields for $j < \mathrm{j}$,

$$\begin{aligned} I_{n,j}(y_0) &= g_j(0)(1-\Phi(\sqrt{n}y_0)) + n^{-1/2}m_j(y_0)\phi(\sqrt{n}y_0) \\ &\quad - \left[\exp(-\frac{n}{2}y^2)[m_j(y)-m_j(y_0)]/n\right]_{y_0}^{\infty} + n^{-1}\sqrt{n}\int_{y_0}^{\infty}\phi(\sqrt{n}y)m_j'(y)\,dy \\ &= g_j(0)(1-\Phi(\sqrt{n}y_0)) + n^{-1/2}m_j(y_0)\phi(\sqrt{n}y_0) + n^{-1}I_{n,j+1}(y_0), \end{aligned}$$

since $m_j'(y) = g_{j+1}(y)$. This integration by parts is valid only in a formal sense; strictly speaking, m_j need not be differentiable outside a neighborhood of y_0. The contribution to the integral outside of this neighborhood, however, converges to 0 geometrically. Inductively, m_l has $\mathrm{j}-l$ continuous derivatives in a neighborhood of y_0, and g_l has $\mathrm{j}-l+1$ continuous derivatives in a neighborhood of y_0. This integration by parts can also be performed in the Riemann-Stieltjes sense for $j = \mathrm{j}$; here $I_{n,\mathrm{j}+1}(y_0) = \sqrt{n}\int_{y_0}^{\infty}\phi(\sqrt{n}y)m_j(dy)$. Since $I_{n,\mathrm{j}+1}(y_0) = \phi(\sqrt{n}y_0)O(1)$, then $I_{n,0}(y_0) = \sum_{j=0}^{\mathrm{j}} g_j(0)n^{-j}(1-\Phi(\sqrt{n}y_0)) + (\sum_{j=0}^{\mathrm{j}} m_j(y_0)n^{-j} + O(n^{-\mathrm{j}-1/2}))\phi(\sqrt{n}y_0)/\sqrt{n}$. This uses the fact that for $s > 0$,

$$1 - \Phi(s) \le \phi(s)/s. \tag{94}$$

Still to be determined are the coefficients $g_j(0)$. Since $g(y) = \sum_{j=0}^{\mathrm{j}} g^{(j)}(0)y^j/j! + o(y^{\mathrm{j}})$, then $g_0(0) = g(0)$, by definition, and by induction

$$m_l(y) = \sum_{j=1+2l}^{\mathrm{j}} g^{(j)}(0)(j+1)\cdots(j+1-2l)y^{j-1-2l}/(j+1)! + o(y^{\mathrm{j}-1-2l})$$

$$g_l(y) = \sum_{j=2l}^{\mathrm{j}} g^{(j)}(0)(j+1)\cdots(j+1-2l)y^j/(j+1)! + o(y^{\mathrm{j}-j}),$$

and hence $g_l(0) = g^{(2l)}(0)2^{-l}/l!$. By Lemma 4.5.1,

$$I_{n,j}(-\infty) = \sum_{j=0}^{\mathrm{j}} g^{(2j)}(0)/((2n)^j j!) + O(n^{-\mathrm{j}-1/2}).$$

Let the series $\sum_{j=0}^{\mathrm{j}} a_l n^{-l}$ be the quotient of $\sum_{j=0}^{\mathrm{j}} m_j(y_0)n^{-j}$ and $\sum_{j=0}^{\mathrm{j}} g^{(2j)}(0) \times (j!)^{-1}(2n)^{-j}$, satisfying $m_k(y_0) = \sum_{j=0}^{k} g^{(2j)}(0)/(2^j j!)a_{k-j}$. Hence (93) holds. Furthermore, under the last conditions above, uniformity holds since none of the error terms depended on y_0 except through values of m_j.

Q.E.D

Now to apply this lemma to the problem at hand:

Theorem 5.5.2: *If X_n has the density $f_n(x) = c_n\sqrt{n}\exp(-np(x))q(x)$ on (a,b), where*

1. *p has a single local minimum at $\hat{x}$, and $p(x) \to \infty$ as $x \to a$ or $x \to b$.*
2. *p and q have expansions $p(x) - \sum_{j=0}^{2\mathrm{j}+1} p_j(x-\hat{x})^j + o((x-\hat{x})^{2\mathrm{j}+1})$ and $q(x) - \sum_{j=0}^{2\mathrm{j}} q_j(x-\hat{x})^j + o((x-\hat{x})^{2\mathrm{j}})$.*

3. $p^{(j+1)}$ *and* $q^{(j)}$ *are continuous in a neighborhood* $(x-\delta, x+\delta)$ *in which* p' *has no zeros.*

4. $I_n(y) = \int_{\hat{x}}^{b} \exp(-np(x))q(x)\,dx$ *converges absolutely*

then $Q_X(x_0) = \sum_{j=0}^{j-1} g_j(0)n^{-j}(1-\Phi(\sqrt{n}y_0)) + \sum_{j=0}^{j} m_j(y_0)n^{-j-1/2}\phi(\sqrt{n}y_0) + O(n^{-j})$. *Furthermore, if the functions* m_l *as defined below, and their derivatives, exist and are bounded on* $\Re$*, then the error term given above is uniform.*

Proof: Apply Lemma 5.5.1 to $g(y) = yq(x(y))/p'(x(y))$. The coefficients $g_j(0)$ and $m_j(y_0)$ are calculated based on the functions $p(x)$ and $q(x)$.

Q.E.D

Without loss of generality $p(x)$ and $q(x)$ may be chosen such that $p(\hat{x}) = 0$, $p''(\hat{x}) = 1$, and $q(\hat{x}) = 1$. Then $g_0(0) = 1$, $m_0(y) = (g(y)-1)/y$, and $a_0 = m_0(y_0) = q(x_0)/p'(x_0) - 1/y_0$. Hence

$$Q_X(x_0) = 1 - \Phi(\sqrt{n}y_0) + \left((q(x_0)/p'(x_0) - 1/y_0) + O(n^{-1/2})\right)\phi(\sqrt{n}y_0).$$

When the cumulant generating function of a distribution exists, Lemma 5.5.1 can also be used to calculate integrals of the form (84). This alternate method also produces (89).

5.6. Saddlepoint Series for Lattice Variables

Now consider saddlepoint cumulative distribution function approximations for lattice distributions. As before, suppose that $\{Y_n\}$ are independent and identically distributed random variables, but explicitly drop the assumption that they have a density. Rather, suppose that they take values on the lattice $\{a + \Delta\mathfrak{Z}\}$, and that $X = \sum_{k=1}^{n} Y_k/n$. Assume further, without loss of generality, that $a = 0$ and $\Delta = 1$. Again an approximation to the distribution function F_X of X is desired. Recall from §2.7 that characteristic functions for lattice distributions are periodic, thus changing the form of the inversion integrals. This periodicity is inherited by cumulant generating functions.

The inversion integral (28) can be rewritten as

$$\begin{aligned}
\mathrm{P}[X \geq x] &= \frac{1}{2\pi i}\int_{c-i\pi}^{c+i\pi} \frac{\exp(n\mathcal{K}_Y(\beta/n) - \beta(x - 1/[2n]))}{\beta}\frac{\beta/[2n]}{\sinh(\beta/[2n])}\,d\beta \\
&= \frac{1}{2\pi i}\int_{c-i\pi}^{c+i\pi} \frac{\exp(n[\mathcal{K}_Y(\beta) - \beta(x - 1/[2n])])}{\beta}\frac{\beta/2}{\sinh(\beta/2)}\,d\beta \\
&= \frac{1}{2\pi i}\int_{c-i\pi}^{c+i\pi} \frac{\exp(-\mathcal{K}_U(\beta) + n[\mathcal{K}_Y(\beta) - \beta(x - 1/[2n])])}{\beta}\,d\beta
\end{aligned}$$

Here $\mathcal{K}_U = \log(\sinh(\beta/2) - \log(\beta/2)$ is the cumulant generating function for the uniform variable on $(-1/2, 1/2)$. One might be tempted to use usual saddlepoint cumulative distribution function methods (86), (87), or (89), with continuity correction and the cumulant generating function corrected by subtracting

$\mathcal{K}_U(\beta)$; however, since this does not have the factor n the behavior of this approximation is an open question. Daniels (1987) includes $\exp(-\mathcal{K}_U(\beta))$ in the n-independent function ϖ in Watson's Lemma. The analogue of (89) is $1-\Phi(\sqrt{n}\hat{\omega})+n^{-1/2}\phi(\sqrt{n}\hat{\omega})\left(1/\tilde{z}-1/\hat{\omega}+O(n^{-3/2})\right)$. Here $\tilde{z}=2\sqrt{\mathcal{K}_Y''(\hat{\beta})}\sinh(\frac{\hat{\beta}}{2})$. The quantity $O(n^{-3/2})$ may also be replaced by $(1/[\hat{\omega}^3 n])+(3\rho_4-5\rho_3^2-12\tilde{\rho}_3\cosh(\hat{\beta}/2)/\tilde{z}-24\cosh(\hat{\beta}/2)^2/\tilde{z}^2+3/\kappa_2)(24\,\tilde{z}n)^{-1}+O(n^{-5/2})$.

5.7. Exercises

1. Numerically calculate the cumulative distribution function for the convolution of 5, 10, and 20 copies of the distribution of Kolassa (1992) whose cumulative distribution function is given by

$$F(x)=\begin{cases}0 & \text{if } x<0\\ \Phi(\sqrt{x}) & \text{otherwise}\end{cases}$$

at its mean and one, two, and three standard deviations above the mean, and compare to the Robinson and Lugannani and Rice approximations.

2. Verify formula (94), by considering the integral representation of $1-\Phi(s)$, and changing variables so that $\phi(s)/s$ may be factored out, leaving an integrand depending on s. This integrand may be bounded by something that integrates to unity.

6
Multivariate Expansions

Edgeworth and saddlepoint expansions also have analogues for distributions of random vectors. As in the univariate case these expansions will be derived with reference to characteristic functions and cumulant generating functions, and hence these will be defined first. Subsequently Edgeworth density approximations will be defined. Just as in the univariate case, the Edgeworth approximation to probabilities that a random vector lies in a set is the integral of the Edgeworth density over that set; however, since sets of interest are usually not rectangular, theorems for the asymptotic accuracy of these approximations are difficult to prove. These proofs are not presented here. Approximation for variables on a multivariate lattice are discussed. Multivariate saddlepoint approximations are also defined, by a multivariate extension of steepest descent methods. These methods are also used to approximate conditional probabilities.

Results presented in this chapter will have their primary applications in justifying techniques presented in the following two chapters.

6.1. Multivariate Generating Functions

Define the multivariate characteristic function of $\boldsymbol{X} = (X^1, ..., X^{\mathfrak{k}})^\top$ as

$$\zeta_{\boldsymbol{X}}(\boldsymbol{\beta}) = \mathrm{E}\left[\exp(i\boldsymbol{\beta}^\top \boldsymbol{X})\right] = \mathrm{E}\left[\exp(i\sum_{j=1}^{\mathfrak{k}} \beta_j X^j)\right].$$

This always exists over $\mathfrak{R}^{\mathfrak{k}}$. Define the multivariate moment generating function of $\boldsymbol{X}$ as $\mathcal{M}_{\boldsymbol{X}}(\boldsymbol{\beta}) = \mathrm{E}\left[\exp(\boldsymbol{\beta}^\top \boldsymbol{X})\right]$. As in the univariate case, $\mathcal{M}_{\boldsymbol{X}}$ exists at $\mathbf{o} = (0, \ldots, 0)$. Saddlepoint expansions will be available when $\mathcal{M}_{\boldsymbol{X}}$ exists in a neighborhood $\mathfrak{Q} \subset \mathfrak{R}^{\mathfrak{k}}$ of $\mathbf{o}$.

Many of the useful properties of univariate moment generating functions hold in the multivariate case as well. For instance, if $\boldsymbol{A}$ and $\boldsymbol{B}$ are matrices and $\boldsymbol{X}$ and $\boldsymbol{Y}$ are independent random vectors such that the combination $\boldsymbol{AX} + \boldsymbol{BY}$ is defined, (that is, $\boldsymbol{A}$ and $\boldsymbol{B}$ have the same number of rows, and the numbers of columns equal the lengths of $\boldsymbol{X}$ and $\boldsymbol{Y}$ respectively), then

$$\begin{aligned}\mathcal{M}_{\boldsymbol{AX}+\boldsymbol{BY}}(\boldsymbol{\beta}) &= \mathrm{E}\left[\exp(\boldsymbol{\beta}^\top(\boldsymbol{AX}+\boldsymbol{BY}))\right] = \mathrm{E}\left[\exp(\boldsymbol{\beta}^\top \boldsymbol{AX})\exp(\boldsymbol{\beta}^\top \boldsymbol{BY}))\right] \\ &= \mathrm{E}\left[\exp(\boldsymbol{\beta}^\top \boldsymbol{AX})\right]\mathrm{E}\left[\exp(\boldsymbol{\beta}^\top \boldsymbol{BY}))\right] = \mathcal{M}_{\boldsymbol{X}}(\boldsymbol{A}^\top\boldsymbol{\beta})\mathcal{M}_{\boldsymbol{Y}}(\boldsymbol{B}^\top\boldsymbol{\beta}) \quad (95)\end{aligned}$$

Expand the moment generating function as

$$\mathcal{M}_{\boldsymbol{X}}(\boldsymbol{\beta}) = \sum_{j=0}^{\infty}\sum_{\boldsymbol{s}\in S(j)} \frac{1}{j!}\mu^{s_1\cdots s_j}\beta_{m_1}\cdots\beta_{m_j}$$

where $S(j) = \{1,\ldots,\mathfrak{k}\}^j$, the set of vectors of integers with j components and all entries between 1 and $\mathfrak{k}$. This series by itself is not enough to uniquely define the μ's, since all coefficients μ with the same set of indices permuted correspond to the same product of elements of $\boldsymbol{\beta}$. Adding the requirement that the μ's have the same value if their indices are permuted, however, does allow the above expansion to uniquely determine the coefficients. As in the univariate case, $\mu^{s_1\cdots s_j} = \mathrm{E}\left[X^{s_1}\cdots X^{s_j}\right]$. For example, $\mu^1 = \mathrm{E}\left[X^1\right]$, $\mu^2 = \mathrm{E}\left[X^2\right]$, $\mu^{11} = \mathrm{E}\left[X^1X^1\right]$, and $\mu^{12} = \mathrm{E}\left[X^1X^2\right]$.

One may also define the multivariate cumulant generating function $\mathcal{K}_{\boldsymbol{X}}(\boldsymbol{\beta})$ using the moment generating function in analogy with the univariate case: $\mathcal{K}_{\boldsymbol{X}}(\boldsymbol{\beta}) = \log(\mathcal{M}_{\boldsymbol{X}}(\boldsymbol{\beta}))$. Again this function will generate the cumulants. Expand $\mathcal{K}_{\boldsymbol{X}}$ as

$$\mathcal{K}_{\boldsymbol{X}}(\boldsymbol{\beta}) = \sum_{j=1}^{\infty}\sum_{\boldsymbol{s}\in S(j)} \frac{1}{j!}\kappa^{s_1\cdots s_j}\beta_{m_1}\cdots\beta_{m_j}$$

with the same definition for the inner sum. As above, require that coefficients κ with the same indices up to permutation have the same value. This is enough to define the multivariate cumulants κ uniquely via the series expansion.

As in the univariate case, these cumulants may be expressed in terms of the moments. For example, $\kappa^1 = \mathrm{E}\left[X^1\right]$, $\kappa^2 = \mathrm{E}\left[X^2\right]$, $\kappa^{11} = \mathrm{Var}\left[X^1\right]$, and $\kappa^{12} = \mathrm{Cov}[X^1, X^2]$. If the components of $\boldsymbol{X}$ are independent, then

$$\begin{aligned}\mathcal{K}_{\boldsymbol{X}}(\boldsymbol{\beta}) &= \log\left(\mathrm{E}\left[\exp\left(\sum_{j=1}^{\mathfrak{k}}\beta_j X^j\right)\right]\right) = \log\left(\prod_{j=1}^{\mathfrak{k}}\mathrm{E}\left[\exp(\beta_j X^j)\right]\right) \\ &= \sum_{j=1}^{\mathfrak{k}}\log\left(\mathrm{E}\left[\exp(\beta_j X^j)\right]\right) = \sum_{j=1}^{\mathfrak{k}}\mathcal{K}_{X^j}(\beta_j)\end{aligned}$$

Hence in the case of independent components mixed cumulants are zero. That is, $\kappa^{s_1\cdots s_j} = 0$ if the indices are not all the same.

Multivariate cumulants have some of the same invariance properties as have univariate cumulants, as described in §2.1. Applying (95) with $\boldsymbol{B} = \mathbf{o}$, and taking logarithms, $\mathcal{K}_{\boldsymbol{AX}}(\boldsymbol{\beta}) = \mathcal{K}_{\boldsymbol{X}}(\boldsymbol{A}^\top\boldsymbol{\beta})$. If λ denotes the cumulants of the linear combination $\boldsymbol{AX}$, and if the element in the i, j element of $\boldsymbol{A}$ is a_i^j, then

$$\begin{aligned}\mathcal{K}_{\boldsymbol{AX}}(\boldsymbol{\beta}) &= \sum_{j=1}^{\infty} \sum_{\boldsymbol{s}\in S(j)} \frac{1}{j!} \kappa^{s_1 \cdots s_j} (\sum_{r_1} a_{s_1}^{r_1} \beta_{r_1}) \cdots (\sum_{r_j} a_{s_j}^{r_j} \beta_{r_j}) \\ &= \sum_{j=1}^{\infty} \sum_{\boldsymbol{r}\in S(j)} \frac{1}{j!} \sum_{\boldsymbol{s}\in S(j)} \left[\kappa^{s_1 \cdots s_j} a_{s_1}^{r_1} \cdots a_{s_j}^{r_j}\right] \beta_{r_1} \cdots \beta_{r_j}.\end{aligned}$$

Hence $\lambda^{r_1 \cdots r_j} = \sum_{\boldsymbol{s}\in S(j)} \left[\kappa^{s_1 \cdots s_j} a_{s_1}^{r_1} \cdots a_{s_j}^{r_j}\right]$.

6.2. Edgeworth Series

In this section the multivariate counterpart of the Edgeworth approximations to densities and distribution functions are presented. These approximations can be expressed as refinements to the normal approximations, and so these normal approximations are presented first. Proofs are deferred until a later section.

Assume that $\boldsymbol{X}$ is a random vector such that $\mathrm{E}[\boldsymbol{X}] = \mathbf{o}$. Let Σ represent the covariance matrix of $\boldsymbol{X}$, and let $\Sigma^{-1} = [\kappa_{ij}]$ represent its inverse. If the distribution of $\boldsymbol{X}$ has a density, define the normal approximation to this density to be

$$\phi(\boldsymbol{x}, \Sigma) = \frac{\exp(-\frac{1}{2} \sum_{j=1}^{\mathfrak{k}} \sum_{i=1}^{\mathfrak{k}} x^j x^i \kappa_{ji})}{(2\pi)^{\mathfrak{k}/2} \sqrt{\det[\Sigma]}}.$$

Define its cumulative distribution function $F_{\boldsymbol{X}}(\boldsymbol{x})$ to be the normal approximation for the cumulative distribution function $\mathrm{P}[\boldsymbol{X} \leq \boldsymbol{x}] = \mathrm{P}[X^j \leq x^j \; \forall j]$ and define the normal approximation for the cumulative distribution function to be

$$\Phi(\boldsymbol{x}, \Sigma) = \int_{-\infty}^{x^1} \cdots \int_{-\infty}^{x^{\mathfrak{k}}} \phi(\boldsymbol{y}, \Sigma)\, d\boldsymbol{y}.$$

Just as in the univariate case, the normal approximation uses information in the first and second order cumulants to approximate the distribution of interest. Also as in the univariate case, higher order cumulants can be used to correct this series, producing a higher-order analogue to the Edgeworth series.

Multivariate Fourier inversion techniques will be used to approximate a density inversion integral. As before, the exponent will be divided into a quadratic term in $\boldsymbol{\beta}$, and a higher-order term in $\boldsymbol{\beta}$. The higher order term will be exponentiated, and the result will be integrated term by term.

Begin by defining higher-order analogues to pseudo-moments. Let the coefficients μ^* be defined from the power series expansion

$$1 + \sum_{j=1}^{\infty} \sum_{\boldsymbol{s}\in S(j)} \frac{1}{j!} \mu^{*s_1 \cdots s_j} \beta_{m_1} \cdots \beta_{m_j} = \exp\left(\sum_{j=3}^{\infty} \sum_{\boldsymbol{s}\in S(j)} \frac{1}{j!} \kappa^{s_1 \cdots s_j} \beta_{m_1} \cdots \beta_{m_j}\right),$$

again with the convention that the μ^*'s have the same value if their indices are permuted, and again where the inner sum is over all j-long vectors of integers $\boldsymbol{s}$

between 1 and $\mathfrak{k}$. For a vector $\boldsymbol{s} = (s_1, \ldots, s_{\mathfrak{k}})$ let $\kappa^{\boldsymbol{s}}$ and $\mu^{*\boldsymbol{s}}$ denote the cumulants and pseudo-moments with superscripts $s_1, \ldots, s_{\mathfrak{k}}$. Although $\boldsymbol{s}$ is a vector these quantities do not depend on the order of the components of $\boldsymbol{s}$. Then, still assuming $\mathrm{E}[\boldsymbol{X}] = \mathbf{o}$, the characteristic function of $f_{\boldsymbol{X}}(\boldsymbol{x})$ is

$$\int_{-\infty}^{\infty} \cdots \int_{-\infty}^{\infty} \exp(-i\boldsymbol{\beta}\boldsymbol{x} + \tfrac{1}{2}\boldsymbol{\beta}^\top \boldsymbol{\Sigma}\boldsymbol{\beta}) \left\{ 1 + \sum_{j=1}^{\infty} \sum_{\boldsymbol{s}\in S(j)} \frac{i^j}{j!} \mu^{*\boldsymbol{s}} \beta_{m_1} \cdots \beta_{m_j} \right\}.$$

Inverting this relationship termwise, and extending (36), the density of $\boldsymbol{X}$ can be approximated by an Edgeworth series derived from the formal series expansion

$$\sum_{j=0}^{\infty} \sum_{\boldsymbol{s}\in S(j)} \frac{1}{j!} \mu^{*s_1 \cdots s_j} (-1)^j \frac{d^j}{dx^{s_1} \cdots dx^{s_j}} \phi(\boldsymbol{x}, \Sigma) \tag{96}$$

where the $\mu^{*s_1 \cdots s_j}$ are sums of products of the joint cumulants. This can also be defined in terms of the generalized Hermite polynomials

$$\frac{d^j}{dx^{s_1} \cdots dx^{s_j}} \phi(\boldsymbol{x}, \Sigma) / \phi(\boldsymbol{x}, \Sigma).$$

Integrating with respect to $\boldsymbol{x}$, the cumulative distribution function of $\boldsymbol{X}$ can be approximated by an Edgeworth series derived from the formal series expansion

$$\sum_{j=1}^{\infty} \sum_{\boldsymbol{s}\in S(j)} \frac{1}{j!} \mu^{*s_1 \cdots s_j} (-1)^j \frac{d^j}{dx^{s_1} \cdots dx^{s_j}} \Phi(\boldsymbol{x}, \Sigma). \tag{97}$$

When $\boldsymbol{X}$ depends on a parameter n, the cumulants $\kappa^{\boldsymbol{s}}_{\boldsymbol{X}}$ then also depend on n. When $\boldsymbol{X}$ is the standardized sum of independent and identically distributed components, $\kappa^{\boldsymbol{s}}_{\boldsymbol{X}} = n^{1-|\boldsymbol{s}|/2}\kappa^{\boldsymbol{s}}_1$, and in this case discarding terms in (96) of size $o(n^{-j/2})$ leaves an Edgeworth series $e_j(\boldsymbol{x}, \boldsymbol{\kappa}^{*n})$ such that

$$f_{\boldsymbol{X}}(\boldsymbol{x}) = e_j(\boldsymbol{x}, \boldsymbol{\kappa}^{*n}) + o(n^{-j/2}). \tag{98}$$

Under other assumptions on the distribution for $\boldsymbol{X}$, $\kappa^{\boldsymbol{s}}_{\boldsymbol{X}} = O(n^{1-|\boldsymbol{s}|/2})$ (Chambers, 1967). Here $|\boldsymbol{s}|$ denotes the number of elements in $\boldsymbol{s}$. McCullagh (1987) treats this multivariate case in detail. Proofs demonstrating the order of the error are provided below.

6.3. Saddlepoint Approximations

As in the univariate case, the first order saddlepoint density approximation can be calculated directly from the normal approximation. Suppose that a random vector $\boldsymbol{X}$ arising as the mean of n independent and identically distributed copies of $\boldsymbol{Y}$ has a density $f_{\boldsymbol{X}}(\boldsymbol{x})$ and a multivariate cumulant generating function $\mathcal{K}_{\boldsymbol{X}}(\boldsymbol{\beta})$. Embed $\boldsymbol{X}$ in an exponential family, to define a density $f_{\boldsymbol{X}}(\boldsymbol{x}, \boldsymbol{\beta}) = \exp(\boldsymbol{\beta}^\top \boldsymbol{x} - \mathcal{K}_{\boldsymbol{X}}(\boldsymbol{\beta})) f_{\boldsymbol{X}}(\boldsymbol{x})$. As before we desire an approximation $g_n(\boldsymbol{x})$ to $f_{\boldsymbol{X}}(\boldsymbol{x})$, and proceed by finding $\hat{\boldsymbol{\beta}}$ dependent on $\boldsymbol{x}$ such that $f_{\boldsymbol{X}}(\boldsymbol{x}, \hat{\boldsymbol{\beta}})$ has mean $\boldsymbol{x}$ and approximating $f_{\boldsymbol{X}}(\boldsymbol{x}, \hat{\boldsymbol{\beta}})$ by a normal density with mean 0. Then $\hat{\boldsymbol{\beta}}$ is defined by the multivariate saddlepoint equation

$$\mathcal{K}'_{\boldsymbol{X}}(\hat{\boldsymbol{\beta}}) = \boldsymbol{x}. \tag{99}$$

This normal approximation is $(2\pi)^{(-\mathfrak{k}/2)} \det \left[\mathcal{K}''_{\boldsymbol{X}}(\hat{\boldsymbol{\beta}})\right]^{-1/2}$. By (98),

$$f_{\boldsymbol{X}}(\boldsymbol{x}) = \exp(\mathcal{K}_{\boldsymbol{X}}(\hat{\boldsymbol{\beta}}) - \hat{\boldsymbol{\beta}}^{\top}\boldsymbol{x})(2\pi)^{(-\mathfrak{k}/2)} \det \left[\mathcal{K}''_{\boldsymbol{X}}(\hat{\boldsymbol{\beta}})\right]^{-1/2} + O(n^{-1/2}).$$

As in the univariate case, when (99) has a solution the only conditions to be checked are the those necessary to demonstrate the order of error of the underlying Edgeworth series. We will see that this requires only the existence of the proper cumulants, which is verified by noting that moments of all orders exist, and the integrability of some power of the characteristic function, which is verified by noting that the density $f_{\boldsymbol{X}}$ is assumed to exist. Also as in the univariate case no uniformity claim is made here, although the same heuristics lead one to expect good uniform error properties.

One might address the question of the existence of solutions to (99) in much the same manner as in the univariate case; this will not be undertaken here. Note, however, that the solution to (99) is once again the maximum likelihood estimator, this time for the multivariate parameter $\boldsymbol{\beta}$ in the exponential family defined by $f_{\boldsymbol{X}}(\boldsymbol{x}, \boldsymbol{\beta})$. Various authors including Albert and Anderson (1984), Barndorff–Nielsen (1978), Clarkson and Jennrich (1991), and Jacobsen (1989), have considered the case when the data $\boldsymbol{x}$ lie outside of but on the boundary of the range of $\mathcal{K}'_{\boldsymbol{X}}(\boldsymbol{\beta})$. Kolassa (1996a) considers the implications of lack of existence of $\hat{\boldsymbol{\beta}}$ on saddlepoint applications.

Predictably, the higher-order saddlepoint approximations can be derived from the Edgeworth approximation. In the multivariate setting, as in the univariate setting, the factor multiplying $1/\sqrt{n}$ in the multiplicative correction factor to the baseline approximation is zero, since the associated Hermite polynomial is evaluated at $\mathbf{o}$. One might expect that $f_{\boldsymbol{X}}(\boldsymbol{x})$ might have an expansion of the form

$$f_{\boldsymbol{X}}(\boldsymbol{x}) = \exp(n[\mathcal{K}_{\boldsymbol{Y}}(\hat{\boldsymbol{\beta}}) - \hat{\boldsymbol{\beta}}^{\top}\boldsymbol{x}]) \left(\frac{n}{2\pi}\right)^{\mathfrak{k}/2} \det \left[\mathcal{K}''_{\boldsymbol{Y}}(\hat{\boldsymbol{\beta}})\right]^{-1/2} \left[1 + \frac{b(\hat{\boldsymbol{\beta}})}{2n} + O\left(n^{-2}\right)\right]. \tag{100}$$

To obtain $b(\hat{\boldsymbol{\beta}})$, approximate the tilted distribution $f_{\boldsymbol{X}}(\boldsymbol{x}, \hat{\boldsymbol{\beta}})$ by

$$\sum_{j=0}^{\infty} \sum_{\boldsymbol{s} \in S(j)} \frac{1}{j!} \mu^{*s_1 \cdots s_j} (-1)^j \frac{d^j}{dx^{s_1} \cdots dx^{s_j}} \phi(0, K''(\hat{\boldsymbol{\beta}})), \tag{101}$$

where the cumulants involved in (101) are derivatives of $\mathcal{K}_{\boldsymbol{X}}$ at $\hat{\boldsymbol{\beta}}$, identify the terms of order $O(1/\sqrt{n})$, and check that they are zero. Then identify the terms of order

$O(n^{-1})$ and evaluate the generalized Hermite polynomials

$$h_{s_1,\dots,s_j}(\boldsymbol{x}; K''(\hat{\boldsymbol{\beta}})) = \left[\frac{d^j}{dx^{s_1}\cdots dx^{s_j}}\phi(\boldsymbol{x}, K''(\hat{\boldsymbol{\beta}}))\right] / \phi(\boldsymbol{x}, K''(\hat{\boldsymbol{\beta}}))$$

at zero. The pseudo-moments in (101) are generated by equating coefficients of the series expansion of

$$\begin{aligned} 1+\sum_{j=3}^{\infty}\sum_{s\in S(j)}\frac{1}{j!}\mu^{*s}\beta_{s_1}\cdots\beta_{s_j} &= \exp\left(\sum_{j=3}^{\infty}\sum_{s\in S(j)}\frac{1}{j!}\kappa^{s}\beta_{s_1}\cdots\beta_{s_j}\right) \\ &= 1+\left(\frac{1}{3!}\kappa^{ijk}\beta_i\beta_j\beta_k+\frac{1}{4!}\kappa^{ijkl}\beta_i\beta_j\beta_k\beta_l+\cdots\right) \\ &\quad +\frac{1}{2}\left(\frac{1}{3!}\kappa^{ijk}\beta_i\beta_j\beta_k+\frac{1}{4!}\kappa^{ijkl}\beta_i\beta_j\beta_k\beta_l+\cdots\right)^2+\cdots \end{aligned}$$

Here and below invoke the convention that an index appearing both as a subscript and as a superscript is summed over. After replacing products of components of $\boldsymbol{\beta}$ by the associated Hermite polynomials, terms of order $O(1/\sqrt{n})$ in (101) are

$$\frac{1}{3!}\mathcal{K}_{\boldsymbol{X}}^{ijk}(\hat{\boldsymbol{\beta}})h_{ijk}(\boldsymbol{x}; K''(\hat{\boldsymbol{\beta}}));$$

the terms of order $O(n^{-1})$ in (101) are

$$\frac{1}{4!}\mathcal{K}_{\boldsymbol{X}}^{ijkl}(\hat{\boldsymbol{\beta}})h_{ijkl}(\boldsymbol{x}; K''(\hat{\boldsymbol{\beta}}))+\frac{1}{2\times 3!3!}\mathcal{K}_{\boldsymbol{X}}^{ijk}(\hat{\boldsymbol{\beta}})\mathcal{K}_{\boldsymbol{X}}^{lmn}(\hat{\boldsymbol{\beta}})h_{ijklmn}(\boldsymbol{x}; K''(\hat{\boldsymbol{\beta}})). \quad (102)$$

Here dependence of $\mathcal{K}_{\boldsymbol{X}}$ on n is suppressed, and the superscripts refer to derivatives with respect to components of $\boldsymbol{\beta}$. For some purposes it is preferable to have the pseudo-moments be symmetric in their indices; the coefficients of products of components of $\boldsymbol{\beta}$ in (102) do not have this property. In particular, the second term of size $O(n^{-1})$ is asymmetric. These terms may be made symmetric by summing over all permutations giving recognizably different index patterns and dividing by the number of such patterns. Were this necessary in the present context we would be required to add up over all $(6!/(3!3!))/2 = 10$ ways to partition $ijklmn$ into two groups of 3 each, and then divide by 10. This is not necessary for what follows. McCullagh (1987) gives formulas for the Hermite polynomials associated with an arbitrary covariance matrix Σ as equation (5.7); for the present purposes it suffices to note that

$$\begin{aligned} h_{ijk}(\mathbf{o};\Sigma) &= 0 \\ h_{ijkl}(\mathbf{o},\Sigma) &= \kappa_{ij}\kappa_{kl}[3] = \kappa_{ij}\kappa_{kl}+\kappa_{ik}\kappa_{jl}+\kappa_{il}\kappa_{jk} \\ h_{ijklmn}(\mathbf{o};\Sigma) &= -\kappa_{ij}\kappa_{kl}\kappa_{mn}[15]. \end{aligned}$$

These polynomials will be evaluated with Σ equal to the covariance matrix $[\hat{\kappa}_{ij}]$ associated with the saddlepoint. Here we use the notation of McCullagh (1987) in denoting by quantity with subscripts followed by an integer in brackets the sum of all similar but distinct terms generated by permuting the indices. After replacing quantities involving $\mathcal{K}_X$ by counterparts involving $\mathcal{K}_Y$,

$$\begin{aligned}2b(\hat{\beta}) &= \frac{1}{4!}\mathcal{K}_Y^{ijkl}(\hat{\beta})h_{ijkl}(\mathbf{o};K''(\hat{\beta})) + \frac{1}{2\times 3!3!}\mathcal{K}_Y^{ijk}(\hat{\beta})\mathcal{K}_Y^{lmo}(\hat{\beta})h_{ijklmo}(\mathbf{o};K''(\hat{\beta}))\\ &= \frac{1}{4!}\hat{\kappa}^{ijkl}(\hat{\kappa}_{ij}\hat{\kappa}_{kl}[3]) - \frac{10}{6!}\hat{\kappa}^{ijk}\hat{\kappa}^{lmo}(\hat{\kappa}_{ij}\hat{\kappa}_{kl}\hat{\kappa}_{mo}[15]).\end{aligned}$$

Sorting the fifteen summands represented by $\hat{\kappa}_{ij}\hat{\kappa}_{kl}\hat{\kappa}_{mo}[15]$ into nine for which the indices for two factors are subsets of the indices of $\hat{\kappa}^{ijk}$ or $\hat{\kappa}^{lmo}$, and six for which indices of all three factors are split between the indices of $\hat{\kappa}^{ijk}$ and $\hat{\kappa}^{lmo}$,

$$\begin{aligned}2b(\hat{\beta}) &= \frac{1}{4!}\hat{\kappa}^{ijkl}(\hat{\kappa}_{ij}\hat{\kappa}_{kl}[3]) - \frac{10}{6!}\left[(\hat{\kappa}^{ijk}\hat{\kappa}^{lmo}\hat{\kappa}_{ij}\hat{\kappa}_{kl}\hat{\kappa}_{mo}[9]) + (\hat{\kappa}^{ijk}\hat{\kappa}^{lmo}\hat{\kappa}_{il}\hat{\kappa}_{jm}\hat{\kappa}_{ko}[6])\right]\\ &= \frac{3}{4!}\hat{\kappa}^{ijkl}\hat{\kappa}_{ij}\hat{\kappa}_{kl} - \left[\frac{90}{6!}\hat{\kappa}^{ijk}\hat{\kappa}^{lmo}\hat{\kappa}_{ij}\hat{\kappa}_{kl}\hat{\kappa}_{mo} + \frac{60}{6!}\hat{\kappa}^{ijk}\hat{\kappa}^{lmo}\hat{\kappa}_{il}\hat{\kappa}_{jm}\hat{\kappa}_{ko}\right]\end{aligned}$$

Then $b(\hat{\beta}) = \frac{1}{4}\hat{\rho}_4 - \frac{1}{4}\hat{\rho}_{13} - \frac{1}{6}\hat{\rho}_{23}$, for

$$\hat{\rho}_4 = \hat{\kappa}^{ijkl}\hat{\kappa}_{ij}\hat{\kappa}_{kl},\ \hat{\rho}_{13} = \hat{\kappa}^{ijk}\hat{\kappa}^{lmo}\hat{\kappa}_{ij}\hat{\kappa}_{kl}\hat{\kappa}_{mo},\ \hat{\rho}_{23} = \hat{\kappa}^{ijk}\hat{\kappa}^{lmo}\hat{\kappa}_{il}\hat{\kappa}_{jm}\hat{\kappa}_{ko}. \tag{103}$$

For the sake of brevity all cumulants $\hat{\kappa}$ denote those of the distribution tilted to $\hat{\beta}$ and are given by derivatives of $\mathcal{K}_Y(\beta) - \beta^\top \mathbf{x}$ at $\hat{\beta}$. For example, $\hat{\kappa}^1 = \partial\mathcal{K}_Y/\partial\beta_1 - x^1$, and $\hat{\kappa}^{12} = \partial^2\mathcal{K}_Y/\partial\beta_1\partial\beta_2$. Let $\hat{\kappa}_{ij}$ refer to the i,j entry of the inverse of the matrix formed by $\hat{\kappa}^{ij}$. The quantities $\hat{\rho}_4$, $\hat{\rho}_{13}$, and $\hat{\rho}_{23}$ are multivariate skewness and kurtosis measures proposed by Mardia (1970) and McCullagh (1986). Hence, as in the univariate case the terms of order $O(n^{-1/2})$ and $O(n^{-3/2})$ are zero, and (100) holds.

Wang (1990b) provides a saddlepoint approximation to multivariate cumulative distribution functions.

6.4. Multivariate Inversion Integrals

We require the following multivariate extension of the univariate theorem on inverting characteristic functions to recover the underlying cumulative distribution function, presented in §2.4. This proof is also found in Billingsley (1986).

Theorem 6.4.1: *If a probability distribution* $\mathrm{P}[\cdot]$ *on* $\mathfrak{R}^{\mathfrak{k}}$ *corresponds to a characteristic function* ζ, *and the rectangle* $\mathfrak{B} = \{\mathbf{x} \in \mathfrak{R}^{\mathfrak{k}} | x^j \in [b^{1j}, b^{2j}]\ \forall j = 1,\ldots,\mathfrak{k}\}$ *has a boundary*

$$\partial\mathfrak{B} = \{\mathbf{x} \in \mathfrak{R}^{\mathfrak{k}} | x^j \in [b^{1j}, b^{2j}]\ \forall j = 1,\ldots,\mathfrak{k}, x^j = b^{1j} \text{ or } b^{2j} \textit{ for some } j\}$$

that is assigned zero probability, then

$$\mathrm{P}[\mathfrak{B}] = \lim_{T\to\infty}\int_{-T}^{T}\cdots\int_{-T}^{T}\prod_{j=1}^{\mathfrak{k}}\frac{\exp(-i\beta_j b^{1j}) - \exp(-i\beta_j b^{2j})}{i\beta_j}\zeta(\beta)\frac{d\beta}{(2\pi)^{\mathfrak{k}}}. \tag{104}$$

The factors in the product over j *in the integrand have a removable singularity at* $\beta_j = 0$; *as before substitute the value* $b^{2j} - b^{1j}$.

Proof: Let

$$I_T = \int_{-T}^{T} \cdots \int_{-T}^{T} \prod_{j=1}^{\mathfrak{k}} \frac{\exp(-i\beta_j b^{1j}) - \exp(-i\beta_j b^{2j})}{i\beta_j} \zeta(\boldsymbol{\beta}) \frac{d\boldsymbol{\beta}}{(2\pi)^{\mathfrak{k}}}$$

be the quantity inside the limit operation in (104). After expressing this characteristic function as an integral over possible values of $\boldsymbol{x}$,

$$I_T = \int_{-T}^{T} \cdots \int_{-T}^{T} \int_{-\infty}^{\infty} \cdots \int_{-\infty}^{\infty} \prod_{j=1}^{\mathfrak{k}} \frac{\exp(i\beta_j(x^j - b^{1j})) - \exp(i\beta_j(x^j - b^{2j}))}{i\beta_j} dF(\boldsymbol{x}) \frac{d\boldsymbol{\beta}}{(2\pi)^{\mathfrak{k}}}.$$

The integrand above is bounded by $\prod_{j=1}^{\mathfrak{k}} |b^{2j} - b^{1j}|$ by Theorem 2.5.3. The order of integration can therefore be interchanged, and the complex exponential can be expressed in terms of sines and cosines:

$$I_T = \int_{-\infty}^{\infty} \cdots \int_{-\infty}^{\infty} \int_{-T}^{T} \cdots \int_{-T}^{T} \prod_{j=1}^{\mathfrak{k}} \left(\frac{\cos(\beta_j(x^j - b^{1j})) + i\sin(\beta_j(x^j - b^{2j}))}{i\beta_j} \right.$$
$$\left. - \frac{\cos(\beta_j(x^j - b^{2j})) + i\sin(\beta_j(x^j - b^{2j}))}{i\beta_j} \right) \frac{d\boldsymbol{\beta}}{(2\pi)^{\mathfrak{k}}} dF(\boldsymbol{x}).$$

Since the cosine terms involve odd order terms in β_j these terms integrate to zero:

$$I_T = \int_{-\infty}^{\infty} \cdots \int_{-\infty}^{\infty} \int_{-T}^{T} \cdots \int_{-T}^{T} \prod_{j=1}^{\mathfrak{k}} \frac{\sin(\beta_j(x^j - b^{1j})) - \sin(\beta_j(x^j - b^{1j}))}{\beta_j} \frac{d\boldsymbol{\beta}}{(2\pi)^{\mathfrak{k}}} dF(\boldsymbol{x})$$
$$= \int_{-T}^{T} \cdots \int_{-T}^{T} \prod_{j=1}^{\mathfrak{k}} \frac{2S(\beta\,|x^j - b^{1j}|)\operatorname{sgn}(x^j - b^{1j}) - 2S(\beta\,|x^j - b^{2j}|)\operatorname{sgn}(x^j - b^{2j})}{2\pi} dF(\boldsymbol{x})$$

where again $S(\theta) = \int_0^\theta \sin(x)/x\, dx$. Since the integrand is bounded, one can pass to the limit:

$$\lim_{T\to\infty} I_T = (S(\infty)/\pi)^{\mathfrak{k}} \int_{-\infty}^{\infty} \cdots \int_{-\infty}^{\infty} \prod_{j=1}^{\mathfrak{k}} (\operatorname{sgn}(x^j - b^{1j}) - \operatorname{sgn}(x^j - b^{2j}))\, dF(\boldsymbol{x})$$
$$- \int_{-\infty}^{\infty} \cdots \int_{-\infty}^{\infty} \prod_{j=1}^{\mathfrak{k}} (\operatorname{sgn}(x^j - b^{1j}) - \operatorname{sgn}(x^j - b^{2j}))\, dF(\boldsymbol{x}),$$

since $S(\infty)/\pi = 1$ as was found in §2.4.

Q.E.D

A theorem for recovering multivariate density functions also exists:

Theorem 6.4.2: *If a random vector has the characteristic function ζ satisfying*

$$\int_{-\infty}^{\infty} \cdots \int_{-\infty}^{\infty} |\zeta(\boldsymbol{\beta})|\, d\boldsymbol{\beta} < \infty \tag{105}$$

then its density is given by

$$f_Y(\boldsymbol{y}) = \int_{-\infty}^{\infty}\cdots\int_{-\infty}^{\infty} \exp(-i\boldsymbol{\beta}\boldsymbol{y})\zeta(\boldsymbol{\beta})\,\frac{d\boldsymbol{\beta}}{(2\pi)^{\mathfrak{k}}}. \tag{106}$$

Proof: By (105) one can replace the limit of proper integrals as $T \to \infty$ by the corresponding improper integral over the real line in (104). As in §2.4 replace the difference in exponentials by an integral. Then

$$F(\boldsymbol{b}^2) - F(\boldsymbol{b}^1) = \int_{-\infty}^{\infty}\cdots\int_{-\infty}^{\infty}\int_{b^{11}}^{b^{21}}\cdots\int_{b^{1\mathfrak{k}}}^{b^{2\mathfrak{k}}} \exp(-i\boldsymbol{\beta}\boldsymbol{y})\,d\boldsymbol{y}\,\zeta(\boldsymbol{\beta})\,\frac{d\boldsymbol{\beta}}{(2\pi)^{\mathfrak{k}}}$$
$$= \int_{b^{11}}^{b^{21}}\cdots\int_{b^{1\mathfrak{k}}}^{b^{2\mathfrak{k}}}\int_{-\infty}^{\infty}\cdots\int_{-\infty}^{\infty} \exp(-i\boldsymbol{\beta}\boldsymbol{y})\zeta(\boldsymbol{\beta})\,\frac{d\boldsymbol{\beta}}{(2\pi)^{\mathfrak{k}}}\,d\boldsymbol{y}$$

Interchange of the order of integration is justified by the absolute convergence of the integral (105). Hence the density is (106).

Q.E.D

The bounded convergence theorem implies that the resulting density is continuous. A condition like integrability of the first derivative of the density implies the integrability condition of the cumulant generating function (105). The presence of probability atoms implies that this condition (105) is violated.

The inversion integrand in (106) is the $\mathfrak{k}$-order derivative of the inversion integrand in (104) with respect to each component of $\boldsymbol{b}$, and evaluated with $b^{1j} = b^{2j}$. This will become important in our discussion of saddlepoint series.

For completeness, and because it will be useful in the later development of multivariate saddlepoint density approximations, the following analog of Lemma 2.4.3 is stated:

Lemma 6.4.3: *The characteristic function ζ of a random vector satisfies*

$$\int_{-\infty}^{\infty}\cdots\int_{-\infty}^{\infty} |\zeta(\boldsymbol{\beta})|^{\mathfrak{r}}\, d\boldsymbol{\beta} < \infty$$

for some $\mathfrak{r} > 1$ if and only if the density of a j-fold convolution of the random variable with itself exists and is bounded, for some integer j.

Proof: Left as an exercise.

Finally, there is also a multivariate inversion counterpart to (26) for the probability mass function for a random vector $\boldsymbol{X}$ confined to a multivariate lattice:

Theorem 6.4.4: *Suppose that a random vector $\boldsymbol{X}$ of length $\mathfrak{k}$ is confined to a generalized lattice in the following sense: The distribution of $X^{\mathfrak{k}}$ is confined to a lattice of form $a^{\mathfrak{k}} + \Delta_{\mathfrak{k}}\mathfrak{Z}$, and for each $j < \mathfrak{k}$ the distribution of X^j is confined to a lattice of form $a^j(x^{j+1}, \ldots, x^{\mathfrak{k}}) + \Delta_j\mathfrak{Z}$. Then*

$$\prod_{j=1}^{\mathfrak{k}} \frac{\Delta_j}{2\pi}\int_{-\pi/\Delta_1}^{\pi/\Delta_1}\cdots\int_{-\pi/\Delta_{\mathfrak{k}}}^{\pi/\Delta_{\mathfrak{k}}} \exp(-i\boldsymbol{\beta}\boldsymbol{x})\zeta(\boldsymbol{\beta})\,d\boldsymbol{\beta} = \mathrm{P}\left[\boldsymbol{X} = \boldsymbol{x}\right]. \tag{107}$$

Proof: By Fubini's theorem,

$$\prod_{j=1}^{\mathfrak{k}} \frac{\Delta_j}{2\pi} \int_{-\pi/\Delta_1}^{\pi/\Delta_1} \cdots \int_{-\pi/\Delta_{\mathfrak{k}}}^{\pi/\Delta_{\mathfrak{k}}} \sum_{l_1=-\infty}^{\infty} \cdots \sum_{l_{\mathfrak{k}}=-\infty}^{\infty} \mathrm{P}\left[\boldsymbol{X} = \boldsymbol{x} + \Delta \boldsymbol{l}\right] \exp(i\boldsymbol{\beta}(\boldsymbol{x} + \Delta \boldsymbol{l}) - i\boldsymbol{\beta}\boldsymbol{x})\, d\boldsymbol{\beta}$$
$$= \prod_{j=1}^{\mathfrak{k}} \frac{\Delta_j}{2\pi} \int_{-\pi/\Delta_1}^{\pi/\Delta_1} \cdots \int_{-\pi/\Delta_{\mathfrak{k}}}^{\pi/\Delta_{\mathfrak{k}}} \mathrm{P}\left[\boldsymbol{X} = \boldsymbol{x}\right]\, d\boldsymbol{\beta}$$

The left hand side of the above equation is equal to the left hand side of (107) after expanding the definition of the characteristic function; the right hand side of the above equation is trivially equal to the right hand side of (107).

Q.E.D

The above generalized lattice definition generalizes the conventional definition as the set

$$\boldsymbol{a} + \Delta_1 \mathfrak{Z} \times \cdots \times \Delta_{\mathfrak{k}} \mathfrak{Z}$$

in that it allows the starting point for the lattice to depend on other components of the vector $\boldsymbol{x}$. This generalization will prove useful in a later chapter on conditional cumulative distribution function approximations for lattice random variables.

6.5. Error Analysis for the Edgeworth Series

Below is a theorem on convergence of Edgeworth series for multivariate densities and cumulative distribution functions.

Theorem 6.5.1: *Suppose F is a cumulative distribution function, with mean $\boldsymbol{\mu}$ and positive definite variance matrix $\boldsymbol{\Sigma}$, and $\mathfrak{j}$ is an integer greater than or equal to two, such that F has all cumulants of order $\mathfrak{j}$. Suppose further that*

$$\int_{-\infty}^{\infty} \cdots \int_{-\infty}^{\infty} |\zeta(\boldsymbol{\beta})|^{\mathfrak{r}}\, d\boldsymbol{\beta} < \infty \text{ for some } \mathfrak{r} \geq 1. \tag{108}$$

Let $\boldsymbol{Y}_j$ be independent and identically distributed random vectors with distribution function F. Let $\boldsymbol{X} = (\sum_{j=1}^{n} \boldsymbol{Y}_j - n\boldsymbol{\mu})/\sqrt{n}$. Then $\boldsymbol{X}$ has a bounded density $f_{\boldsymbol{X}}$ for sufficiently large n, and when (96) is calculated using only the first $\mathfrak{j}$ cumulants, terms of order $o(n^{1-\mathfrak{j}/2})$ dropped, and the result $e_{\mathfrak{j}}(\boldsymbol{x}, \boldsymbol{\kappa}^{*n})$ is used to approximate the density $f_{\boldsymbol{X}}$, the absolute error is uniformly of order $o(n^{1-\mathfrak{j}/2})$.

Proof: The following proof is almost identical to the univariate proof. Without loss of generality assume $\Sigma = \boldsymbol{I}$. Since the cumulative distribution functions can be recovered from their characteristic functions using (15), the difference between $f_{\boldsymbol{X}}(\boldsymbol{x})$ and $e_{\mathfrak{j}}(\boldsymbol{x}, \boldsymbol{\kappa}^*_{\boldsymbol{X}})$ can be bounded by the integral

$$\left(\frac{1}{2\pi}\right)^{\mathfrak{k}} \int_{-\infty}^{\infty} \cdots \int_{-\infty}^{\infty} 2\left|\zeta_{\boldsymbol{X}}(\boldsymbol{\beta}) - \xi_{\boldsymbol{X}}(\boldsymbol{\beta})\right|\, d\boldsymbol{\beta}, \tag{109}$$

the analogue of (44). Here $\xi_{\boldsymbol{X}}$ is the Fourier transform of the approximate density $e_{\mathfrak{j}}(\boldsymbol{x}, \boldsymbol{\kappa}^*_{\boldsymbol{X}})$. Because the integral (109) converges absolutely, this argument will not need the Smoothing Theorem. The Series Theorem is still a key idea here.

The range of integration will now be split in two parts. The area very close to the origin, and the area farther out will be handled separately.

a. Very near the origin ($\|\boldsymbol{\beta}\| < \delta\sqrt{n}$ for δ to be determined below), set $\alpha^*(\boldsymbol{\beta}) = \log(\zeta(\boldsymbol{\beta})) + \sum_j (\beta_j)^2/2$, and $v^*(\boldsymbol{\beta}) = \sum_{j=3}^{\mathrm{j}} \sum_{s\in S(j)} i^j \kappa^{s_1\cdots s_t} \beta_{s_1}\cdots\beta_{s_t}/j!$. Set $\alpha(\boldsymbol{\beta}) = n\alpha^*(\boldsymbol{\beta}/\sqrt{n})$, and $v(\boldsymbol{\beta}) = nv^*(\boldsymbol{\beta}/\sqrt{n})$. Then $\alpha(\boldsymbol{\beta}) - v(\boldsymbol{\beta})$ has j continuous derivatives at 0 and all are 0. Hence there exists $\delta > 0$ such that if $\|\boldsymbol{\beta}\| < \delta$ then $\alpha^*(\boldsymbol{\beta}) - v^*(\boldsymbol{\beta}) < \epsilon \|\boldsymbol{\beta}\|^{\mathrm{j}}$, and if $\|\boldsymbol{\beta}\| < \sqrt{n}\delta$ then $\alpha(\boldsymbol{\beta}) - v(\boldsymbol{\beta}) < \epsilon \|\boldsymbol{\beta}\|^{\mathrm{j}} n^{1-\mathrm{j}/2}$. Furthermore require that $\|\boldsymbol{\beta}\| < \delta$ imply that $\left|\zeta(\boldsymbol{\beta}) + \sum_j \beta_j^2/2\right| < \|\boldsymbol{\beta}\|^2/4$; hence $\|\boldsymbol{\beta}\| < \delta\sqrt{n}$ implies that

$$|\alpha(\boldsymbol{\beta})| = \left| n\zeta(\boldsymbol{\beta}/\sqrt{n}) + \sum_j \beta_j^2/2 \right| < \|\boldsymbol{\beta}\|^2/4.$$

Also, since $v^*(\boldsymbol{\beta})$ has a third derivative at 0, there exists C such that $\|\boldsymbol{\beta}\| < \delta\sqrt{n}$ implies

$$v(\boldsymbol{\beta}) = nv^*(\boldsymbol{\beta}/\sqrt{n}) < C\|\boldsymbol{\beta}\|^3/\sqrt{n}.$$

Hence the integrand is bounded by

$$\exp(-\|\boldsymbol{\beta}\|^2/4)\left[\frac{\epsilon\|\boldsymbol{\beta}\|^{\mathrm{j}}}{n^{\mathrm{j}/2-1}} + \frac{C^{\mathrm{j}-1}\|\boldsymbol{\beta}\|^{3(\mathrm{j}-1)}}{(\mathrm{j}-1)!n^{\mathrm{j}/2-1/2}}\right],$$

for $\boldsymbol{\beta}$ such that $\|\boldsymbol{\beta}\| < \delta\sqrt{n}$. When integrated over $(-\delta\sqrt{n}, \delta\sqrt{n})^{\mathfrak{k}}$ the result is still of order $o(n^{1-\mathrm{j}/2})$.

b. For more extreme values of $\boldsymbol{\beta}$ ($\|\boldsymbol{\beta}\| > \delta\sqrt{n}$), the integrability condition on the characteristic function is used to bound the integrand. There exists $q < 1$ such that if $\|\boldsymbol{\beta}\| > \delta\sqrt{n}$ then $|\zeta(\boldsymbol{\beta}/\sqrt{n})| < q$. Hence the contribution of (109) from this part of the range of integration can be bounded by

$$q^{n-p}\int_{-\infty}^{\infty}\cdots\int_{-\infty}^{\infty} |\zeta(\boldsymbol{\beta})|^p \, d\boldsymbol{\beta} + \int\cdots\int_{\|\boldsymbol{\beta}\|>\delta\sqrt{n}} |\xi_{\mathbf{X}}(\boldsymbol{\beta})| \, d\boldsymbol{\beta},$$

and approaches 0 geometrically.

Q.E.D

The following theorem is the analogue of the univariate Edgeworth cumulative distribution function theorem. As the flavor of its proof deviates from that of the rest of this volume, and as it will not be needed to prove later results, it is cited without complete proof.

Theorem 6.5.2: *Suppose F is a cumulative distribution function, with mean $\boldsymbol{\mu}$ and positive definite variance matrix $\boldsymbol{\Sigma}$, and* j *is an integer greater than or equal to two, such that F has all cumulants of order* j*. Suppose further that*

$$\begin{cases} |\zeta(\boldsymbol{\beta})| < 1 \ \forall \boldsymbol{\beta} \neq 0 & \text{if } \mathrm{j} = 3, \text{ or} \\ \limsup_{\|\boldsymbol{\beta}\|\to\infty} |\zeta(\boldsymbol{\beta})| < 1 & \text{if } \mathrm{j} > 3. \end{cases} \tag{110}$$

The condition when $\mathrm{j} > 3$ *is a multivariate version of Cramér's condition. Let* $\boldsymbol{Y}_j$ *be independent and identically distributed random vectors with distribution function* F*. Let* $\boldsymbol{X} = (\sum_{j=1}^{n} \boldsymbol{Y}_j - n\boldsymbol{\mu})/\sqrt{n}$*, and let* $F_{\boldsymbol{X}}$ *be the cumulative distribution function for* $\boldsymbol{X}$*. When (97) is calculated using only the first* j *cumulants, terms of order* $o(n^{1-\mathrm{j}/2})$ *dropped, and the result* $E_{\mathrm{j}}(\boldsymbol{x}, \boldsymbol{\kappa}^{*n})$ *is used to approximate the distribution given by* $F_{\boldsymbol{X}}$*, the absolute error is uniformly of order* $o(n^{1-\mathrm{j}/2})$*, in the sense that not only*

$$\sup_{\boldsymbol{x}} |E_{\mathrm{j}}(\boldsymbol{x}, \boldsymbol{\kappa}^{*n}) - F_{\boldsymbol{X}}(\boldsymbol{x})| = o(n^{1-\mathrm{j}/2}), \tag{111}$$

but furthermore, that $\sup_C |\int_C dE_{\mathrm{j}}(\boldsymbol{x}, \boldsymbol{\kappa}^{*n}) - \int_C dF_{\boldsymbol{X}}(\boldsymbol{x})| = o(n^{1-\mathrm{j}/2})$*, where the supremum is taken over all convex Borel sets* C*.*

Proof: The precise definition of a Borel set is unimportant for the present purposes; it suffices to note that this class of sets contains all open sets and all closed sets.

Note that (111) provides a result applicable when approximating probabilities of multivariate rectangles and their unions. Proof of this result is a simple extension of univariate results. A multivariate version of the Smoothing Theorem could be constructed to prove that the difference between $F_{\boldsymbol{X}}(\boldsymbol{x})$ and $E_{\mathrm{j}}(\boldsymbol{x}, \boldsymbol{\kappa}^{*n})$ can be bounded by the principal value of the integral

$$\lim_{T\to\infty} \frac{1}{2\pi} \int_{-T}^{T} \cdots \int_{-T}^{T} \frac{2}{\|\boldsymbol{\beta}\|} |\zeta_{\boldsymbol{X}}(\boldsymbol{\beta}) - \xi_{\boldsymbol{X}}(\boldsymbol{\beta})| \, d\boldsymbol{\beta}, \tag{112}$$

where ξ is the Fourier transform of $E_{\mathrm{j}}(\boldsymbol{x}, \boldsymbol{\kappa}^{*n})$. Remember that ξ is not really a characteristic function since $E_{\mathrm{j}}(\boldsymbol{x}, \boldsymbol{\kappa}^{*n})$ is not really a cumulative distribution function. Nevertheless, just as in the univariate case the cumulative distribution function inversion theorem arguments hold for ξ as well, justifying (112). As in the univariate case, the range of integration is divided into three parts: within a multivariate cube with edges of length δ, outside a multivariate cube with edges of length T, and in the region between these. The outer region again is eliminated through the use of the Smoothing Theorem. The inner region can be bounded just as in the multivariate density proof, and the intermediate region bounded using conditions (110). Proving the final result is more difficult. See Bhattacharya and Rao (1976).

Q.E.D

Result (111) is, however, too weak to be useful in the sequel, since of interest will be primarily approximately elliptically-shaped regions. Bhattacharya and Rao (1976) also present a direct analogy to Esseen's series, given in Theorem 3.14.2, for lattice variables. These theorems in general are more complicated, since of interest are probabilities of sets in $\boldsymbol{Y}$-space with complicated shapes. As in the univariate case, this series evaluated at continuity corrected points is equivalent to the multivariate Edgeworth series, to the proper order in n (Kolassa, 1989).

6.6. Multivariate Integral Expansion Theorems

Multivariate saddlepoint approximations may also be derived via a multivariate version of Watson's Lemma due to Temme (1982). The aim is to calculate integrals whose integrand is the exponential of a quadratic term times sample size, times a factor of the form

$$f_n(\omega) = \sum_{j=0}^{\mathfrak{j}} \theta_j(\omega)/n^j. \tag{113}$$

Theorem 6.6.1: *If $\omega \in \mathfrak{C}^{\mathfrak{k}}$, if $f_n(\omega)$ is as in (113), with $\hat{\omega} \in Q$, and*

$$A_s = \sum_{\substack{v_1,\cdots,v_{\mathfrak{k}} \geq 0 \\ \sum v_j \leq s}} \frac{(-2)^{-\sum_{j=1}^{\mathfrak{k}} v_j}}{v_1! \cdots v_{\mathfrak{k}}!} \left[\frac{\partial^{2v_1+\cdots+2v_{\mathfrak{k}}}}{\partial^{2v_1}\omega_1 \cdots \partial^{2v_{\mathfrak{k}}}\omega_{\mathfrak{k}}} \theta_{s-\sum v_j} \right](\hat{\omega}) \tag{114}$$

then

$$\left(\frac{n}{2\pi}\right)^{\mathfrak{k}/2} i^{-\mathfrak{k}} \int_{\hat{\omega}_1-i\infty}^{\hat{\omega}_1+i\infty} \cdots \int_{\hat{\omega}_{\mathfrak{k}}-i\infty}^{\hat{\omega}_{\mathfrak{k}}+i\infty} \exp(\frac{n}{2}\sum_j(\omega_j - \hat{\omega}_j)^2) f_n(\omega)\, d\omega = \sum_{s=0}^{\mathfrak{j}-1} A_s n^{-s} + O(n^{-\mathfrak{j}}). \tag{115}$$

Proof: For $\mathfrak{k} = 1$ this is just Lemma 4.5.2, applied termwise. Suppose that this holds for some integer $\mathfrak{k}$, and choose f a function of $\mathfrak{k}+1$ complex variables. By the induction hypothesis, with $\omega_{\mathfrak{k}+1}$ held fixed,

$$\begin{aligned}\left(\frac{n}{2\pi}\right)^{\mathfrak{k}/2} i^{-\mathfrak{k}} \int_{\hat{\omega}_1-i\infty}^{\hat{\omega}_1+i\infty} \cdots \int_{\hat{\omega}_{\mathfrak{k}}+1-i\infty}^{\hat{\omega}_{\mathfrak{k}}+1+i\infty} \exp(-\frac{n}{2}\sum_{j=1}^{\mathfrak{k}+1}(\omega_j - \hat{\omega}_j)^2) f_n(\omega)\, d\omega \\ = \int_{-\infty}^{\infty} \exp(-\frac{n}{2}(\omega_{\mathfrak{k}+1} - \hat{\omega}_{\mathfrak{k}+1})^2) \sum_{s=0}^{\mathfrak{j}-1} \frac{\psi_s(\omega_{\mathfrak{k}+1})}{n^s}\, d\omega_{\mathfrak{k}+1} + O(n^{-\mathfrak{j}})\end{aligned}$$

where

$$\psi_s(\omega_{\mathfrak{k}+1}) = \sum_{\mathfrak{S}(s,\mathfrak{k})} \frac{(-2)^{-\sum_{j=1}^{\mathfrak{k}} v_j}}{v_1! \cdots v_{\mathfrak{k}}!} \left[\frac{\partial^{2v_1+\cdots+2v_{\mathfrak{k}}}}{\partial^{2v_1}\omega_1 \cdots \partial^{2v_{\mathfrak{k}}}\omega_{\mathfrak{k}}} \theta_{s-\sum v_j} \right](\mathbf{o}, \omega_{\mathfrak{k}+1}),$$

and $\mathfrak{S}(s,\mathfrak{k}) = \{v_1, \cdots, v_{\mathfrak{k}} \geq 0 | \sum_{j=1}^{\mathfrak{k}} v_j \leq s\}$. Applying Lemma 4.5.2,

$$\left(\frac{n}{2\pi}\right)^{(\mathfrak{k}+1)/2} i^{-\mathfrak{k}-1} \int_{\hat{\omega}_1-i\infty}^{\hat{\omega}_1+i\infty} \cdots \int_{\hat{\omega}_{\mathfrak{k}+1}-i\infty}^{\hat{\omega}_{\mathfrak{k}+1}+i\infty} \exp(-\frac{n}{2}\sum_j(\omega_j - \hat{\omega}_j)^2) f_n(\omega)\, d\omega = \sum_{s=0}^{\mathfrak{j}-1} A_s n^{-s} + O(n^{-\mathfrak{j}}),$$

where

$$A_s = \sum_{v_{\mathfrak{k}+1}=0}^{s} \frac{(-2)^{-v_{\mathfrak{k}+1}}}{v_{\mathfrak{k}+1}!}$$

$$\left[\frac{\partial^{2v_{\mathfrak{k}+1}}}{\partial^{2v_{\mathfrak{k}+1}}\omega_{\mathfrak{k}+1}} \sum_{\mathfrak{S}(s-v_{\mathfrak{k}+1},\mathfrak{k})} \frac{(-2)^{-\sum_{j=1}^{\mathfrak{k}} v_j}}{v_1!\cdots v_{\mathfrak{k}}!}\left[\frac{\partial^{2v_1+\cdots+2v_{\mathfrak{k}}}}{\partial^{2v_1}\omega_1\cdots\partial^{2v_{\mathfrak{k}}}\omega_{\mathfrak{k}}}\theta_{s-\sum_{j=0}^{\mathfrak{k}+1} v_j}\right]\right](\mathbf{o})$$

$$= \sum_{\mathfrak{S}(s,\mathfrak{k}+1)} \frac{(-2)^{-\sum_{j=1}^{\mathfrak{k}+1} v_j}}{v_1!\cdots v_{\mathfrak{k}+1}!}\left[\frac{\partial^{2v_1+\cdots+2v_{\mathfrak{k}+1}}}{\partial^{2v_1}\omega_1\cdots\partial^{2v_{\mathfrak{k}+1}}\omega_{\mathfrak{k}+1}}\theta_{s-\sum v_j}\right](\mathbf{o}).$$

Q.E.D

The following theorem will allow some terms in the series expansions to be dropped:

Theorem 6.6.2: *Let $\sum_{\boldsymbol{v}\in\mathcal{N}} c_{\boldsymbol{v}}\boldsymbol{\omega}^{\boldsymbol{v}}$ be the power series representation of the function θ in (113). Here $\mathcal{N}$ represents the collection of vectors of $\mathfrak{k}$ non-negative integers. Then the coefficients (114) can be computed from the function θ^{*d} whose power series expansion is $\sum_{\boldsymbol{v}\in\mathcal{N}_l} c_{\boldsymbol{v}}\boldsymbol{\omega}^{\boldsymbol{v}}$, where $\mathcal{N}_l$ is the set of all vectors of positive integers in which $v_l,\ldots,v_{\mathfrak{k}}$ are even.*

Proof: Note that only even–order derivatives are present in the expansion.

Q.E.D

If $\boldsymbol{\omega}$ is a $\mathfrak{k}$-long vector $(\omega_1,\cdots,\omega_{\mathfrak{k}})$, $\boldsymbol{v}$ is a $\mathfrak{k}$-long vector of integers $(v_1,\cdots,v_{\mathfrak{k}})$, $\mathbf{1}^\top\boldsymbol{v} = \sum_{j=1}^{\mathfrak{k}} v_j$, $\boldsymbol{v}! = v_1!\times\cdots\times v_{\mathfrak{k}}!$, and $\left[\frac{\partial^{\boldsymbol{v}}}{\partial^{\boldsymbol{v}}\boldsymbol{\omega}}\right] = \left[\frac{\partial^{v_1+\cdots+v_{\mathfrak{k}}}}{\partial^{v_1}\omega_1\cdots\partial^{v_{\mathfrak{k}}}\omega_{\mathfrak{k}}}\right]$, then (114) can be reexpressed as

$$A_s = \sum_{\mathfrak{S}(s,\mathfrak{k})} \frac{(-2)^{-\mathbf{1}^\top\boldsymbol{v}}}{\boldsymbol{v}!}\left[\frac{\partial^{2\mathbf{1}^\top\boldsymbol{v}}}{\partial^{2\mathbf{1}^\top\boldsymbol{v}}\boldsymbol{\omega}}\theta_{s-\mathbf{1}^\top\boldsymbol{v}}\right](\mathbf{o}).$$

6.7. A Direct Derivation of the Multivariate Saddlepoint Approximation

The multivariate saddlepoint density might be derived directly from (106), using the results of the previous section. Consider specifically the case in which $f_{\boldsymbol{X}}$ represents the distribution of the mean of n independent and identically distributed copies of $\boldsymbol{Y}$. After changing variables,

$$f_{\boldsymbol{X}}(\boldsymbol{x}) = \left(\frac{n}{2\pi i}\right)^{\mathfrak{k}} \int_{-i\infty}^{+i\infty}\cdots\int_{-i\infty}^{+i\infty} \exp(n[\mathcal{K}_{\boldsymbol{Y}}(\boldsymbol{\beta}) - \boldsymbol{\beta}\boldsymbol{x}])\, d\boldsymbol{\beta}.$$

Choose a differentiable function $\boldsymbol{\omega}(\boldsymbol{\beta})$ so that

$$\tfrac{1}{2}(\boldsymbol{\omega}-\hat{\boldsymbol{\omega}})^\top(\boldsymbol{\omega}-\hat{\boldsymbol{\omega}}) = \mathcal{K}_{\boldsymbol{Y}}(\boldsymbol{\beta}) - \boldsymbol{\beta}^\top\boldsymbol{x} - \mathcal{K}_{\boldsymbol{Y}}(\hat{\boldsymbol{\beta}}) + \hat{\boldsymbol{\beta}}^\top\boldsymbol{x}. \qquad (116)$$

A specific choice for $\boldsymbol{\omega}(\boldsymbol{\beta})$ that will be both useful and easily shown to be analytic will be constructed in this section.

As these results will be used not only in the present chapter but also in a subsequent chapter on approximating conditional distributions, the function $\boldsymbol{\omega}$ will be constructed such that for each integer l, $\omega_1, \ldots, \omega_l$ depend only on $\beta_1, \ldots, \beta_l$. In essence, ω_l becomes the signed likelihood ratio statistic for the hypothesis that the canonical parameter associated with X^l is zero in the exponential family into which $(X^1, \cdots, X^l)$ is embedded, as in §6.3.

For real $\boldsymbol{\beta}$, and $1 \leq l \leq \mathfrak{k}$, let

$$\left.\begin{aligned} -\tfrac{1}{2}(\omega_l - \hat{\omega}_l)^2 &= \min(\mathcal{K}_{\boldsymbol{Y}}(\boldsymbol{\gamma}) - \gamma_j x^j | \gamma_j = \beta_j \,\forall j > l) - \\ &\quad \min(\mathcal{K}_{\boldsymbol{Y}}(\boldsymbol{\gamma}) - \gamma_j x^j | \gamma_j = \beta_j \,\forall j \geq l) \\ -\tfrac{1}{2}\hat{\omega}_l^2 &= \min(\mathcal{K}_{\boldsymbol{Y}}(\boldsymbol{\gamma}) - \gamma_j x^j | \gamma_j = 0 \,\forall j > l) - \\ &\quad \min(\mathcal{K}_{\boldsymbol{Y}}(\boldsymbol{\gamma}) - \gamma_j x^j | \gamma_j = 0 \,\forall j \geq l). \end{aligned}\right\} \tag{117}$$

We adopt the notation that indices appearing both as superscripts and subscripts are to be summed over. This parameterization makes proving the following lemmas easy. Recall that a function of one or more complex variables having a power series representation about a point is called analytic at that point.

Lemma 6.7.1: *There exists a function $\boldsymbol{\beta}(\boldsymbol{\omega})$ satisfying (117) for $\boldsymbol{\beta} \in \mathfrak{Q}$, that is analytic for $\boldsymbol{\omega}$ in a neighborhood of $\hat{\boldsymbol{\omega}}$. Furthermore, for $j \in \{1, \ldots, \mathfrak{k}\}$, if the first j components of $\boldsymbol{\beta}$ are zero, so are the corresponding first j components of $\boldsymbol{\omega}$.*

Proof: For each l the components $\gamma_{l+1}, \ldots, \gamma_{\mathfrak{k}}$ at which the minima in (117) occur satisfy $(\partial/\partial\beta_j)\mathcal{K}_{\boldsymbol{Y}}(\beta_1, \ldots, \beta_l, \gamma_{l+1}, \ldots, \gamma_{\mathfrak{k}}) = x^j$ for $j = 1, \ldots, l$, and hence by the inverse function theorem for complex variables (Bochner and Martin, 1948) is a differentiable function of $\beta_1, \ldots, \beta_l$. The right hand side of (117) has zero constant and first order terms when viewed as a function of $\beta_1, \ldots, \beta_l$ at $\hat{\boldsymbol{\beta}}$, and so it can be factored as $(\beta_l - \hat{\beta}_l)^2$ times a function $h_l(\boldsymbol{\beta})$ of $\boldsymbol{\beta}$ analytic at $\hat{\boldsymbol{\beta}}$ and which does not vanish at $\hat{\boldsymbol{\beta}}$. Then $\omega_l - \hat{\omega}_l = (\beta_l - \hat{\beta}_l)\sqrt{2h_l(\boldsymbol{\beta})}$ describes the function ω_l of (117). The quantity $h_l(\hat{\boldsymbol{\beta}})$ is easily shown to be positive, and the branch of the square root function assigning a positive result may be used. Since the equation for ω_l does not contain $\beta_1, \ldots, \beta_{l-1}$ the determinant of the matrix of derivatives of the right hand sides with respect to $\boldsymbol{\beta}$ is the product of the diagonals; one can easily show that these derivatives, evaluated at $\hat{\boldsymbol{\beta}}$, are non-zero. Hence the requirements of the inverse function theorem for complex variables are satisfied, and the lemma follows. Kolassa (1996d) demonstrates that this parameterization may be constructed in a ball whose size is bounded away from zero in an appropriate metric, as $\hat{\boldsymbol{\beta}}$ varies.

Q.E.D

Then

$$f_{\boldsymbol{X}}(\boldsymbol{x}) = \left(\frac{n}{2\pi i}\right)^{\mathfrak{k}} \exp(n[\mathcal{K}_{\boldsymbol{Y}}(\hat{\boldsymbol{\beta}}) - \hat{\boldsymbol{\beta}}^\top \boldsymbol{x}]) \int_{-i\infty}^{+i\infty} \cdots \int_{-i\infty}^{+i\infty} \exp(\frac{n}{2}(\boldsymbol{\omega} - \hat{\boldsymbol{\omega}})^\top(\boldsymbol{\omega} - \hat{\boldsymbol{\omega}})) \left|\frac{\partial\boldsymbol{\beta}}{\partial\boldsymbol{w}}\right| d\boldsymbol{\omega}. \tag{118}$$

By differentiating both sides of (116) with respect to $\boldsymbol{\beta}$, $\sum_l (\omega_l - \hat{\omega}_l)(\partial\omega_l/\partial\beta_j) = \mathcal{K}_{\mathbf{Y}}^j(\boldsymbol{\beta}) - x^j$ and $\sum_l (\partial\omega_l/\partial\beta_j)(\partial\omega_l/\partial\beta_k) + (\omega_l - \hat{\omega}_l)(\partial^2\omega_l/\partial\beta_j\partial\beta_k) = \mathcal{K}_{\mathbf{Y}}^{jk}(\boldsymbol{\beta})$. Hence

$$\det\left[\frac{\partial\omega}{\partial\boldsymbol{\beta}}(\hat{\omega})\right] = \sqrt{\det\left[\mathcal{K}_{\mathbf{Y}}''(\hat{\boldsymbol{\beta}})\right]}. \tag{119}$$

The application of Theorem 6.6.1 yields the first order terms in (100). Higher-order terms will be calculated by evaluating derivatives of $d\boldsymbol{\beta}/d\boldsymbol{\omega}$; as (100) has already been completely derived in §6.3, and as these calculations are lengthy, the calculations will be deferred until the next chapter.

6.8. Exercises

1. Prove Lemma 6.4.3.

7
Conditional Distribution Approximations

Often inference for a subset of model parameters is desired, and the others are treated as nuisance parameters. Among the many methods for attacking this problem is conditional inference, in which sufficient statistics for nuisance parameters are conditioned on. Calculations involving these conditional distributions are often quite difficult. This chapter will develop methods for approximating densities and distribution functions for conditional distributions.

Consider full canonical exponential family distributions consisting of distributions with densities of the form

$$f_Y(\boldsymbol{y};\boldsymbol{\upsilon}) = \exp(\boldsymbol{y}^\top\boldsymbol{\upsilon} - \mathcal{H}_Y(\boldsymbol{\upsilon}) - \mathcal{G}_Y(\boldsymbol{y})). \tag{120}$$

Here the cumulant generating function of $\boldsymbol{Y}$ is $\mathcal{K}_Y(\boldsymbol{\beta};\boldsymbol{\upsilon}) = \mathcal{H}_Y(\boldsymbol{\beta}+\boldsymbol{\upsilon}) - \mathcal{H}_Y(\boldsymbol{\upsilon})$. In such families the conditional distribution of a subset of sufficient statistics conditional on the rest is also a canonical exponential family, depending only on the corresponding canonical parameters:

$$f_Y(Y^1 = y^1,\cdots,Y^{\mathrm{m}} = y^{\mathrm{m}}|Y^{\mathrm{m}+1} = y^{\mathrm{m}+1},\cdots,Y^{\mathfrak{k}} = y^{\mathfrak{k}};\boldsymbol{\upsilon})$$
$$= \frac{\exp(y^1\upsilon_1 + \cdots + y^{\mathfrak{k}}\upsilon_{\mathfrak{k}} - \mathcal{G}_Y(\boldsymbol{Y}))}{\int_{-\infty}^{\infty}\cdots\int_{-\infty}^{\infty}\exp(y^1\upsilon_1 + \cdots + y^{\mathfrak{k}}\upsilon_{\mathfrak{k}} - \mathcal{G}_Y(\boldsymbol{Y}))\,dy^1\cdots dy^{\mathrm{m}}}.$$

This conditional distribution may be used for inference on a subset of model parameters, while treating the remaining model parameters as nuisance parameters.

Also note that if $\boldsymbol{X}$ is is the mean of n independent and identically distributed random vectors with density (120), then the density for $\boldsymbol{X}$ may be expressed as

$$f_X(\boldsymbol{x};\theta) = \exp(\boldsymbol{x}^\top\boldsymbol{\theta} - \mathcal{H}_X(\boldsymbol{\theta}) - \mathcal{G}_X(\boldsymbol{x})), \tag{121}$$

and so a similar conditioning argument can be used for inference

This chapter will consider approximations that aid in conditional inference for some components of a parameter vector that is a linear transformation of the canonical parameter in an exponential family as in (121), especially those arising under repeated sampling from a family like (120). The question of inference on a subset of model parameters that cannot be expressed as canonical parameters in an exponential family will be addressed in the next chapter.

7.1. Double Saddlepoint Density Approximations

In this section an approximation to the conditional density of the first $\mathfrak{m}$ components of a random vector $\boldsymbol{X} = (X^1, \ldots, X^{\mathfrak{d}})$ arising as the mean of n independent and identically distributed random vectors $\boldsymbol{Y}_j$, each with cumulant generating function $\mathcal{K}_{\boldsymbol{Y}}$. is derived. Recall the expression (118) for the density of a random variable in terms of a multivariate line integral over the paths of steepest descent:

$$f_{\boldsymbol{X}}(\boldsymbol{x}) = \left(\frac{n}{2\pi i}\right)^{\mathfrak{d}} \exp(n[\mathcal{K}_{\boldsymbol{Y}}(\hat{\boldsymbol{\beta}}) - \hat{\boldsymbol{\beta}}^{\top}\boldsymbol{x}]) \int_{-i\infty}^{+i\infty} \cdots \int_{-i\infty}^{+i\infty} \exp(\frac{n}{2}(\omega - \hat{\omega})^{\top}(\omega - \hat{\omega})) \left|\frac{\partial \boldsymbol{\beta}}{\partial \boldsymbol{w}}\right| \, d\omega.$$

Here again $\hat{\boldsymbol{\beta}}$ is defined by $\mathcal{K}'_{\boldsymbol{Y}}(\hat{\boldsymbol{\beta}}) = \boldsymbol{x}$. The approximation to $f_{\boldsymbol{X}}(\boldsymbol{x})$ will be factored into two parts, one of which will be identified with the density of the conditioning random variables. The remainder will then be the conditional density of interest. The particular choice of $\boldsymbol{\beta}(\boldsymbol{\omega})$ of §6.7 implies that for each j, ω_j is a function of $\beta_j, \ldots, \beta_{\mathfrak{d}}$, demonstrating that $\partial\boldsymbol{\beta}/\partial\boldsymbol{w}$ is triangular, and hence the Jacobian is the product of the derivatives; specifically, for any $j \in \{2, \ldots, \mathfrak{d}\}$,

$$\left|\frac{\partial \boldsymbol{\beta}}{\partial \boldsymbol{w}}\right| = \det\left[\frac{\partial \beta_1, \ldots, \beta_{j-1}}{\partial \omega_1, \ldots, \omega_{j-1}}\right](\boldsymbol{\omega}) \det\left[\frac{\partial \beta_j, \ldots, \beta_{\mathfrak{d}}}{\partial \omega_j, \ldots, \omega_{\mathfrak{d}}}\right](\omega_j, \ldots, \omega_{\mathfrak{d}}). \tag{122}$$

By (100), (119), and (122),

$$f_{\boldsymbol{X}}(\boldsymbol{x}) = \left[\frac{\exp(n[\mathcal{K}_{\boldsymbol{Y}}(\tilde{\boldsymbol{\beta}}) - \tilde{\boldsymbol{\beta}}^{\top}\boldsymbol{x}])}{(2\pi n)^{(\mathfrak{d}-\mathfrak{m})/2}} \det\left[\frac{\partial \beta_{\mathfrak{m}+1}, \ldots, \beta_{\mathfrak{d}}}{\partial \omega_{\mathfrak{m}+1}, \ldots, \omega_{\mathfrak{d}}}\right](\hat{\omega}_{\mathfrak{m}+1}, \ldots, \hat{\omega}_{\mathfrak{d}})\right] \times$$
$$\left[\frac{\exp(n[\mathcal{K}_{\boldsymbol{Y}}(\hat{\boldsymbol{\beta}}) - \mathcal{K}_{\boldsymbol{Y}}(\tilde{\boldsymbol{\beta}}) - (\hat{\boldsymbol{\beta}} - \tilde{\boldsymbol{\beta}})^{\top}\boldsymbol{x}])}{(2\pi n)^{\mathfrak{m}/2}} \det\left[\frac{\partial \beta_1, \ldots, \beta_{\mathfrak{m}}}{\partial \omega_1, \ldots, \omega_{\mathfrak{m}}}\right](\hat{\boldsymbol{\omega}})\right](1 + O(n^{-1})),$$

where $\tilde{\boldsymbol{\beta}}$ satisfies

$$\tilde{\beta}_1, \ldots, \tilde{\beta}_{\mathfrak{m}} = 0,\ \mathcal{K}^j_{\boldsymbol{Y}}(\tilde{\boldsymbol{\beta}}) = x^j \text{ for } j > \mathfrak{m}. \tag{123}$$

The first factor above is the approximation for the density of $(X^{\mathfrak{m}+1}, \ldots, X^{\mathfrak{d}})$, to $O(n^{-1})$. Hence the second factor is the approximation for the conditional density of $(X^1, \ldots, X^{\mathfrak{m}})$ conditional on $(X^{\mathfrak{m}+1}, \ldots, X^{\mathfrak{d}})$, to $O(n^{-1})$. By (122) and (119),

$$\det\left[\frac{\partial \beta_1, \ldots, \beta_{\mathfrak{m}}}{\partial \omega_1, \ldots, \omega_{\mathfrak{m}}}\right](\hat{\boldsymbol{\omega}}) = \left|\frac{\partial \boldsymbol{\beta}}{\partial \boldsymbol{w}}\right|(\hat{\boldsymbol{\omega}}) / \det\left[\frac{\partial \beta_{\mathfrak{m}+1}, \ldots, \beta_{\mathfrak{d}}}{\partial \omega_{\mathfrak{m}+1}, \ldots, \omega_{\mathfrak{d}}}\right](\hat{\omega}_{\mathfrak{m}+1}, \ldots, \hat{\omega}_{\mathfrak{d}})$$
$$= \det\left[\mathcal{K}''_{Y^{\mathfrak{m}+1}, \ldots, Y^{\mathfrak{d}}}(\tilde{\boldsymbol{\beta}})\right] / \det\left[\mathcal{K}''_{\boldsymbol{Y}}(\hat{\boldsymbol{\beta}})\right],$$

where $\mathcal{K}''_{Y^{\mathfrak{m}+1}, \ldots, Y^{\mathfrak{d}}}$ is the submatrix of $\mathcal{K}''_{\boldsymbol{Y}}$ associated with components $\mathfrak{m}+1$ through $\mathfrak{d}$ of the argument $\boldsymbol{\beta}$. The double saddlepoint approximation to the conditional density is now

$$f_{X^1,\dots,X^{\mathfrak{m}}|X^{\mathfrak{m}+1},\dots,X^{\mathfrak{k}}}(x^1,\dots,x^{\mathfrak{m}}|x^{\mathfrak{m}+1},\dots,x^{\mathfrak{k}}) = \det\left[\mathcal{K}''_{Y^{\mathfrak{m}+1},\dots,Y^{\mathfrak{k}}}(\tilde{\boldsymbol{\beta}})\right]/\det\left[\mathcal{K}''_{\boldsymbol{Y}}(\hat{\boldsymbol{\beta}})\right]\times$$
$$\frac{\exp(n[\mathcal{K}_{\boldsymbol{Y}}(\hat{\boldsymbol{\beta}}) - \mathcal{K}_{\boldsymbol{Y}}(\tilde{\boldsymbol{\beta}}) - (\hat{\boldsymbol{\beta}}-\tilde{\boldsymbol{\beta}})^\top \boldsymbol{x}])}{(2\pi n)^{\mathfrak{m}/2}}(1+O(n^{-1}))$$

(Barndorff–Nielsen and Cox, 1979). Its name derives from the fact that the saddlepoint equations are solved twice, once in (99) to obtain $\hat{\boldsymbol{\beta}}$ and once in (123) to obtain $\tilde{\boldsymbol{\beta}}$. This approximation might also be derived from two applications of (100); the ratio of (100) calculated for $(X^{\mathfrak{m}+1},\dots,X^{\mathfrak{k}})$ and for $\boldsymbol{X}$ might be calculated; their ratio, with terms of size $o(1/n)$ deleted, is valid to $O(n^{-2})$, and is

$$\frac{\det\left[\mathcal{K}''_{Y^{\mathfrak{m}+1},\dots,Y^{\mathfrak{k}}}(\tilde{\boldsymbol{\beta}})\right]}{\det\left[\mathcal{K}''_{\boldsymbol{Y}}(\hat{\boldsymbol{\beta}})\right]}\frac{\exp(n[\mathcal{K}_{\boldsymbol{Y}}(\hat{\boldsymbol{\beta}}) - \mathcal{K}_{\boldsymbol{Y}}(\tilde{\boldsymbol{\beta}}) - (\hat{\boldsymbol{\beta}}-\tilde{\boldsymbol{\beta}})^\top \boldsymbol{x}])}{(2\pi n)^{\mathfrak{m}/2}}$$
$$\times\left(1+n^{-1}[\tfrac{1}{8}(\hat{\rho}_{13}-\tilde{\rho}_{13})+\tfrac{1}{12}(\hat{\rho}_{23}-\tilde{\rho}_{23})-\tfrac{1}{8}(\hat{\rho}_{13}-\tilde{\rho}_{13})]+O(n^{-2})\right). \quad (124)$$

Here the invariants $\hat{\rho}$ are exactly as in (103), and the invariants $\tilde{\rho}$ are as in (103), calculated from the derivatives of the cumulant generating function of $(X^{\mathfrak{m}+1},\dots,X^{\mathfrak{k}})$, evaluated at $\tilde{\boldsymbol{\beta}}$.

7.2. The Sequential Saddlepoint Approximation

Perhaps the most direct approach to calculating conditional densities involves calculating the cumulant generating function for the conditional distribution. Specifically, when desiring the distribution of $(X^1,\dots,X^{\mathfrak{m}})$ conditional on $(X^{\mathfrak{m}+1},\dots,X^{\mathfrak{k}})$, one could calculate

$$\mathcal{K}_{X^1,\dots,X^{\mathfrak{m}}|X^{\mathfrak{m}+1},\dots,X^{\mathfrak{k}}}(\beta_1,\dots,\beta_{\mathfrak{m}}) = \frac{\int_{-\infty}^{\infty}\cdots\int_{-\infty}^{\infty}\exp(\sum_{j=1}^{\mathfrak{m}} x^j\beta_j)f_{\boldsymbol{X}}(\boldsymbol{x})\,dx^1\cdots dx^{\mathfrak{m}}}{\int_{-\infty}^{\infty}\cdots\int_{-\infty}^{\infty}f_{\boldsymbol{X}}(\boldsymbol{x})\,dx^1\cdots dx^{\mathfrak{m}}}. \quad (125)$$

Usually, this is often not obtainable in closed form even when the unconditional cumulant generating function is available.

In this case cumulant generating function derivatives may be obtained from log likelihood derivatives. This relationship will be further explored in the next chapter. Fraser, Reid, and Wong (1991) exploit this fact, and in the case where $\mathfrak{k} = \mathfrak{m}+1$ suggest approximating the marginal density in the denominator of (125) by its saddlepoint approximation, and approximating numerically the derivatives of this approximate conditional likelihood. The result of this method is called the sequential saddlepoint approximation. Consider the density (121), arising from (120) under repeated sampling. The cumulant generating function for the marginal distribution of $\boldsymbol{X}_{-1} = (X^2,\dots,X^{\mathfrak{k}})$ is $\mathcal{H}_{\boldsymbol{X}}(\boldsymbol{\theta}+\boldsymbol{\beta}) - \mathcal{H}_{\boldsymbol{X}}(\boldsymbol{\theta})$, with $\beta_1 = 0$, and the multivariate saddlepoint satisfies $\mathcal{H}^i_{\boldsymbol{X}}(\boldsymbol{\theta}+\tilde{\boldsymbol{\beta}}) = x^i$ for $i \geq 2$ and $\tilde{\beta}_1 = 0$, where $\mathcal{H}^i_{\boldsymbol{X}}$

is the partial derivative of $\mathcal{H}_{\boldsymbol{X}}$ with respect to argument i. Let $\tilde{\boldsymbol{\theta}} = \boldsymbol{\theta} + \tilde{\boldsymbol{\beta}}$. Then $\tilde{\boldsymbol{\theta}}$ satisfies

$$\mathcal{H}^{i}_{\boldsymbol{X}}(\tilde{\boldsymbol{\theta}}) = x^{i} \text{ for } i \geq 2,\ \tilde{\theta}_1 = \theta_1. \tag{126}$$

The saddlepoint approximation to this marginal density $f_{X_{-1}}(\boldsymbol{x}_{-1})$, treating $\boldsymbol{X}$ as the mean of 1 independent and identically distributed copy of $\boldsymbol{X}$, is proportional to

$$\exp\left(\sum_{j=2}^{\mathfrak{t}}(\theta_j - \tilde{\theta}_j)x^j - \mathcal{H}_{\boldsymbol{X}}(\boldsymbol{\theta}) + \mathcal{H}_{\boldsymbol{X}}(\tilde{\boldsymbol{\theta}})\right) / \sqrt{\det\left[\mathcal{H}''_{\boldsymbol{X}_{-1}}(\tilde{\boldsymbol{\theta}})\right]},$$

where $\mathcal{H}_{\boldsymbol{X}_{-1}}(\tilde{\boldsymbol{\theta}})$ is the matrix of second derivatives of $\mathcal{H}$ with respect to all arguments except the first. Dependence in sample size is implicit in the relationship between $\mathcal{H}_{\boldsymbol{X}}$ and $\mathcal{H}_{\boldsymbol{Y}}$. An approximation to the conditional density $f_{X^1|\boldsymbol{X}_{-1}}(x^1|\boldsymbol{x}_{-1}) = f_{\boldsymbol{X}}(\boldsymbol{x})/f_{\boldsymbol{X}_{-1}}(\boldsymbol{x}_{-1})$ is

$$\exp\left(x^1\theta_1 + \sum_{j=2}^{\mathfrak{t}}\tilde{\theta}_j x^j - \mathcal{H}(\tilde{\boldsymbol{\theta}}) - \mathcal{G}_{X^1|\boldsymbol{X}_{-1}}(\boldsymbol{X})\right) / \sqrt{\det\left[\mathcal{H}''_{\boldsymbol{X}_{-1}}(\tilde{\boldsymbol{\theta}})\right]},$$

where $\mathcal{G}_{X^1|\boldsymbol{X}_{-1}}(\boldsymbol{X})$ is a function that might be quite difficult to evaluate, but fortunately will not be needed in the sequel. This corresponds to the conditional profile log likelihood of Cox and Reid (1987) after adjustment for non-orthogonality of parameters; also see Levin (1990). This thread, with more references, will be picked up later in §8.5. Approximately, then, the cumulant generating function for this conditional distribution is

$$-\tilde{\theta}_2 x^2 - \cdots - \tilde{\theta}_{\mathfrak{t}} x^{\mathfrak{t}} + \mathcal{H}_{\boldsymbol{Y}}(\tilde{\boldsymbol{\theta}}) - \tfrac{1}{2}\log\left(\det\left[\mathcal{H}''_{\boldsymbol{Y}_{-1}}(\tilde{\boldsymbol{\theta}})\right]\right). \tag{127}$$

Derivatives of this cumulant generating function can be calculated numerically, and be used in conjunction with univariate saddlepoint formulae such as (57) and (89).

As an example, let X^1 and X^2 be independent exponential random variables, from the family $f_X(x;\theta) = \exp(x\theta - x + \log(1-\theta))$ for $x > 0$ and $\theta < 1$. Consider the distribution of $T^1 = X^1$ conditional on $T^2 = t^2$, where $T^2 = X^1 + X^2$. The joint cumulant generating function of $\boldsymbol{T} = (T^1, T^2)$ is $-\log(1 - \theta_1 - 2\,\theta_2 + \theta_2\ (\theta_1 + \theta_2))$. The saddlepoint equation for θ_1 fixed is $(2-\theta_1-2\,\theta_2)/(1-\theta_1-2\,\theta_2+\theta_2\,(\theta_1+\theta_2)) = t^2$; this equation has two solutions for θ_2:

$$\theta_2 = 1 - \tfrac{1}{2}\theta_1 - \left(1 \pm \sqrt{1 + \tfrac{1}{4}\theta_1^2\,(t^2)^2}\right)/t^2.$$

The solution taking $\pm$ as $-$ reduces to $\theta_2 = 1$ when $\theta_1 = 0$, and as this lies outside the range of definition for the full cumulant generating function, the solution with $\pm$ taken as $+$ is chosen. The approximate conditional profile log likelihood is

$$\tfrac{1}{2}\log(2 + \sqrt{4 + \theta_1^2\,(t^2)^2}) + \tfrac{1}{2}\log(\sqrt{4 + \theta_1^2(t^2)^2}) + \theta_1\, t^1 - \tfrac{1}{2}\theta_1\, t^2 - \tfrac{1}{2}\sqrt{4 + \theta_1^2\,(t^2)^2},$$

and hence $\mathcal{K}^{\dagger}_{T^1|T^2}(\theta_1) = 2 - \frac{1}{2}\log(4 + 2\sqrt{4 + {\theta_1}^2\,(t^2)^2}) - \frac{1}{4}\log(4 + \theta_1^2\,(t^2)^2) + \frac{1}{2}\theta_1\,t^2 + \frac{1}{2}\sqrt{4 + {\theta_1}^2\,(t^2)^2}$ represents the cumulant generating function approximation (127). The true conditional distribution of T^1 is uniform over $(0, t^2)$, and if $t^2 = 1$, the true cumulant generating function of this conditional distribution is $\mathcal{K}_{T^1|T^2}(\theta_1) = \log(2\sinh(\theta_1/2)/\theta_1) + \frac{1}{2}\theta_1$, the cumulant generating function of a uniform random variable on $(0, 1)$. Fig. 14 illustrates how well $\mathcal{K}^{\dagger}_{T^1|T^2}(\theta_1)$ approximates $\mathcal{K}_{T^1|T^2}(\theta_1)$.

This example illustrates an interesting aspect of joint cumulant generating functions. The domain of $\mathcal{K}_{\boldsymbol{T}}$ is not the product of the domains of $\mathcal{K}_{T^1}$ and $\mathcal{K}_{T^2}$. Thus, while the domain of $\mathcal{K}_{T^1}$ is $(-\infty, 1)$, the domain of $\mathcal{K}^{\dagger}_{T^1|T^2}(\theta_1)$ is $\mathfrak{R}$, as is the domain of $\mathcal{K}_{T^1|T^2}$.

An alternate expression for (127) may be derived by considering the problem of conditional inference based on a $\boldsymbol{T}$, a linear transformation of $\boldsymbol{X}$. Then the likelihood equations (126) can be reexpressed in terms of $\mathcal{H}_{\boldsymbol{X}}$ rather than $\mathcal{H}_{\boldsymbol{T}}$, which is an advantage when $\mathcal{H}_{\boldsymbol{X}}$ has a simple form.

Lemma 7.2.1: *Suppose that $\boldsymbol{X}$ has the density (121), and that $\boldsymbol{T} = \boldsymbol{B}\boldsymbol{X}$, with $\boldsymbol{B}$ invertible. Express $\boldsymbol{B}$ as a partitioned matrix, with the first row, giving T^1, denoted by $\mathbf{c}$, and the remaining rows denoted by the $(\mathfrak{k} - 1) \times \mathfrak{k}$ matrix $\boldsymbol{A}$. Suppose that $\boldsymbol{A}\mathbf{c}^\top = \mathbf{o}$, a vector of all zeros. Then $\boldsymbol{T}$ has a density of the form*

$$f_{\boldsymbol{T}}(\boldsymbol{t}) = \exp(\boldsymbol{t}^\top\boldsymbol{\eta} - \mathcal{H}_{\boldsymbol{T}}(\boldsymbol{\eta}) - \mathcal{G}_{\boldsymbol{T}}(\boldsymbol{t})),$$

with $\mathcal{H}_{\boldsymbol{T}}(\boldsymbol{\eta}) = \mathcal{H}_{\boldsymbol{X}}(\boldsymbol{B}^\top\boldsymbol{\eta})$. Suppose $\boldsymbol{t}$ is the observed value of $\boldsymbol{T}$, and $\boldsymbol{x}$ is the associated value for $\boldsymbol{X}$. Let $\hat{\boldsymbol{\eta}}$ and $\tilde{\boldsymbol{\eta}}$ satisfy $\mathcal{H}'_{\boldsymbol{T}}(\hat{\boldsymbol{\eta}}) = \boldsymbol{t}^\top$ and $\mathcal{H}'_{\boldsymbol{T}_{-1}}(\tilde{\boldsymbol{\eta}}) = \boldsymbol{t}^\top_{-1}$ with $\tilde{\eta}_1 = \eta_1$. Let $\hat{\boldsymbol{\theta}}$ and $\tilde{\boldsymbol{\theta}}$ satisfy $\mathcal{H}'_{\boldsymbol{X}}(\hat{\boldsymbol{\theta}}) = \boldsymbol{x}^\top$ and $\mathcal{H}'_{\boldsymbol{X}_{-1}}(\tilde{\boldsymbol{\theta}})\boldsymbol{A} = \boldsymbol{x}^\top_{-1}\boldsymbol{A}$ with $\mathbf{c}^\top\tilde{\boldsymbol{\theta}}/(\mathbf{c}^\top\mathbf{c}) = \eta_1$. Then $\boldsymbol{B}^\top\hat{\boldsymbol{\theta}} = \hat{\boldsymbol{\eta}}$ and $\boldsymbol{B}^\top\tilde{\boldsymbol{\theta}} = \tilde{\boldsymbol{\eta}}$.

Proof: Follows from straight-forward calculation.

Q.E.D

The following lemma will also be useful:

Lemma 7.2.2: *Suppose that $\boldsymbol{V}$ is a symmetric invertible matrix of size $\mathfrak{k} \times \mathfrak{k}$, that $\boldsymbol{A}$ is a $\mathfrak{k} \times (\mathfrak{k} - 1)$ matrix of full rank, and $\mathbf{c}$ is a vector of length $\mathfrak{k}$ such that $\boldsymbol{A}^\top\mathbf{c} = \mathbf{o}$. Let $\boldsymbol{B}$ be the $\mathfrak{k} \times \mathfrak{k}$ matrix whose first row is $\mathbf{c}$ and whose remaining rows are $\boldsymbol{A}$. Then* $\det\left[\boldsymbol{B}\boldsymbol{V}\boldsymbol{B}^\top\right] = (\mathbf{c}^\top\mathbf{c})^2 \det\left[\boldsymbol{A}\boldsymbol{V}\boldsymbol{A}^\top\right]/(\mathbf{c}^\top\boldsymbol{V}^{-1}\mathbf{c})$.

Proof: Expanding the product of partitioned matrices, and using standard determinant formulae for partitioned matrices (Hocking, 1985, Appendix A.II.1.2),

$$\begin{aligned}\det\left[(\mathbf{c}\,\boldsymbol{A}^\top)^\top\boldsymbol{V}(\mathbf{c}\,\boldsymbol{A}^\top)\right] &= \det\left[\begin{pmatrix}\mathbf{c}^\top\boldsymbol{V}\mathbf{c} & \mathbf{c}^\top\boldsymbol{V}\boldsymbol{A}^\top\\ \boldsymbol{A}\boldsymbol{V}\mathbf{c} & \boldsymbol{A}\boldsymbol{V}\boldsymbol{A}^\top\end{pmatrix}\right]\\ &= \det\left[\boldsymbol{A}\boldsymbol{V}\boldsymbol{A}^\top\right]\left|\mathbf{c}^\top\boldsymbol{V}\mathbf{c} - \mathbf{c}^\top\boldsymbol{V}\boldsymbol{A}^\top(\boldsymbol{A}\boldsymbol{V}\boldsymbol{A}^\top)^{-1}\boldsymbol{A}\boldsymbol{V}\mathbf{c}\right|.\end{aligned}$$

Sequential Saddlepoint Approximation to the Uniform Cumulant Generating Function

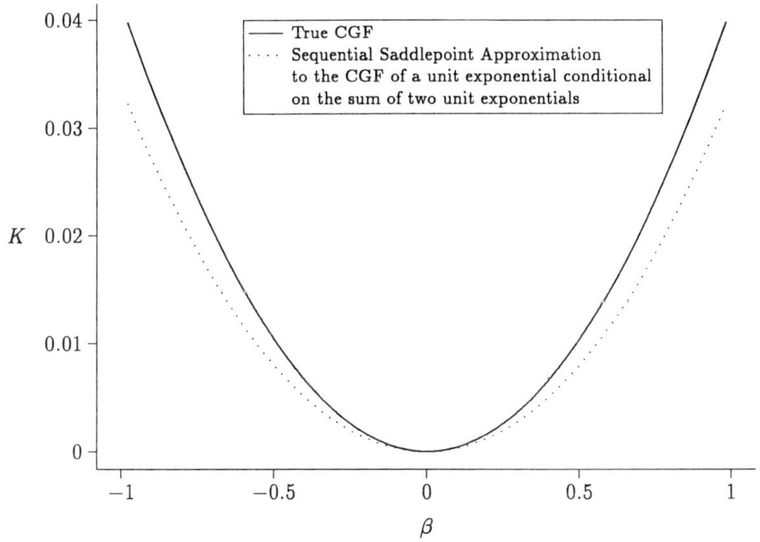

Fig. 14a.

Derivative of Sequential Saddlepoint Approximation to the Uniform Cumulant Generating Function

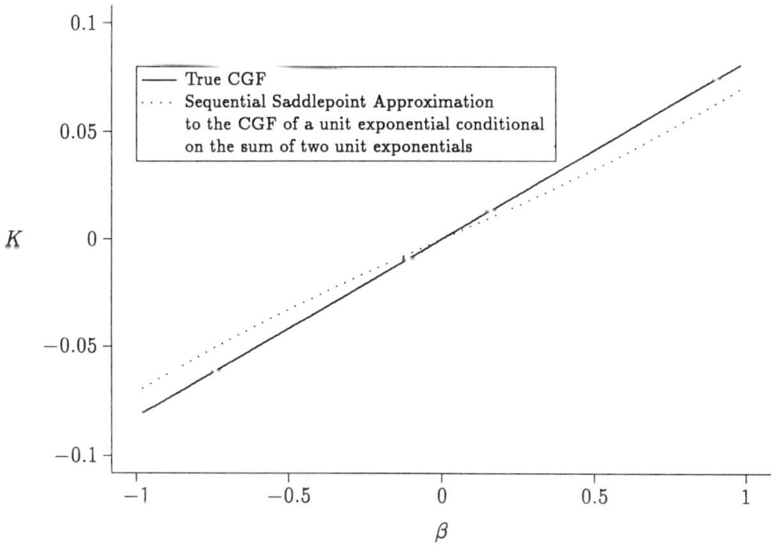

Fig. 14b.

Letting $\boldsymbol{U}$ be a matrix such that $\boldsymbol{U}\boldsymbol{U}^\top = \boldsymbol{V}$, this last factor can be reexpressed as

$$\left|\boldsymbol{c}^\top \boldsymbol{U}(\boldsymbol{I} - (\boldsymbol{A}\boldsymbol{U})^\top(\boldsymbol{A}\boldsymbol{U}(\boldsymbol{A}\boldsymbol{U})^\top)^{-1}\boldsymbol{A}\boldsymbol{U})\boldsymbol{U}^\top \boldsymbol{c}\right|.$$

The inner matrix is the projection matrix onto the space perpendicular to the space spanned by $\boldsymbol{A}\boldsymbol{U}$; this space is spanned by $\boldsymbol{c}^\top \boldsymbol{U}^{-1\top}$, and the projection matrix onto this space can be expressed as

$$\boldsymbol{U}^{-1}\boldsymbol{c}(\boldsymbol{c}^\top \boldsymbol{V}^{-1}\boldsymbol{c})^{-1}\boldsymbol{c}^\top \boldsymbol{U}^{-1\top}.$$

Hence $\left|\boldsymbol{c}^\top \boldsymbol{V}\boldsymbol{c} - \boldsymbol{c}^\top \boldsymbol{V}\boldsymbol{A}^\top(\boldsymbol{A}\boldsymbol{V}\boldsymbol{A}^\top)^{-1}\boldsymbol{A}\boldsymbol{V}\boldsymbol{c}\right| = (\boldsymbol{c}^\top\boldsymbol{c})^2(\boldsymbol{c}^\top \boldsymbol{V}^{-1}\boldsymbol{c})^{-1}.$

Q.E.D

Then assuming the setup of Lemma 7.2.1, (127) can be expressed as

$$-\tilde{\boldsymbol{\theta}}^\top \boldsymbol{x} + \mathcal{H}_{\boldsymbol{X}}(\tilde{\boldsymbol{\theta}}) - \tfrac{1}{2}\log\left(\det\left[\mathcal{H}''_{\boldsymbol{X}}(\tilde{\boldsymbol{\theta}})\right]\right) - \tfrac{1}{2}\log(\boldsymbol{c}^\top \mathcal{H}''_{\boldsymbol{X}}(\tilde{\boldsymbol{\theta}})^{-1}\boldsymbol{c}).$$

Alternatively, McCullagh and Tibshirani (1990) attack the problem of inference in the presence of nuisance parameters by approximating marginal rather than conditional distributions, and derive an adjusted profile log likelihood. The adjusted profile log likelihood is

7.3. Double Saddlepoint Distribution Function Approximations

Skovgaard (1987) applies double saddlepoint techniques to the problem of approximating tail probabilities for conditional distributions, by approximating the saddlepoint inversion integral for the quantity $D(x^1|\boldsymbol{x}_{-1}) = \int_{x^1}^{\infty} f_n(y, x^2, \ldots, x^{\mathfrak{k}})\,dy$; the ratio

$$D(x^1|\boldsymbol{x}_{-1})/f_{n,X^2,\ldots,X^{\mathfrak{k}}}(x^2, \ldots, x^{\mathfrak{k}})$$

is by definition the conditional cumulative distribution function. Here $\boldsymbol{x}_{-1} = (x^2, \ldots, x^{\mathfrak{k}})$. An inversion integral representation for D is derived by replacing x^1 by a dummy integration variable y in (100), and then integrating with respect to y between x^1 and ∞, to yield, for $C_{n,\mathfrak{k}} = (n/[2\pi i])^{\mathfrak{k}}\exp(n[\mathcal{K}_{\boldsymbol{Y}}(\hat{\boldsymbol{\beta}}) - \hat{\boldsymbol{\beta}}^\top \boldsymbol{x}])$,

$$\begin{aligned}
D(x^1|\boldsymbol{x}_{-1}) &= \left(\frac{n}{2\pi i}\right)^{\mathfrak{k}} \int_{x^1}^{\infty}\int_{-i\infty}^{+i\infty}\cdots\int_{-i\infty}^{+i\infty} \exp\left(n[\mathcal{K}_{\boldsymbol{Y}}(\boldsymbol{\beta}) - \beta_1(y - x^1) - \boldsymbol{\beta}^\top\boldsymbol{x}]\right)\,d\boldsymbol{\beta}\,dy \\
&= \left(\frac{n}{2\pi i}\right)^{\mathfrak{k}} \int_{-i\infty}^{+i\infty}\cdots\int_{-i\infty}^{+i\infty}\int_{x^1}^{\infty} \exp\left(n[\mathcal{K}_{\boldsymbol{Y}}(\boldsymbol{\beta}) - \beta_1(y - x^1) - \boldsymbol{\beta}^\top\boldsymbol{x}]\right)\,dy\,d\boldsymbol{\beta} \\
&= C_{n,\mathfrak{k}} \int_{-i\infty}^{+i\infty}\cdots\int_{-i\infty}^{+i\infty} \exp\left(n[\mathcal{K}_{\boldsymbol{Y}}(\boldsymbol{\beta}) - \mathcal{K}_{\boldsymbol{Y}}(\hat{\boldsymbol{\beta}}) + (\boldsymbol{\beta}^\top - \hat{\boldsymbol{\beta}})\boldsymbol{x}]\right)\,d\boldsymbol{\beta}/\beta_1. \quad (128)
\end{aligned}$$

The last equality in (128) follows only if the path of integration for the integral with respect to β_1 is deformed so that the real part of the path is bounded below by some positive number. After changing the variable of integration from $\boldsymbol{\beta}$ to $\boldsymbol{\omega}$ in the same way that generated (118), we obtain,

$$D(x^1|\boldsymbol{x}_{-1}) = C_{n,\mathfrak{k}} \int_{-i\infty}^{+i\infty}\cdots\int_{-i\infty}^{+i\infty} \exp\left(\frac{n}{2}(\boldsymbol{\omega} - \hat{\boldsymbol{\omega}})^\top(\boldsymbol{\omega} - \hat{\boldsymbol{\omega}})\right)\left|\frac{\partial\boldsymbol{\beta}}{\partial\boldsymbol{w}}\right| \frac{d\boldsymbol{\omega}}{\beta_1(\boldsymbol{\omega})}. \quad (129)$$

Skovgaard expands D in terms of the multivariate saddlepoints $\hat{\boldsymbol{\beta}}$ for the full distribution of $\boldsymbol{X}$ and $\tilde{\boldsymbol{\beta}}$ for the distribution of the shorter random vector $(X^2, \ldots, X^{\mathfrak{k}})$, and in terms of derivatives of $\mathcal{K}_Y$ at these saddlepoints $\hat{\boldsymbol{\beta}}$ and $\tilde{\boldsymbol{\beta}}$. Analytic difficulties in expanding (129) are similar to those that arose in expanding (85), in that the integrand has a simple pole at $\beta_1 = 0$. As with the univariate cumulative distribution function approximation of Lugannani and Rice (1980), presented in §5.3, the singularity will be isolated into a term that can be integrated exactly, and the remaining integral will be approximated. As with the saddlepoint density, the expansion for D can be factored into two components, with one factor identical to the saddlepoint approximation to the density of $X^2, \ldots, X^{\mathfrak{k}}$. The second factor is then the conditional tail probability approximation of interest.

Let $g(\boldsymbol{\omega}) = \det[d\boldsymbol{\beta}/d\boldsymbol{\omega}]\ \omega_1/\beta_1(\boldsymbol{\omega})$. Expressing $g(\boldsymbol{\omega})$ as $g(0, \boldsymbol{w}_{-1}) + (g(\boldsymbol{\omega}) - g(0, \boldsymbol{w}_{-1}))$ separates out the singularity into a term that can be integrated exactly. Here $\boldsymbol{w}_{-1}$ denotes the vector $(\omega_2, \ldots, \omega_{\mathfrak{k}})$. These two terms will be integrated separately. That is, D is expressed as $D_1 + D_2$, where

$$
\begin{aligned}
D_1 &= C_{n,\mathfrak{k}} \int_{-i\infty}^{+i\infty} \cdots \int_{-i\infty}^{+i\infty} \exp(\frac{n}{2}(\omega - \hat{\omega})^\top(\omega - \hat{\omega}))\frac{g(0, \boldsymbol{w}_{-1})}{n\omega_1}]\ d\omega \\
D_2 &= C_{n,\mathfrak{k}} \int_{-i\infty}^{+i\infty} \cdots \int_{-i\infty}^{+i\infty} \exp(\frac{n}{2}(\omega - \hat{\omega})^\top(\omega - \hat{\omega}))\frac{g(\boldsymbol{\omega}) - g(0, \boldsymbol{w}_{-1})}{n\omega_1}\ d\omega.
\end{aligned}
\tag{130}
$$

Since $\lim_{\omega_1 \to 0} \beta_1/\omega_1 = \partial\beta_1/\partial\omega_1$, equation (122) implies that

$$
\lim_{\omega_1 \to 0} \frac{\beta_1}{\omega_1}\left|\frac{\partial\boldsymbol{\beta}}{\partial\boldsymbol{w}}\right| = \frac{\partial\boldsymbol{\beta}_{-1}}{\partial\boldsymbol{w}_{-1}}(0, \boldsymbol{w}_{-1}).
$$

Since $g(0, \boldsymbol{w}_{-1})/\omega_1$ factors as the product of ω_1 times a function of the other variables, the resulting integral D_1 may be factored the same way, and may be expressed as

$$
D_1 = \frac{\phi(n\hat{\omega}_1)}{i\sqrt{2\pi}} \int_{c-i\infty}^{c+i\infty} \exp\left(-\frac{n}{2}(\omega_1 - \hat{\omega}_1)^2\right)\frac{d\omega_1}{\omega_1}\left\{\left(\frac{n}{2\pi i}\right)^{\mathfrak{k}-1}\exp(n[\mathcal{K}_Y(\tilde{\boldsymbol{\beta}}) - \tilde{\boldsymbol{\beta}}^\top\boldsymbol{x}])\times\right.
$$
$$
\left.\int_{c-i\infty}^{c+i\infty} \cdots \int_{c-i\infty}^{c+i\infty} \exp\left(-\frac{n}{2}\sum_{j=2}^{\mathfrak{k}}(\omega_j - \hat{\omega}_j)^2\right)\det\left[\frac{\partial\boldsymbol{\beta}_{-1}}{\partial\boldsymbol{w}_{-1}}(0, \boldsymbol{w}_{-1})\right] d\boldsymbol{w}_{-1}\right\}.
$$

By (118) the quantity in brackets is exactly the unconditional density of $(x^2, \ldots, x^{\mathfrak{k}})$; the first factor is $1 - \Phi(\sqrt{n}\hat{\omega}_1)$, by (88).

Since $\partial\boldsymbol{\beta}_{-1}/\partial\boldsymbol{\omega}_{-1}$ does not depend on ω_1, then $(g(\boldsymbol{\omega}) - g(0, \boldsymbol{w}_{-1}))/\omega_1$ can be expressed as

$$
\frac{\partial\boldsymbol{\beta}_{-1}}{\partial\boldsymbol{w}_{-1}}(0, \boldsymbol{w}_{-1}) \times \left(\frac{\partial\beta_1}{\partial\omega_1}/\beta_1 - 1/\omega_1\right).
$$

Since the second factor depends on $\boldsymbol{w}_{-1}$ the integrals will not factor as those in the preceding paragraph did. However, the first order Watson's Lemma approximation to this integral does factor into

$$D_2 = \left[\frac{n^{(\mathfrak{k}-1)/2}\exp(n[\mathcal{K}_{\boldsymbol{Y}}(\tilde{\boldsymbol{\beta}}) - \sum_{j=2}^{\mathfrak{k}}\tilde{\beta}_j x^j])}{(2\pi)^{(\mathfrak{k}-1)/2}} \det\left[\frac{\partial\beta_2,\ldots,\beta_{\mathfrak{k}}}{\partial\omega_2,\ldots,\omega_{\mathfrak{k}}}\right](\hat{\omega}_2,\ldots,\hat{\omega}_{\mathfrak{k}})\right]\times$$
$$\left[\frac{\exp(n[\mathcal{K}_{\boldsymbol{Y}}(\hat{\boldsymbol{\beta}}) - \mathcal{K}_{\boldsymbol{Y}}(\tilde{\boldsymbol{\beta}}) - (\hat{\boldsymbol{\beta}}-\tilde{\boldsymbol{\beta}})^\top\boldsymbol{x}])}{(2\pi)^{1/2}\sqrt{n}}\left(\det\left[\frac{\partial\beta_1}{\partial\omega_1}\right](\hat{\boldsymbol{\omega}})/\hat{\beta}_1 - \frac{1}{\hat{\omega}_1}\right)\right](1+O(n^{-1})).$$

Again the first factor is the density approximation for $(X^2,\ldots,X^{\mathfrak{k}})$ to $O(n^{-1})$. This implies that

$$1 - F_{x^1|\boldsymbol{x}_{-1}}(X^1|\boldsymbol{X}_{-1}) = 1 - \Phi(\sqrt{n}\hat{\omega}_1) + \frac{\phi(\sqrt{n}\hat{\omega}_1)}{\sqrt{n}}\left(\frac{\vartheta}{\hat{u}_1} - \frac{1}{\hat{\omega}_1} + O(n^{-1})\right), \quad (131)$$

where

$$\hat{u}_1 = \hat{\beta}_1,\ \hat{\omega}_1 = \operatorname{sgn}(\hat{\beta}_1)\sqrt{2[\hat{\boldsymbol{\beta}}^\top\boldsymbol{x} - \mathcal{K}_{\boldsymbol{Y}}(\hat{\boldsymbol{\beta}})] - 2[\tilde{\boldsymbol{\beta}}^\top\boldsymbol{x} - \mathcal{K}_{\boldsymbol{Y}}(\tilde{\boldsymbol{\beta}})]},$$
$$\vartheta = \det\left[\mathcal{K}''_{\boldsymbol{Y}_{-1}}(\tilde{\boldsymbol{\beta}})\right]^{1/2}\det\left[\mathcal{K}''_{\boldsymbol{Y}}(\hat{\boldsymbol{\beta}})\right]^{-1/2}, \quad (132)$$

$\mathcal{K}_{\boldsymbol{Y}_{-1}}$ is the $(\mathfrak{k}-1)\times(\mathfrak{k}-1)$ submatrix of the matrix of second derivatives of $\mathcal{K}$, corresponding to all components of $\boldsymbol{\beta}$ except the first, and Φ and ϕ are the normal distribution function and density respectively, since D is (131) times the saddlepoint density approximation for $(X^2,\ldots,X^{\mathfrak{k}})$ to $O(n^{-1})$. This is the double saddlepoint approximation to the conditional distribution function.

Also of interest are inversion techniques random vectors $\boldsymbol{X}$ confined to a generalized lattice in the sense of Lemma 6.4.4. Let $D(x^1|\boldsymbol{x}_{-1}) = \sum_{x^1}^{\infty} p_n(y, x^2,\ldots,x^{\mathfrak{k}})$ where p_n is the probability mass function for $\boldsymbol{X}$. Using the multivariate lattice probability inversion theorem (107), the counterpart of equation (128) is

$$D(x^1|\boldsymbol{x}_{-1}) = \sum_{y^1=x^1}^{\infty}\left(\frac{n}{2\pi i}\right)^{\mathfrak{k}}\int_{-\pi/\Delta_1}^{\pi/\Delta_1}\cdots\int_{-\pi/\Delta_{\mathfrak{k}}}^{\pi/\Delta_{\mathfrak{k}}}\exp(n[\mathcal{K}_{\boldsymbol{Y}}(\boldsymbol{\beta}) - y^1\beta_1 - \boldsymbol{\beta}_{-1}^\top\boldsymbol{x}_{-1}])\,d\boldsymbol{\beta}$$
$$= \left(\frac{n}{2\pi i}\right)^{\mathfrak{k}}\int_{-\pi/\Delta_1}^{\pi/\Delta_1}\cdots\int_{-\pi/\Delta_{\mathfrak{k}}}^{\pi/\Delta_{\mathfrak{k}}}\frac{\exp(n[\mathcal{K}_{\boldsymbol{Y}}(\boldsymbol{\beta}) - (x^1 - \frac{1}{2}\Delta_1)\beta_1 - \boldsymbol{\beta}_{-1}^\top\boldsymbol{x}_{-1}])}{2\sinh(\Delta_1\beta_1/2)}\,d\boldsymbol{\beta}.$$

The last equality above follows only if the path of integration for the integral with respect to β_1 is deformed so that the real part of the path is bounded below by some positive number. The same analysis as above is performed with $g(\boldsymbol{\omega}) = \det[d\boldsymbol{\beta}/d\boldsymbol{\omega}]\ \omega_1/2\sinh(\beta_1(\boldsymbol{\omega})/2)$. In the lattice case (131) holds, with

$$\hat{u}_1 = 2\sinh(\tfrac{1}{2}\Delta_1\hat{\beta}_1)/\Delta_1,\ \hat{\omega}_1 = \operatorname{sgn}(\hat{\beta}_1)\sqrt{2[\hat{\boldsymbol{\beta}}^\top\boldsymbol{x} - \mathcal{K}_{\boldsymbol{Y}}(\hat{\boldsymbol{\beta}})] - 2[\tilde{\boldsymbol{\beta}}^\top\boldsymbol{x} - \mathcal{K}_{\boldsymbol{Y}}(\tilde{\boldsymbol{\beta}})]},$$
$$\vartheta = \det\left[\mathcal{K}''_{\boldsymbol{Y}_{-1}}(\tilde{\boldsymbol{\beta}})\right]^{1/2}\det\left[\mathcal{K}''_{\boldsymbol{Y}}(\hat{\boldsymbol{\beta}})\right]^{-1/2}, \quad (133)$$

and x^1 is corrected for continuity when calculating $\hat{\boldsymbol{\beta}}$. That is, if possible values for X^1 are Δ_1 units apart, $\hat{\boldsymbol{\beta}}$ solves $\mathcal{K}'_{\boldsymbol{Y}}(\hat{\boldsymbol{\beta}}) = \tilde{\boldsymbol{x}}$ where $\tilde{x}^j = x^j$ if $j \neq 1$ and $\tilde{x}^1 = x^1 - \frac{1}{2}\Delta_1$.

When $\hat{u}_1 = 0$ (131) is undefined; these can be replaced by their limiting values, which are given by Skovgaard (1987).

These methods have been applied to distributions that are either wholly continuous or wholly lattice. Results in mixed cases are also not only possible but easily derived.

Reconsider the case in which $\boldsymbol{X}$ arises from the full exponential family (120). In this case the double saddlepoint distribution function approximation can be reexpressed in terms of the quantities in (120). Skovgaard's approximation (131) holds, with

$$\hat{\omega}_1 = \operatorname{sgn}(\hat{\theta}_1 - \theta_1)\sqrt{2[\hat{\boldsymbol{\theta}}^\top \boldsymbol{x} - \mathcal{H}_{\boldsymbol{Y}}(\hat{\boldsymbol{\theta}}) - \tilde{\boldsymbol{\theta}}^\top \boldsymbol{x} + \mathcal{H}_{\boldsymbol{Y}}(\tilde{\boldsymbol{\theta}})]},\ \hat{u}_1 = \hat{\theta}_1 - \theta_1,$$
$$\vartheta = \det\left[\mathcal{H}''_{\boldsymbol{Y}_{-1}}(\tilde{\boldsymbol{\theta}})\right]^{1/2} \det\left[\mathcal{H}''_{\boldsymbol{Y}}(\hat{\boldsymbol{\theta}})\right]^{-1/2} \tag{134}$$

and depends on the parameter vector $\boldsymbol{\theta}$ only through $\hat{\omega}_1$, the restricted maximum likelihood estimator $\tilde{\boldsymbol{\theta}}$, and the second derivative matrix of $\mathcal{H}_{\boldsymbol{Y}}$ at $\tilde{\boldsymbol{\theta}}$. The saddlepoint $\tilde{\boldsymbol{\theta}}$ is chosen to satisfy (126), and the saddlepoint $\hat{\boldsymbol{\theta}}$ is chosen to satisfy $\mathcal{H}^i_{\boldsymbol{Y}}(\hat{\boldsymbol{\theta}}) = X^i$ for all i, where the superscript denotes partial differentiation. Hence Skovgaard's approximation also depends only on the first component of the parameter vector and not on any of the other components. When X^1 has a lattice distribution approximation (131) holds with

$$\hat{\omega}_1 = \operatorname{sgn}(\hat{\theta}_1 - \theta_1)\sqrt{2[\hat{\boldsymbol{\theta}}^\top \boldsymbol{x} - \mathcal{H}_{\boldsymbol{Y}}(\hat{\boldsymbol{\theta}}) - \tilde{\boldsymbol{\theta}}^\top \boldsymbol{x} + \mathcal{H}_{\boldsymbol{Y}}(\tilde{\boldsymbol{\theta}})]},\ \hat{u}_1 = 2\sinh(\Delta_1(\hat{\theta}_1 - \theta_1)/2)/\Delta_1,$$
$$\vartheta = \det\left[\mathcal{H}''_{\boldsymbol{Y}_{-1}}(\tilde{\boldsymbol{\theta}})\right]^{1/2} \det\left[\mathcal{H}''_{\boldsymbol{Y}}(\hat{\boldsymbol{\theta}})\right]^{-1/2} \tag{135}$$

where $\hat{\boldsymbol{\theta}}$ is calculated from continuity-corrected data.

Alternate versions of (134) and (135) for conditional calculations involving random vectors that are linear transformations of more simple vectors are also desirable. Assume the definitions of Lemma 7.2.1. Suppose further that the distribution of T^1 conditional on all other components of $\boldsymbol{T}$ is desired. Then $\mathcal{H}_{\boldsymbol{T}}(\hat{\boldsymbol{\eta}}) - \hat{\boldsymbol{\eta}}^\top \boldsymbol{t} - \mathcal{H}_{\boldsymbol{T}}(\tilde{\boldsymbol{\eta}}) + \tilde{\boldsymbol{\eta}}^\top \boldsymbol{t} = \mathcal{H}_{\boldsymbol{X}}(\hat{\boldsymbol{\theta}}) - \hat{\boldsymbol{\theta}}^\top \boldsymbol{x} - \mathcal{H}_{\boldsymbol{X}}(\tilde{\boldsymbol{\theta}}) + \tilde{\boldsymbol{\theta}}^\top \boldsymbol{x}$, and so the magnitude of $\hat{\omega}_1$ is unchanged by substituting $\mathcal{H}_{\boldsymbol{X}}$, $\hat{\boldsymbol{\theta}}$, and $\tilde{\boldsymbol{\theta}}$ for $\mathcal{H}_{\boldsymbol{T}}$, $\hat{\boldsymbol{\eta}}$, and $\tilde{\boldsymbol{\eta}}$. Furthermore, $\hat{\eta}_1 = (\boldsymbol{c}^\top \boldsymbol{c})^{-1} \boldsymbol{c}^\top \hat{\boldsymbol{\theta}}$, and $\det[\mathcal{H}''_{\boldsymbol{T}}(\hat{\boldsymbol{\eta}})] = \det\left[\mathcal{H}''_{\boldsymbol{X}}(\hat{\boldsymbol{\theta}})\right] \det[\boldsymbol{B}]^2$. By Lemma 7.2.2, $\det\left[\mathcal{H}''_{\boldsymbol{T}_{-1}}(\tilde{\boldsymbol{\eta}})\right] = \det\left[\boldsymbol{A}\mathcal{H}''_{\boldsymbol{X}}(\tilde{\boldsymbol{\theta}})\boldsymbol{A}^\top\right] = \det[\mathcal{H}''_{\boldsymbol{X}}(\tilde{\boldsymbol{\eta}})]\,(\boldsymbol{c}^\top \left[\mathcal{H}''_{\boldsymbol{X}}(\tilde{\boldsymbol{\theta}})\right]^{-1} \boldsymbol{c}) \det[\boldsymbol{B}]^2 (\boldsymbol{c}^\top \boldsymbol{c})^{-2}$. After accounting for n by replacing $\mathcal{H}_{\boldsymbol{X}}$ by its expression in $\mathcal{H}_{\boldsymbol{Y}}$, (131) holds, with T substituted for X, and

$$\vartheta = \sqrt{\det\left[\mathcal{H}''_{\boldsymbol{Y}}(\tilde{\boldsymbol{\theta}})\right](\boldsymbol{c}^\top \left[\mathcal{H}''_{\boldsymbol{Y}}(\tilde{\boldsymbol{\theta}})\right]^{-1} \boldsymbol{c})} \Big/ \left(\boldsymbol{c}^\top \boldsymbol{c}\sqrt{\det\left[\mathcal{H}''_{\boldsymbol{Y}}(\hat{\boldsymbol{\theta}})\right]}\right),\ \hat{u}_1 = \boldsymbol{c}^\top \hat{\boldsymbol{\theta}}/\boldsymbol{c}^\top \boldsymbol{c} - \eta_1$$
$$\hat{\omega}_1 \text{ is as in (135)}, \tag{136}$$

when $\boldsymbol{Y}$ has a continuous distribution, and

$$\vartheta = \sqrt{\det\left[\mathcal{H}''_{\boldsymbol{Y}}(\tilde{\boldsymbol{\theta}})\right](\boldsymbol{c}^{\top}\left[\mathcal{H}''_{\boldsymbol{Y}}(\tilde{\boldsymbol{\theta}})\right]^{-1}\boldsymbol{c})} / \left(\boldsymbol{c}^{\top}\boldsymbol{c}\sqrt{\det\left[\mathcal{H}''_{\boldsymbol{Y}}(\hat{\boldsymbol{\theta}})\right]}\right),$$

$$u_1 = 2\sinh\left(\Delta_1[(\boldsymbol{c}^{\top}\hat{\boldsymbol{\theta}})/(\boldsymbol{c}^{\top}\boldsymbol{c}) - \boldsymbol{\eta}_1]/2\right)/\Delta_1,\ \hat{\omega}_1 \text{ is as in (135)}, \tag{137}$$

when $\boldsymbol{Y}$ has a lattice distribution.

The following theorem summarizes the results of this this section:

Theorem 7.3.1: *Suppose that the random vector $\boldsymbol{X}$ is the mean of n independent and identically distributed random vectors, each of which has a cumulant generating function defined on an open neighborhood about $\mathbf{o}$. Take a potential value $\boldsymbol{x}$ for the mean of n independent and identically distributed copies of such vectors. If one of the following approximations is desired:*

a. *An expansion for the cumulative distribution function of X^1 conditional on $X^2, \ldots, X^{\mathfrak{t}}$ in terms of the cumulant generating function*
b. *An expansion for the cumulative distribution function of X^1 conditional on $X^2, \ldots, X^{\mathfrak{t}}$ in terms of the exponential family expression*
c. *An expansion for the cumulative distribution function of T^1 conditional on $T^2, \ldots, T^{\mathfrak{t}}$, where $\boldsymbol{T} = \boldsymbol{BX}$ with $\boldsymbol{B}$ as above, in terms of the cumulant generating function of $\boldsymbol{X}$*

and one of the following conditions holds:

1. *A joint density exists for $\boldsymbol{X}$,*
2. *$\boldsymbol{X}$ is confined to a generalized lattice as in Lemma 6.4.4.*

Then (131), with components defined as indicated in Table 5, gives valid a asymptotic expansion.

Table 5: Appropriate Double Saddlepoint Distribution Function Results

Condition	*Objective*		
	a.	*b.*	*c.*
1.	*(132)*	*(134)*	*(136)*
2.	*(133)*	*(135)*	*(137)*

Proof: See the above discussion in this section.

Q.E.D

When conditioning on variables taking values at the ends of their ranges, (123) may not have a finite solution. Additionally, when limits are taken in (124) as components of this vector converge to plus or minus infinity, the matrices whose determinants are taken become of less than full rank. Kolassa (1996a) discusses modifications to (124) in this case.

Note the formal equivalence between (131) and (89). The adjusted r^* approximation of §5.4 may also be used here, following Lemma 5.4.1.

The calculations of this chapter are applied to conditional inference in one of the exercises; the application is due to Davison (1988).

7.4. Higher-Order Double Saddlepoint Approximations

Kolassa (1996b) derives a higher-order double saddlepoint distribution function approximation from (130). Recall that the first-order approximation to D_1 was exact; higher order terms will all come from D_2. Skovgaard (1987) suggests performing the integrals with respect to $\omega_2, \ldots, \omega_{\mathfrak{k}}$ to obtain the standard higher order saddlepoint approximation to the density of the conditioning variables, and then applying Temme's Theorem to the remaining univariate integral. The present approach of using the extended version of Temme's Theorem is analytically more straight-forward. Theorem 6.6.1 is applied once to D_2. The series expansion for D_2 will be expressed as the product of two factors. The first will be the series expansion for the density of $\boldsymbol{X}_{-1}$; the second will then be what is added to $1 - \Phi(\hat{\omega}_1)$ to produce the desired expansion for the conditional cumulative distribution function.

Use of Theorem 6.6.1 requires evaluation of the quantities A_s defined by (114). Since the integrand is free of n except in the exponent, the functions θ_s vanish unless $s = 0$. The inequality in the description of the set over which summation is taken in (114) becomes equality, yielding, after combining factors of i in (115) and in the leading factor of D_2, $D_2 = (n/2\pi)^{\mathfrak{k}/2} \exp(-\frac{n}{2} \sum_{j=1}^{\mathfrak{k}} \hat{\omega}_j^2) \sum_{s=0}^{\infty} A_s n^{-s}$, where

$$A_s = \sum_{v_1,\cdots,v_{\mathfrak{k}} \geq 0,\ \sum v_j = s} \frac{(-2)^{-s}}{\boldsymbol{v}!} \left[\frac{\partial^{2\boldsymbol{v}}}{\partial^{2\boldsymbol{v}} \omega} \theta_0 \right] (\hat{\omega}),$$

and $\theta_0(\boldsymbol{\omega}) = \det\left[\partial \boldsymbol{\beta} / \partial \boldsymbol{\omega}\right] / \beta^1 - \det\left[\partial \boldsymbol{\beta}_{-1} / \partial \boldsymbol{w}_{-1}(0, \boldsymbol{w}_{-1})\right] / \omega_1$. Again employ the convention that indices appearing as superscripts and subscripts in a term are summed over. Let $h(\boldsymbol{\omega})$ be the function of $\boldsymbol{\omega}$ whose power series about $\hat{\boldsymbol{\omega}}$ is constructed from that of $\det\left[(\partial \boldsymbol{\beta}_{-1} / \partial \boldsymbol{w}_{-1})(0, \boldsymbol{w}_{-1})\right]$, with all terms involving components of $\boldsymbol{\omega} - \hat{\boldsymbol{\omega}}$ to odd powers omitted. Let

$$g(\boldsymbol{\omega}) = \theta_0(\boldsymbol{\omega})/h(\boldsymbol{\omega}) = \det\left[\frac{\partial \boldsymbol{\beta}}{\partial \boldsymbol{w}}(\boldsymbol{\omega})\right] / (h(\boldsymbol{\omega})\beta_1) - \det\left[\frac{\partial \boldsymbol{\beta}_{-1}}{\partial \boldsymbol{w}_{-1}}(0, \boldsymbol{w}_{-1})\right] / (h(\boldsymbol{\omega})\omega_1). \tag{138}$$

Then

$$\frac{(-2)^{-s}}{\boldsymbol{v}!} \left[\frac{\partial^{2\boldsymbol{v}}}{\partial^{2\boldsymbol{v}} \omega} \theta_0 \right] (\hat{\omega}) = \frac{1}{\boldsymbol{v}!} \sum_{0 \leq u_j < 2v_j} \frac{2^{-s}(2\boldsymbol{v})!}{\boldsymbol{u}!(2\boldsymbol{v} - \boldsymbol{u})!} \left[\frac{\partial^{\boldsymbol{u}}}{\partial^{\boldsymbol{u}} \omega} h \right] (\hat{\omega}) \left[\frac{\partial^{2\boldsymbol{v} - \boldsymbol{u}}}{\partial^{2\boldsymbol{v} - \boldsymbol{u}} \omega} g \right] (\hat{\omega}).$$

Applying Theorem 6.6.2, and recalling that h has only even order terms,

$$A_s = \sum_{\substack{v_1,\cdots,v_{\mathfrak{k}} \geq 0 \\ \sum v_j = s}} \frac{(2\boldsymbol{v})!}{\boldsymbol{v}!} \sum_{0 \leq u_j \leq v_j} \frac{(-2)^{-s}}{(2\boldsymbol{u})!(2\boldsymbol{v} - 2\boldsymbol{u})!} \left[\frac{\partial^{2\boldsymbol{u}}}{\partial^{2\boldsymbol{u}} \omega} h \right] (\hat{\omega}) \left[\frac{\partial^{2\boldsymbol{v} - 2\boldsymbol{u}}}{\partial^{2\boldsymbol{v} - 2\boldsymbol{u}} \omega} g \right] (\hat{\omega}). \tag{139}$$

The aim here, then, is to express the expansion $\sum_{s=0}^{j-1} A_s n^{-s}$ as

$$\sum_{s=0}^{j-1} B_s n^{-s} \times \sum_{s=0}^{j-1} C_s n^{-s} + O(n^{-j}),$$

where by Theorem 6.6.2, B_s are coefficients in the asymptotic expansion of the density for the conditioning variables:

$$B_s = \sum_{\substack{v_2,\cdots,v_{\mathfrak{k}}\geq 0 \\ \sum_{j=2}^{\mathfrak{k}} v_j = s}} \frac{(-2)^{-\sum_{j=2}^{\mathfrak{k}} v_j}}{v_2!\cdots v_{\mathfrak{k}}!}\left[\frac{\partial^{2v_2+\cdots+2v_{\mathfrak{k}}}}{\partial^{2v_2}\omega_2\cdots\partial^{2v_{\mathfrak{k}}}\omega_{\mathfrak{k}}} h\right](\hat{\omega})$$

such that $f_{X^2,\dots,X^{\mathfrak{k}}} = \sum_{s=0}^{j-1} B_s n^{-s} + O(n^{-j})$. Since

$$\left(1-\Phi(\sqrt{n}\hat{\omega}_1)+\phi(\sqrt{n}\hat{\omega}_1)n^{-1/2}\sum_{s=0}^{j-1} C_s n^{-s}\right)\left(\phi(\sqrt{n}\hat{\omega}_{-1})n^{(1-\mathfrak{k})/2}\sum_{s=0}^{j-1} B_s n^{-s}\right)$$

is an asymptotic expansion for D, $1-\Phi(\sqrt{n}\hat{\omega}_1)+\phi(\sqrt{n}\hat{\omega}_1)n^{-1/2}\sum_{s=0}^{j-1} C_s n^{-s}$ is the required expansion for the conditional tail probability. Skovgaard (1987) provides such a decomposition for $j = 1$; that is, he factors the lead term $A_0 = \det\left[\frac{\partial \beta_{-1}}{\partial w_{-1}}(0, w_{-1})\right] \times g(\hat{\omega})$ into the lead term $B_0 = \det\left[\frac{\partial \beta_{-1}}{\partial w_{-1}}(0, w_{-1})\right]$ in the asymptotic expansion for the density of the conditioning density times a factor $C_0 = g(\hat{\omega}) = \sqrt{\det\left[\mathcal{K}''_{X_{-1}}(0,\hat{\beta}_{-1})\right]}/(\hat{\beta}_1\sqrt{\det\left[\mathcal{K}''_{X}(\hat{\beta})\right]}) - 1/\hat{\omega}_1$ which consequently contributes to the lead term in the asymptotic expansion for the conditional tail probability desired.

When $j = 2$, of concern are A_0 and A_1. When $s = 0$ or $s = 1$, $(2\boldsymbol{v})!$ in (139) is equal either to $(2\boldsymbol{u})!$ or $(2\boldsymbol{v}-2\boldsymbol{u})!$, or to both. The other quantity is 1. Also, $\boldsymbol{u}!$, $\boldsymbol{v}!$, and $(\boldsymbol{v}-\boldsymbol{u})!$ are all 1. Hence when $s = 0$ or $s = 1$,

$$A_s = \sum_{\substack{v_1,\cdots,v_{\mathfrak{k}}\geq 0 \\ \sum v_j = s}} \sum_{0\leq u_j\leq v_j} \frac{(-2)^{-s}}{(\boldsymbol{u})!(\boldsymbol{v}-\boldsymbol{u})!}\left[\frac{\partial^{2\boldsymbol{u}}}{\partial^{2\boldsymbol{u}}\omega} h\right](0, w_{-1})\left[\frac{\partial^{2\boldsymbol{v}-2\boldsymbol{u}}}{\partial^{2\boldsymbol{v}-2\boldsymbol{u}}\omega} g\right](\hat{\omega}),$$

indicating that $C_1 = \sum_{j=1}^{\mathfrak{k}} \frac{(-2)^{-1}}{1!}\frac{\partial^2}{\partial\omega_j^2} g(\hat{\omega}) = -\frac{1}{2}\sum_{k=1}^{\mathfrak{k}} g^{kk}(\hat{\omega})$, where g is given by (138), and superscripts on g indicate derivatives with respect to components of $\boldsymbol{\omega}$.

The next section shows that

$$C_1 = -\frac{1}{\hat{z}}\left(\tfrac{1}{8}(\hat{\rho}_{13}-\tilde{\rho}_{13}-\hat{\rho}_4+\tilde{\rho}_4)+\tfrac{1}{12}(\hat{\rho}_{23}-\tilde{\rho}_{23})+\frac{1}{2}\frac{\hat{\kappa}_{1j}\hat{\kappa}^{ijk}\hat{\kappa}_{ik}}{\hat{\beta}_1}+\frac{\hat{\kappa}_{11}}{(\hat{\beta}_1)^2}\right)-\frac{1}{(\hat{\omega}_1)^3}, \tag{140}$$

where $\hat{z} = \hat{\beta}_1 \det\left[\frac{\partial\beta}{\partial w}(\hat{\omega})\right] / \det\left[\frac{\partial\beta_{-1}}{\partial w_{-1}}(0,\hat{\omega}_{-1})\right]$, and $\hat{\rho}_4$, $\hat{\rho}_{13}$, $\hat{\rho}_{23}$, $\tilde{\rho}_4$, $\tilde{\rho}_{13}$, and $\tilde{\rho}_{23}$ are as in §7.1. The second-order saddlepoint approximation to the conditional cumulative distribution function is now, to $O(n^{-5/2})$,

$$1 - \tilde{F}(x^1|\boldsymbol{x}_{-1}) = 1 - \Phi(\sqrt{n}\hat{\omega}_1) + \frac{\phi(\sqrt{n}\hat{\omega}_1)}{\sqrt{n}}\Bigg(\frac{1}{n\hat{\omega}_1^3} - \frac{1}{\hat{\omega}_1} + \frac{1}{\hat{z}}\Big(1 + \frac{1}{n}[\tfrac{1}{8}(\hat{\rho}_4 - \tilde{\rho}_4) - \tfrac{1}{8}(\hat{\rho}_{13} - \tilde{\rho}_{13}) - \tfrac{1}{12}(\hat{\rho}_{23} - \tilde{\rho}_{23}) - \tfrac{1}{2}\frac{\hat{\kappa}_{1k}\hat{\kappa}^{ijk}\hat{\kappa}_{ij}}{\hat{\beta}_1} - \frac{\hat{\kappa}^{11}}{(\hat{\beta}_1)^2}]\Big)\Bigg).$$

In the lattice case,

$$1 - \tilde{F}(x^1|\boldsymbol{x}_{-1}) = 1 - \Phi(\sqrt{n}\hat{\omega}_1) + \frac{\phi(\sqrt{n}\hat{\omega}_1)}{\sqrt{n}}\Bigg(- \frac{1}{\hat{\omega}_1} + \frac{1}{\hat{z}}\Big(1 + \frac{1}{n}\Big[\tfrac{1}{8}(\hat{\rho}_4 - \tilde{\rho}_4) - \tfrac{1}{8}(\hat{\rho}_{13} - \tilde{\rho}_{13}) - \tfrac{1}{12}(\hat{\rho}_{23} - \tilde{\rho}_{23}) - \tfrac{1}{4}\Delta\hat{\kappa}_{1k}\hat{\kappa}^{ijk}\hat{\kappa}_{ij}\coth(\tfrac{1}{2}\hat{\beta}_1\Delta_1) - \left(\tfrac{1}{4}\coth(\tfrac{1}{2}\hat{\beta}_1\Delta_1)^2 - \tfrac{1}{8}\right)\Delta^2\hat{\kappa}_{11}\Big]\Big) + \frac{1}{n\hat{\omega}_1^3}\Bigg) \quad (141)$$

where $\hat{z} = 2\sinh(\frac{1}{2}\hat{\beta}_1\Delta_1)\det\left[\frac{\partial\beta}{\partial w}(\hat{\omega})\right] / \det\left[\frac{\partial\beta_{-1}}{\partial w_{-1}}(0, \hat{\omega}_{-1})\right]$.

7.5. Derivatives of the Jacobian

In this section are presented calculations of derivatives of functions of $\partial\boldsymbol{\beta}/\partial\boldsymbol{\omega}$ necessary for for deriving via Temme's Theorem higher-order terms in the series expansions given in §6.3, §7.1, and §7.4. In this section again we employ the convention that an index appearing in a term both as a subscript and as a superscript is summed over.

Let $\hat{\beta}_r^{m\ldots o}$ be the derivative of component r of $\boldsymbol{\beta}$ with respect to $m, \ldots, o$, evaluated at $\hat{\boldsymbol{\omega}}$. Implicitly differentiating (116),

$$\left.\begin{aligned} \delta_{m,n} &= \hat{\beta}_l^m\hat{\beta}_j^n\hat{\kappa}^{jl} + \hat{\kappa}^j\hat{\beta}_j^{mn} \\ 0 &= \hat{\kappa}^{jli}\hat{\beta}_j^o\hat{\beta}_l^m\hat{\beta}_i^n + \hat{\kappa}^{ji}\hat{\beta}_j^{mo}\hat{\beta}_i^n[3] + \hat{\kappa}^j\hat{\beta}_j^{mno} \\ 0 &= \hat{\kappa}^{ijkl}\hat{\beta}_j^o\hat{\beta}_l^m\hat{\beta}_i^n\hat{\beta}_k^p + \hat{\kappa}^{ijl}\hat{\beta}_j^o\hat{\beta}_l^m\hat{\beta}_i^{np}[6] + \hat{\kappa}_{ij}\hat{\beta}_j^{mo}\hat{\beta}_i^{np}[3] + \\ &\quad \hat{\kappa}_{ij}\hat{\beta}_j^{mop}\hat{\beta}_i^n[4] + \hat{\kappa}^j\hat{\beta}_j^{mnop}. \end{aligned}\right\} \quad (142)$$

Evaluating these at $\boldsymbol{\beta} = \ddot{\boldsymbol{\beta}}$, note that $-\frac{1}{3}\hat{\beta}_l^m\hat{\beta}_j^n\hat{\kappa}^{jlp} = \hat{\kappa}^{pj}\hat{\beta}_j^{mn}$ satisfies the second equality of (142), substituting into the third equality of (142),

$$\left[-\hat{\kappa}^{ijkl} + \tfrac{5}{9}\hat{\kappa}^{gij}\hat{\kappa}_{gh}\hat{\kappa}^{hkl}[3]\right]\hat{\beta}_j^o\hat{\beta}_l^m\hat{\beta}_i^n\hat{\beta}_k^p = \hat{\kappa}^{ij}\hat{\beta}_j^{mop}\hat{\beta}_i^n[4] + \hat{\kappa}^j\hat{\beta}_j^{mnop}.$$

Then

$$\tfrac{1}{4}\left[\tfrac{5}{9}\hat{\kappa}^{gij}\hat{\kappa}_{gh}\hat{\kappa}^{hkl}[3] - \hat{\kappa}^{ijkl}\right]\hat{\beta}_j^o\hat{\beta}_l^m\hat{\beta}_k^n = \hat{\kappa}^{ij}\hat{\beta}_j^{mon}$$

satisfies this equation. Hence

$$\begin{aligned} \hat{\beta}_r^{mn} &= -\tfrac{1}{3}\hat{\kappa}_{ir}\hat{\kappa}^{ijk}\hat{\beta}_j^m\hat{\beta}_k^n \\ \hat{\beta}_r^{mno} &= \tfrac{1}{4}\hat{\kappa}_{ir}\left[\tfrac{5}{9}\hat{\kappa}^{gij}\hat{\kappa}_{gh}\hat{\kappa}^{hkl}[3] - \hat{\kappa}^{ijkl}\right]\hat{\beta}_j^o\hat{\beta}_l^m\hat{\beta}_k^n. \end{aligned}$$

Let α_q^r represent the matrix inverse of $\hat{\beta}_r^q$. By (142) $\alpha_q^r \hat{\kappa}_{ri} = \hat{\beta}_i^r \delta_{qr}$ and $\hat{\kappa}_{ij} = \hat{\beta}_i^g \delta_{gh} \hat{\beta}_i^h$. Also

$$\begin{aligned}
\alpha_q^r \hat{\beta}_r^{mn} &= -\tfrac{1}{3}\alpha_q^r \hat{\kappa}_{ri}\hat{\kappa}^{ijk}\hat{\beta}_j^m \hat{\beta}_k^n = -\tfrac{1}{3}\delta_{rq}\hat{\kappa}^{ijk}\hat{\beta}_i^r \hat{\beta}_j^m \hat{\beta}_k^n \\
\alpha_q^r \hat{\beta}_r^{mt}\alpha_t^s \hat{\beta}_s^{no} &= \tfrac{1}{9}\delta_{rq}\hat{\kappa}^{gij}\hat{\kappa}_{gh}\hat{\kappa}^{hkl}\hat{\beta}_l^o \hat{\beta}_j^m \hat{\beta}_k^n \hat{\beta}_i^r \\
\alpha_q^r \hat{\beta}_r^{mno} &= \tfrac{1}{4}\left[\tfrac{5}{9}\hat{\kappa}^{gij}\hat{\kappa}_{gh}\hat{\kappa}^{hkl}[3] - \hat{\kappa}^{ijkl}\right]\hat{\kappa}_{ij}\hat{\beta}_l^m \hat{\beta}_k^n. \\
\alpha_q^r \hat{\beta}_r^{mq}\alpha_s^p \hat{\beta}_p^{ns} &= \tfrac{1}{9}\hat{\kappa}^{ijg}\hat{\kappa}_{ig}\hat{\beta}_j^m \hat{\kappa}^{klh}\hat{\beta}_l^m \hat{\kappa}^{kh}.
\end{aligned}$$

Standard matrix determinant differentiation formulae (Hocking, 1985, Appendix A.II.1.2) show that

$$\begin{aligned}
\frac{\partial \boldsymbol{\beta}}{\partial \boldsymbol{w}}(\omega) &= \frac{\partial \boldsymbol{\beta}}{\partial \boldsymbol{w}}(\hat{\omega})\Big(1 + \alpha_q^r \hat{\beta}_r^{qm}(\omega_m - \hat{\omega}_m) + \\
&\tfrac{1}{2}\left(\alpha_q^r \hat{\beta}_r^{qm}\alpha_p^s \hat{\beta}_s^{pn} + \alpha_q^r \hat{\beta}_r^{qmn} - \alpha_q^r \hat{\beta}_r^{sn}\alpha_s^t \hat{\beta}_t^{qm}\right)(\omega_m - \hat{\omega}_m)(\omega_n - \hat{\omega}_n)\Big) + O(\|\omega - \hat{\omega}\|^3) \\
&= \frac{\partial \boldsymbol{\beta}}{\partial \boldsymbol{w}}(\hat{\omega})\left(1 - \tfrac{1}{3}\hat{\kappa}^{ijk}\hat{\kappa}_{ik}\hat{\beta}_j^m(\omega_m - \hat{\omega}_m) + \tfrac{1}{2}\left[\tfrac{1}{4}\hat{\kappa}^{gij}\hat{\kappa}_{gh}\hat{\kappa}^{hkl}\hat{\kappa}_{ij} + \tfrac{1}{6}\hat{\kappa}^{gil}\hat{\kappa}_{gh}\hat{\kappa}^{hjk}\hat{\kappa}_{ij}\right.\right. \\
&\left.\left. - \tfrac{1}{4}\hat{\kappa}^{ijkl}\hat{\kappa}_{ij}\right]\hat{\beta}_k^m \hat{\beta}_l^n(\omega_m - \hat{\omega}_m)(\omega_n - \hat{\omega}_n)\right) + O(\|\omega - \hat{\omega}\|^3). \qquad (143)
\end{aligned}$$

Similarly,

$$\begin{aligned}
\frac{1}{h(\omega)} &= \det\left[\frac{\partial \boldsymbol{\beta}_{-1}}{\partial \boldsymbol{w}_{-1}}(\hat{\omega})\right]^{-1}\left(1 + \tfrac{1}{2}\left[-\tfrac{1}{4}\tilde{\kappa}^{gij}\tilde{\kappa}_{gh}\tilde{\kappa}^{hkl}\tilde{\kappa}_{ij} - \tfrac{1}{6}\tilde{\kappa}^{gil}\tilde{\kappa}_{gh}\tilde{\kappa}^{hjk}\tilde{\kappa}_{ij}\right.\right. \\
&\left.\left. + \tfrac{1}{4}\tilde{\kappa}^{ijkl}\tilde{\kappa}_{ij}\right]\tilde{\beta}_k^m \tilde{\beta}_l^n(\omega_m - \hat{\omega}_m)(\omega_n - \hat{\omega}_n)\right) + O(\|\omega - \hat{\omega}\|^3). \qquad (144)
\end{aligned}$$

Furthermore,

$$\begin{aligned}
\frac{\hat{\beta}_1}{\beta_1} &= 1 - \frac{\hat{\beta}_1^m(\omega_m - \hat{\omega}_m)}{\hat{\beta}_1} + \left[\frac{\hat{\beta}_1^m \hat{\beta}_1^n}{(\hat{\beta}_1)^2} - \frac{\tfrac{1}{2}\hat{\beta}_1^{mn}}{\hat{\beta}_1}\right](\omega_m - \hat{\omega}_m)(\omega_n - \hat{\omega}_n) + O(\|\omega - \hat{\omega}\|^3) \\
&= 1 - \frac{\hat{\beta}_1^m}{\hat{\beta}_1}(\omega_m - \hat{\omega}_m) + \left[\frac{\hat{\beta}_1^m \hat{\beta}_1^n}{(\hat{\beta}_1)^2} + \frac{\hat{\kappa}_{1i}\hat{\kappa}^{ijk}\hat{\beta}_j^m \hat{\beta}_k^n}{6\hat{\beta}_1}\right](\omega_m - \hat{\omega}_m)(\omega_n - \hat{\omega}_n) \\
&+ O(\|\omega - \hat{\omega}\|^3). \qquad (145)
\end{aligned}$$

Also needed for the expansion in the lattice case is:

$$\begin{aligned}
\frac{1}{\sinh(\tfrac{1}{2}\beta^1\Delta)} &= \frac{1}{\sinh(\tfrac{1}{2}\hat{\beta}^1\Delta)}\Big(1 - \tfrac{1}{2}\hat{\beta}_1^m\Delta\coth(\tfrac{1}{2}\hat{\beta}^1\Delta)(\omega_m - \hat{\omega}_m) + \\
&\left(\hat{\beta}_1^m \hat{\beta}_1^n \Delta^2\left(\tfrac{1}{4}\coth(\tfrac{1}{2}\hat{\beta}^1\Delta)^2 - \tfrac{1}{8}\right)\right. \\
&\left. - \tfrac{1}{4}\hat{\beta}_1^{mn}\Delta\coth(\tfrac{1}{2}\hat{\beta}^1\Delta)\right)(\omega_m - \hat{\omega}_m)(\omega_n - \hat{\omega}_n)\Big) + O(\|\omega - \hat{\omega}\|)^3. \qquad (146)
\end{aligned}$$

Then (140) follows by multiplying (143), (144), and (145), and summing over all coefficients of squared components of $\boldsymbol{\omega} - \hat{\boldsymbol{\omega}}$. Also, (141) follows by multiplying (143), (144), and (146), and summing the appropriate coefficients. Furthermore, the unconditional multivariate saddlepoint density formula (100) follows from summing over the same coefficients in (143), and the conditional double saddlepoint density approximation (124) follows by multiplying (144) and (143) and summing over the appropriate coefficients.

7.6. An Example

Skovgaard illustrates (135) in the context of the hypergeometric distribution, which has applications to testing independence in 2×2 contingency tables. Suppose that X^i are independent random variables Poisson variables with intensities $\boldsymbol{\eta}$, for $i = 1, \ldots, 4$, and arranged into the following table:

X^1	X^2	T^2
X^3	X^4	
T^3		T^4 .

Of interest is whether $\eta_1\eta_4 = \eta_2\eta_3$; that is, whether the data come from a hypergeometric distribution. A test will be performed conditional on row and column marginals. This test will be performed by determining whether X^1, or equivalently, $\boldsymbol{c}^\top \boldsymbol{X}$, is too high or low to be consistent with the null hypotheses. Here $\boldsymbol{c} = (1, -1, 1, -1)^\top$. Approximation (137) will be used to approximate tail probabilities. Here $\boldsymbol{T} = \boldsymbol{B}\boldsymbol{X}$, where

$$\boldsymbol{B} = \begin{pmatrix} 1 & -1 & -1 & 1 \\ 1 & 1 & 0 & 0 \\ 1 & 0 & 1 & 0 \\ 1 & 1 & 1 & 1 \end{pmatrix}.$$

Then $\hat{\boldsymbol{\eta}}$, defined by (99), is $\log(\tilde{\boldsymbol{X}})$ where $\tilde{\boldsymbol{X}}$ is $\boldsymbol{X}$ corrected for continuity. Let $\boldsymbol{X}^*$ be the vector of fitted values, that is, the vector of cell values given by the product of the appropriate row and column mean, divided by the grand mean. Then $\boldsymbol{c}^\top \log(\boldsymbol{X}^*) = 0$ and $\boldsymbol{A}\boldsymbol{X}^* = \boldsymbol{A}\boldsymbol{X}$, and $\tilde{\boldsymbol{\eta}} = \log(\boldsymbol{X}^*)$. The cumulant generating function second derivative matrices are diagonal with elements $\tilde{\boldsymbol{X}}$ and $\boldsymbol{X}^*$, and hence their determinants and inverses are simple to calculate. The lattice spacing for $T^1 = \boldsymbol{c}^\top \boldsymbol{X}$ and $\boldsymbol{c}^\top \boldsymbol{c}$ are both 4, and hence cancel. Skovgaard (1987) then simplifies the double saddlepoint approximation to:

$$1 - \Phi(\omega) + \phi(\omega)\left\{\left(\frac{x_{1.}x_{.1}x_{2.}x_{.2}}{x_{..}\tilde{x}_{11}\tilde{x}_{12}\tilde{x}_{21}\tilde{x}_{22}}\right)^{1/2}(2\sinh(\mu))^{-1} - \hat{\omega}^{-1}\right\}$$

where, in an abuse of notation, $\boldsymbol{x}$ is now the matrix of cell counts, and $\tilde{\boldsymbol{x}}$ is the continuity-corrected vector of cell counts. An index replaced with a . denotes summation. Also, $\mu = \log(\tilde{x}_{11}\tilde{x}_{22}/(\tilde{x}_{21}\tilde{x}_{12}))$ is the log of the observed continuity-corrected odds ratio, and

$$\hat{\omega} = \operatorname{sgn}(\mu)\left\{2\left[\sum_{\substack{l=1,2\\ j=1,2}} \tilde{x}_{lj}\log(\tilde{x}_{lj}) - \sum_{l=1,2} x_{l.}\log(x_{l.}) - \sum_{j=1,2} x_{.j}\log(x_{.j}) + x_{..}\log(x_{..})\right]\right\}^{\frac{1}{2}}.$$

Alternatively one might proceed directly from (133), rather than taking advantage of the simplification offered by the methods of Lemma 7.2.1. If $\boldsymbol{\theta}$ are the canonical parameters associated with $\boldsymbol{T}$, then

$$\eta_1 = \theta_1 + \theta_2 + \theta_2 + \theta_4, \eta_2 = \theta_2 + \theta_4, \eta_3 = \theta_3 + \theta_4, \eta_4 = \theta_4.$$

The joint probability mass function of $\boldsymbol{T}$ is given by

$$\frac{\exp(-\exp(\theta_1 + \theta_2 + \theta_3 + \theta_4) - \exp(\theta_2 + \theta_4) - \exp(\theta_3 + \theta_4) - \exp(\theta_4) + \boldsymbol{\theta}^\top \boldsymbol{t})}{t^1!(t^2 - t^1)!(t^3 - t^1)!(t^4 - t^2 - t^3 + t^1)!}.$$

In this case $\mathcal{H}(\boldsymbol{\theta}) = \exp(\theta_1 + \theta_2 + \theta_3 + \theta_4) + \exp(\theta_2 + \theta_4) + \exp(\theta_3 + \theta_4) + \exp(\theta_4)$. When testing whether $\theta_1 = 0$, $\tilde{\boldsymbol{\theta}}$ satisfies

$$\begin{aligned} \exp(\tilde{\theta}_2 + \tilde{\theta}_3 + \tilde{\theta}_4) + \exp(\tilde{\theta}_2 + \tilde{\theta}_4) + \exp(\tilde{\theta}_3 + \tilde{\theta}_4) + \exp(\tilde{\theta}_4) &= t^4 \\ \exp(\tilde{\theta}_2 + \tilde{\theta}_3 + \tilde{\theta}_4) + \exp(\tilde{\theta}_3 + \tilde{\theta}_4) &= t^3 \\ \exp(\tilde{\theta}_2 + \tilde{\theta}_3 + \tilde{\theta}_4) + \exp(\tilde{\theta}_2 + \tilde{\theta}_4) &= t^2. \end{aligned}$$

These equations have the solutions

$$\tilde{\theta}_2 = \log\left(\frac{t^2}{t^4 - t^2}\right), \; \tilde{\theta}_3 = \log\left(\frac{t^3}{t^4 - t^3}\right) \; \tilde{\theta}_4 = \log\left(\left(t^4 - t^2\right)\left(t^4 - t^3\right)/t^4\right).$$

From knowledge of $\hat{\boldsymbol{\theta}}$ and $\tilde{\boldsymbol{\theta}}$ approximation (133) may easily be calculated.

Exact probabilities for atoms for the conditional distribution are given by

$$\mathrm{P}\left[T^1 = t^1 | T^2 = t^2, T^3 = t^3, T^4 = t^4\right] = \binom{t^2}{t^1}\binom{t^4 - t^2}{t^3 - t^1}\Big/\binom{t^4}{t^3}.$$

Exact tail probabilities are quite difficult to compute when t^4, t^2, t^3, t^1, and $\min(t^2, t^3) - t^1$ are large. The double saddlepoint distribution function approximation is very accurate and is a simple function of the x^{ij} (Fig. 15).

Skovgaard (1987) shows that the double saddlepoint approximation performs very well for large tables that are not extreme, but not so well for extreme tables.

The sequential saddlepoint approximation will also be applied to these data. As above, let $\boldsymbol{\theta}$ be the canonical parameter vector for $\boldsymbol{T}$, and let $\boldsymbol{\theta} = \boldsymbol{B}^\top \boldsymbol{\eta}$. Then if $\tilde{\boldsymbol{X}}$

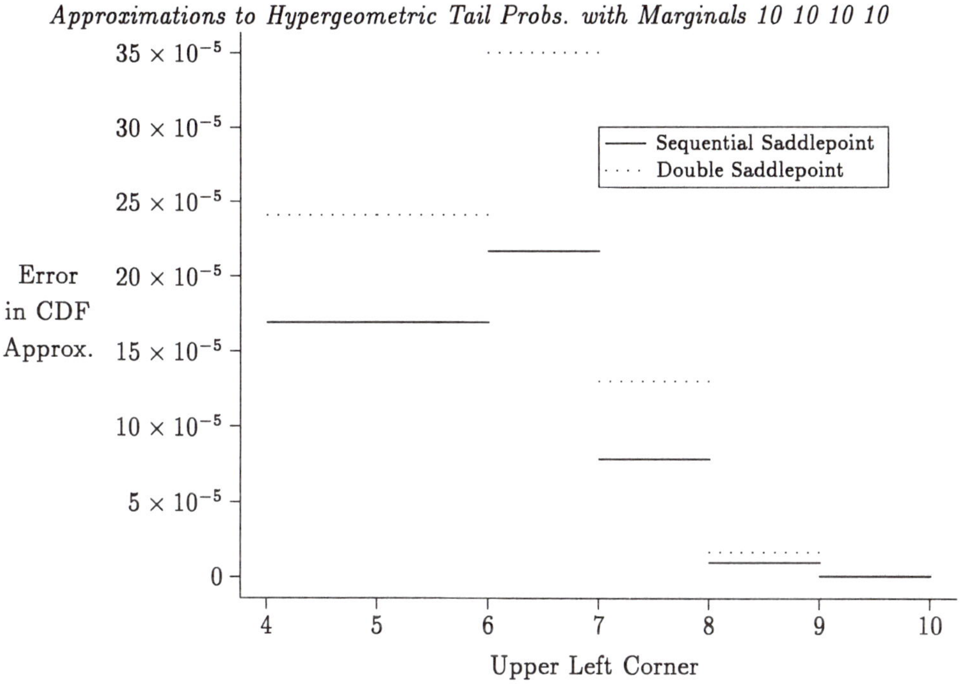

Fig. 15.

is once again the vector of fitted table entries, this time under the null hypothesis that the log odds ratio is $4\theta_1$ rather than 0, then

$$\tilde{X}^1 = \frac{1}{2}\Bigg(t^2 + t^3 - t^4(1 - \exp(4\theta_1))^{-1} + \frac{\sqrt{({t^2}^2 + {t^3}^2)(1 - \exp(4\theta_1))^2 + 2t^2t^3(1 - \exp(4\theta_1)^2) + 2t^4(t^2 + t^3)(\exp(4\theta_1) - 1) + {t^4}^2}}{1 - \exp(4\theta_1)}\Bigg).$$

Other elements of $\boldsymbol{X}$ can be calculated from $\tilde{X}^1$, T^2, T^3, and T^4. The sequential saddlepoint approximation performs even better than the double saddlepoint approximation, but is not so easy to compute.

7.7. Multivariate Conditional Approximations

The preceding sections described methods for approximating one-dimensional conditional probabilities. To approximate multidimensional conditional probabilities, one might integrate with respect to other components of the random vector in (128), to obtain approximations to probabilities of rectangular sets. Kolassa and Tanner (1994) introduce a Markov chain Monte Carlo method for approximating probabilities of less regular sets, including rejection regions for frequentist hypothesis tests. They call this method the Gibbs Skovgaard algorithm. Further details are given by Kolassa (1994, 1995).

The Gibbs sampler is a popular Markov chain method useful for yielding a sample from a posterior or likelihood density. It was first introduced by Geman and Geman (1984) in the context of image reconstruction. See Tanner (1996) for background details and important references.

Let the symbol $p(\cdots|\cdots)$ denote the distribution of those random variables listed before the vertical line conditional on those listed after. To obtain a sample from the joint distribution $p(T^1, \cdots, T^j | T_{j+1}, \ldots, T_{\mathfrak{t}})$, the systematic scan Gibbs sampler iterates the following loop:

1) Sample $T^1_{(i+1)}$ from $p(T^1 | T^2_{(i)}, \cdots, T^j_{(i)}, T_{j+1}, \ldots, T_{\mathfrak{t}})$.
2) Sample $T^2_{(i+1)}$ from $p(T^2 | T^1_{(i+1)}, T^3_{(i)}, \cdots, T^j_{(i)}, T_{j+1}, \ldots, T_{\mathfrak{t}})$.

$\vdots$

j) Sample $T^j_{(i+1)}$ from $p(T^j | T^1_{(i+1)}, \cdots, T^{j-1}_{(i+1)}, T_{j+1}, \ldots, T_{\mathfrak{t}})$.

For a sufficiently large value of I we can take $T^1_{(I)}, \cdots, T^j_{(I)}$ as a simulated observation from the equilibrium distribution $p(T^1, \cdots, T^j | T_{j+1}, \ldots, T_{\mathfrak{t}})$ of the Markov chain. Independently replicating this Markov chain produces an independent and identically distributed sample from the distribution of interest. Kolassa and Tanner (1994) sample instead from approximate conditional distributions of Theorem 7.3.1, to give a sample drawn approximately from the conditional distribution $p(T^1, \cdots, T^j | T_{j+1}, \ldots, T_{\mathfrak{t}})$, in order to perform conditional inference in certain generalized linear models (McCullagh and Nelder, 1989).

Other authors, including Waterman and Lindsay (1996), present an analytic asymptotic approximation to continuous conditional distributions, resulting in accurate estimates and standard errors. The results of Strawderman, Cassella, and Wells (1996) might also be extended to examine these conditional distributions. DiCiccio and Martin (1991) and DiCiccio, Martin, and Young (1993) present double saddlepoint approximations to the distribution of the signed root of the likelihood ratio statistic for conditional inference in continuous distributions.

7.8. Exercises

1. Consider the hypergeometric examples given by Skovgaard (1987):

a.	85	2	b.	14	6	c.	5	3	d.	5	1	e.	6	0
	75	35		8	12		1	9		1	5		0	6

   a. Calculate the tail probability approximation.
   b. Compare these with the tail probability approximations ignoring continuity corrections.
   c. Using this approximation or otherwise, determine values for x_{11} giving approximately the lower 2.5% point of the distribution, and compare the accuracy of the approximation embodying the continuity correction here. Do

not put much effort into determining this point; a normal approximation will probably be good enough.

d. Calculate the higher–order expansion of §7.4.

2. Consider the logistic regression model:

$$y^j \sim B(\pi_j, n_j), \quad \pi^j = \text{alogit}(\lambda_j), \quad \lambda^j = \sum_{i=1}^{2} z_i^j \theta_j,$$

and the data set

Variable	Observation 1	2	3	4	5	6	7	8	9	10	11	12
z_1^j	1	1	1	1	1	1	1	1	1	1	1	1
z_2^j	−1	1	−1	1	−1	1	−1	1	−1	1	−1	1
n^j	8	10	9	9	10	10	9	10	10	9	10	10
y^j	3	4	2	3	2	2	1	1	2	1	1	1 ,

where the first two rows contain covariates and the last two the number of trials and number of successes. This data is a subset of that collected by Berman, *et. al.* (1990) as part of an experiment to measure the effect of magnetic radiation on chicken embryos. Sufficient statistics for the parameter vector are $\boldsymbol{T} = \boldsymbol{Z}^\top \boldsymbol{Y}$, where $\boldsymbol{Z}$ is the transpose of the matrix formed by the first two rows of the table above, and $\boldsymbol{Y}$ is the column vector whose transpose is given in the final row.

a. Graph the approximate conditional probability that T_2 meets or exceeds the observed value, as a function of θ_2, using (136).

b. Graph the approximate conditional probability that T_2 fails to exceed the observed value, as a function of θ_2, using (136).

c. From parts a and b construct an approximate 95% confidence interval for θ_2.

d. Repeat a – c using (89) in conjunction with (127).

8

Applications to Wald, Likelihood Ratio, and Maximum Likelihood Statistics

The breath of this material is similar to that of Reid (1988). Of primary concern will be approximation to densities and distribution functions of maximum likelihood estimators, likelihood ratio statistics, and Wald statistics. Bartlett's correction for the distribution of likelihood ratio statistics is derived. Approximate ancillarity is also discussed.

8.1. The Distribution of the Wald Statistic

Consider the problem of testing a hypothesis specifying a parameter $\boldsymbol{\theta}$ in the model having the sufficient statistic vector $\boldsymbol{T} \in \mathfrak{R}^{\mathfrak{k}}$ using the test statistic $V(\boldsymbol{T}) = (\boldsymbol{T} - \boldsymbol{\mu})^\top \boldsymbol{\Sigma}^{-1}(\boldsymbol{T} - \boldsymbol{\mu})$, when

$$\mathrm{E}\left[\|\boldsymbol{T}\|^s\right] < \infty, \tag{147}$$

for some $s \geq 2$. This statistic a trivial example of a Wald statistic, and in a full exponential family is the score statistic. Suppose that $\boldsymbol{\mu} = \mathrm{E}_{\boldsymbol{\theta}}[\boldsymbol{T}]$ and $\boldsymbol{\Sigma} = \mathrm{Var}_{\boldsymbol{\theta}}[\boldsymbol{T}]$, the expectation and variance of $\boldsymbol{T}$ calculated at the postulated parameter value $\boldsymbol{\theta}$, exist and are finite. Suppose further that $\boldsymbol{\Sigma}$ is invertible. When $\boldsymbol{T}$ is approximately multivariate normal, V has a distribution that is approximately χ^2 on $\mathfrak{k}$ degrees of freedom. This may be seen by expressing $\boldsymbol{\Sigma}^{-1} = \boldsymbol{\Omega}^\top \boldsymbol{\Omega}$, and noting that $\boldsymbol{\Omega}(\boldsymbol{T} - \mathrm{E}_{\boldsymbol{\theta}}[\boldsymbol{T}])$ is approximately multivariate normal with mean $\mathbf{o}$ and all components independent with unit variance.

When $\boldsymbol{T}$ has a distribution dependent on n, such as when $\boldsymbol{T}$ is the standardized sum of independent and identically distributed random vectors $\boldsymbol{X}_i$, each with characteristic function ζ, then the vague notions of approximation above can be refined. Rigorous asymptotic approximation for tail probabilities of V can be derived with reference to the underlying distribution of $\boldsymbol{T}$. Let $\mathfrak{E}'_n = \{\boldsymbol{t} | (\boldsymbol{t} - \boldsymbol{\mu}_n)^\top \boldsymbol{\Sigma}_n^{-1}(\boldsymbol{t} - \boldsymbol{\mu}_n) \leq v\}$ be the elliptical set of $\boldsymbol{t}$ giving rise to $V \leq v$. Theorem 6.5.1 concerning multivariate Edgeworth series, or its analogues in non-independent and identically distributed cases, allows the approximation of probabilities of sets like $\mathfrak{E}'_n$, as long as $\boldsymbol{T}$ satisfies regularity conditions discussed below. These are satisfied, for instance, if $\boldsymbol{T}$ arises from the likelihood for independent and identically distributed components, each of whose distribution is given by a density with the proper moments. The first order approximation is $\int \cdots \int \phi(\boldsymbol{t}; \boldsymbol{\mu}_n, \boldsymbol{\Sigma}_n)\, d\boldsymbol{t}$ where the integral is taken over $\mathfrak{E}'$. After a change of variables this is reexpressed as $F_{\mathfrak{k}}(v)$, where $F_{\mathfrak{k}}$ is the χ^2 cumulative distribution function on $\mathfrak{k}$ degrees of freedom. The χ^2 approximation is accurate to $o(1)$, by the Central Limit Theorem. Under the hypothesis of an integrable characteristic function (108), and with the existence of a third moment as in (147) with $s = 3$ the χ^2 approximation is accurate to $O(1/\sqrt{n})$. If fourth order moments exist, that is, (147) holds with $s = 4$, the next term in the expansion can be added, resulting in an approximation valid to $O(n^{-1})$. However, by the symmetry of $\mathfrak{E}'_n$, the term of $O(1/\sqrt{n})$ is zero, implying that $\mathrm{P}\left[V \leq v\right] = F_{\mathfrak{k}}(v) + O(n^{-1})$.

This expansion holds when $\boldsymbol{T}$ is continuous. Esseen (1945) remarks at the end of §7.4 that under a multivariate version of Cramér's condition of (110),

$$\limsup_{\|\boldsymbol{\beta}\|\to\infty} |\zeta(\boldsymbol{\beta})| < 1$$

, the next term in the asymptotic expansions for the probability of the ellipse is of order $O(n^{-1})$ in this case as well. These conditions allow the expansion with error of order $O(n^{-1})$ in the multivariate Edgeworth series.

Bhattacharya and Rao (1976) prove, under Cramér's condition, crediting Rao (1960, 1961), convergence for the multivariate Edgeworth series for general convex sets, and get error terms of size $o(n^{-s/2})$ when terms of size $O(n^{-s/2})$ are included. This is their Corollary 20.4, and this paraphrased as Theorem 6.5.2 in §6.5. Earlier versions of these theorems included powers of $\log(n)$ in the error term; see their book, p. 222, for more bibliographic information. In the case of ellipses symmetric about the origin, the term of size $O(1/\sqrt{n})$ in the integral of the multivariate case over this region is zero, proving that the χ^2 approximation is valid to $O(n^{-1})$.

The most important class of distributions for which Cramér's condition fails is the class of lattice distributions. A random vector has a lattice distribution if it takes on values in a translation and rotation of $\mathfrak{Z}^{\mathfrak{k}} = \{(z_1, ..., z_{\mathfrak{k}}) | z_j \in \mathfrak{Z} \forall j\}$. When instead $\boldsymbol{T}$ has a lattice distribution approximating probabilities of elliptical regions becomes trickier. Kolassa and McCullagh (1990) show that in the unidimensional lattice case, the Edgeworth series gives approximations valid to $O(n^{-1})$ if evaluated at continuity corrected points and using the third cumulant. The Edgeworth series gives higher order approximations if the cumulants are adjusted and fourth and higher order cumulants are used. In the multivariate case Esseen (1945) showed as Theorem 1 of §7 that $\mathrm{P}\left[\boldsymbol{T}_n \in \mathfrak{E}'_n\right] = F_{\mathfrak{k}}(v) + O(n^{-\mathfrak{k}/(\mathfrak{k}+1)})$ assuming only finite fourth moments for general $\boldsymbol{T}$. Esseen observed (Theorem 1 of §8) that one can approximate atoms of the lattice by the Edgeworth series density approximation using the third cumulants, times the volume of the lattice cell, to order $O(n^{-(\mathfrak{k}+2)/2})$. This isn't as good as it sounds, since estimating tho probability of any set of constant volume involves adding $O(n^{\mathfrak{k}/2})$ of these errors, leaving an error of $O(n^{-1})$.

Sharper results are based on careful expansions for the distribution function of $\boldsymbol{T}$. Rao (1960, 1961) develops an analogue to the Esseen's series for the cumulative distribution function of a lattice distribution in the multivariate case. Error bounds here contain factors of $\log(n)$ later proved unnecessary by Bhattacharya and Rao (1976). Evaluation of this series is, however, difficult for non-rectangular and non-elliptical sets. Kolassa (1989) shows that this Rao series, when evaluated at midpoints of lattice cubes, is equivalent to the Edgeworth series at the same points, with cumulants adjusted by Sheppard's corrections, to the same order of error. Yarnold (1972) addresses the problem of evaluating the Rao series for convex sets, and in particular for standardized ellipses. The Yarnold approximation is the χ^2 approximation plus the difference between the actual number of points in the ellipse and the volume of the ellipse divided by the volume of a unit cube of the lattice, times the normal approximation to the density at each point on the ellipse boundary. Specifically, suppose $\boldsymbol{T}_n$ is the sum of independent and identically distributed distributed vectors $\boldsymbol{X}_1, \ldots, \boldsymbol{X}_n$, divided by $\sqrt{n}$, and suppose that $\boldsymbol{X}_i$ are confined to a lattice with unit spacings, with zero mean and $\mathrm{Var}\left[\boldsymbol{X}_i\right] = \Sigma$. Choose a point $\boldsymbol{a}$ on the lattice. Then

$$\mathrm{P}\left[V(\boldsymbol{T}) \leq v\right] = F_{\mathfrak{k}}(v) + \left(N(nv) - \frac{(\pi n v)^{\mathfrak{k}/2} \det\left[\Sigma\right]^{1/2}}{\Gamma(\mathfrak{k}/2 - 1)}\right) \frac{\exp(-v/2)}{(2\pi n)^{\mathfrak{k}/2} \det\left[\Sigma\right]^{1/2}} + O(n^{-1}),$$

where $N(nv)$ is the number of vectors of integers $\boldsymbol{\mu}$ such that $(\boldsymbol{\mu} + n\boldsymbol{a})^{\top} \Sigma^{-1} (\boldsymbol{\mu} + n\boldsymbol{a}) < v$. In general calculating $N(nv)$ is computationally intensive. Furthermore, finding an order $O(n^{-1})$ correction term independent of v for use in the development of a Bartlett's correction, to be described below, seems very difficult.

Bartlett (1953, 1955) considers asymptotic approximations to the distributions of score statistics in non-exponential-family settings; other authors including Levin and Kong (1990) have addressed the same question.

8.2. Distributions of Maximum Likelihood Estimators

As in §7.2, consider the exponential family model for a random vector $\boldsymbol{T}$ whose density evaluated at $\boldsymbol{t}$ is: $f_T(\boldsymbol{t};\boldsymbol{\theta}) = \exp(\boldsymbol{\theta}^{\top}\boldsymbol{t} - \mathcal{H}_T(\boldsymbol{\theta}) - \mathcal{G}(\boldsymbol{t}))$, such that $\mathcal{H}_T$ is twice differentiable and $\mathcal{H}_T'$ is one to one. The associated cumulant generating function is $\mathcal{K}_T(\boldsymbol{\beta},\boldsymbol{\theta}) = \mathcal{H}_T(\boldsymbol{\beta}+\boldsymbol{\theta}) - \mathcal{H}_T(\boldsymbol{\theta})$, and hence the saddlepoint $\hat{\boldsymbol{\beta}}$ is defined by $\mathcal{H}_T'(\hat{\boldsymbol{\beta}} + \boldsymbol{\theta}) = \boldsymbol{t}$. The log likelihood is $\ell(\boldsymbol{\theta};\boldsymbol{t}) = \boldsymbol{\theta}^{\top}\boldsymbol{t} - \mathcal{H}_T(\boldsymbol{\theta})$, and the maximum likelihood estimator is defined by $\mathcal{H}_T'(\hat{\boldsymbol{\theta}}) = \boldsymbol{t}$. Hence $\hat{\boldsymbol{\beta}} + \boldsymbol{\theta} = \hat{\boldsymbol{\theta}}$. Then by (100),

$$\begin{aligned} f_T(\boldsymbol{t};\boldsymbol{\theta}) &= (2\pi)^{-\mathfrak{k}/2} \det\left[\mathcal{K}_T''(\hat{\boldsymbol{\beta}};\boldsymbol{\theta})\right]^{-1/2} \exp(\mathcal{K}_T(\hat{\boldsymbol{\beta}};\boldsymbol{\theta}) - \hat{\boldsymbol{\beta}}^{\top}\boldsymbol{t})(1 + \tfrac{1}{2}b(\hat{\boldsymbol{\theta}})n^{-1} + O(n^{-2})) \\ &= (2\pi)^{-\mathfrak{k}/2} \det\left[\mathcal{H}_T''(\hat{\boldsymbol{\theta}})\right]^{-1/2} \exp(\mathcal{H}_T(\hat{\boldsymbol{\theta}}) - \mathcal{H}_T(\boldsymbol{\theta}) + (\boldsymbol{\theta} - \hat{\boldsymbol{\theta}})\boldsymbol{t})\times \\ &\quad (1 + \tfrac{1}{2}b(\hat{\boldsymbol{\theta}})n^{-1} + O(n^{-2})) \\ &= (2\pi)^{-\mathfrak{k}/2} \det\left[j(\hat{\boldsymbol{\theta}})\right]^{-1/2} (\mathcal{L}(\boldsymbol{\theta};\boldsymbol{t})/\mathcal{L}(\hat{\boldsymbol{\theta}};\boldsymbol{t}))\left(1 + \tfrac{1}{2}b(\hat{\boldsymbol{\theta}})n^{-1} + O(n^{-2})\right). \qquad (148) \end{aligned}$$

Here j is the observed information, and $\mathcal{L}$ is the likelihood. In these full exponential families, j is equal to i, the Fisher information $\mathrm{E}_{\boldsymbol{\theta}}\left[-\ell''\right]$. Since $\mathcal{H}'_T(\hat{\boldsymbol{\theta}}) = \boldsymbol{t}$, $\mathcal{H}''_T(\hat{\boldsymbol{\theta}})(d\hat{\boldsymbol{\theta}}/d\boldsymbol{t}) = \boldsymbol{I}$, and $(d\hat{\boldsymbol{\theta}}/d\boldsymbol{t}) = \mathcal{H}''_T(\hat{\boldsymbol{\theta}})^{-1} = i(\hat{\boldsymbol{\theta}})^{-1}$. Let $\hat{\boldsymbol{\Theta}}$ be the random variable arising from solutions to $\mathcal{H}'_T(\hat{\boldsymbol{\Theta}}) = \boldsymbol{T}$. Using standard change of variables techniques,

$$f_{\hat{\Theta}}(\hat{\boldsymbol{\theta}};\boldsymbol{\theta}) = (2\pi)^{-\mathfrak{k}/2}\det\left[i(\hat{\boldsymbol{\theta}})\right]^{1/2}\mathcal{L}(\boldsymbol{\theta};\boldsymbol{t})/\mathcal{L}(\hat{\boldsymbol{\theta}};\boldsymbol{t}) + O(n^{-1}). \tag{149}$$

Approximation (149), first noted by Daniels (1958), is a case of Barndorff–Nielsen's formula, and derived in various cases by Barndorff–Nielsen (1980, 1983) and Durbin (1980), who also also noted that arguments analogous to those in §4.8 indicate that after renormalization, the error term in (149) can be replaced by $O(n^{-3/2})$. Formula (149) is also known as the p^* formula, from a series of papers primarily by Barndorff–Nielsen in which the density of the maximum likelihood estimator is denoted by p and the approximation is denoted by adding the star.

Reid (1988) considers the example of a gamma distribution with an unknown shape parameter. Suppose that X_i are independently distributed with density

$$f(x;\mu,\nu) = (\nu/\mu)^{\nu}x^{\nu-1}\exp(-x\nu/\mu)\Gamma^{-1}(\nu).$$

Then the sample $X_1,\ldots,X_n$ gives rise to the log likelihood

$$\ell_n(\mu,\nu;\boldsymbol{X}) = n\nu\log(\nu) - n\nu\log(\mu) + (\nu-1)\sum\log(X_i) - (\nu/\mu)\sum X_i - n\log(\Gamma(\nu)),$$

$\boldsymbol{T} = ((\sum\log X_i)/n,\ (\sum X_i)/n)$ is sufficient, and the maximum likelihood estimators satisfy

$$\nu\,(-\mu + T_1)/\mu^2 = 0 \text{ and } 1 - T_1/\mu + T_2 - \log(\mu) + \log(\nu) - \varphi(\nu) = 0,$$

where φ is the di–gamma function defined to be $\varphi(\nu) = (d/d\nu)\log(\Gamma(\nu))$, and

$$\hat{\mu} = T^1,\ \varphi(\hat{\nu}) - \log(\hat{\nu}) = T^2 - \log(T^1). \tag{150}$$

The matrix of log likelihood second derivatives is

$$\frac{\partial^2 l}{\partial(\mu,\nu)\partial(\mu,\nu)^{\top}} = n\begin{pmatrix}\nu(\mu - 2T^2)/\mu^3 & (-\mu + T^1)/\mu^2\\ (-\mu + T^1)/\mu^2 & 1/\nu - \varphi'(\nu)\end{pmatrix}.$$

Its determinant is

$$\frac{n^2}{\mu^3}\left\{\left(1 - \frac{2T^2}{\mu}\right)(1 - \nu\varphi'(\nu)) - \left(\frac{T^2}{\mu} - 1\right)^2\right\} = \frac{n^2}{\mu^2}\left\{\frac{2T^2\nu\varphi'(\nu)}{\mu} - \nu\varphi'(\nu)\ \frac{(T^2)^2}{\mu^2}\right\}.$$

After using (150) to eliminate $\boldsymbol{T}$, $\det\left[j(\hat{\theta})\right]^{-1/2} = n\{\hat{\nu}\varphi'(\hat{\nu}) - 1\}^{1/2}/\mu$.

Also, $\ell_n(\mu,\nu;\boldsymbol{X})-\ell_n(\hat\mu,\hat\nu;\boldsymbol{X})=n(\hat\nu-\hat\mu\nu/\mu+\nu\log(\hat\mu/\mu)+\log(\Gamma(\hat\nu)/\Gamma(\nu))+\nu\log(\nu/\hat\nu)+(\nu-\hat\nu)\varphi'(\hat\nu))$. The joint maximum likelihood estimator density approximation (149) is

$$n\exp\left(n\left(\hat\nu-\hat\mu\nu/\mu+\nu\log\left(\hat\mu/\mu\right)+\nu\log\left(\nu/\hat\nu\right)+\log\left(\Gamma(\hat\nu)/\Gamma(\nu)\right)+(\nu-\hat\nu)\varphi'(\hat\nu)\right)\right)\times \sqrt{(\hat\nu\varphi'(\hat\nu)-1)/\hat\mu^2}$$

Let

$$g_1(\hat\mu;\mu,\nu)=\exp(n(\nu\log(\hat\mu/\mu)-\hat\mu\nu/\mu-\log(\Gamma(\nu)))(n/\hat\mu)$$

and

$$g_2(\hat\nu;\nu)=\exp(n(\hat\nu+\nu\log(\nu/\hat\nu)+(\nu-\hat\nu)\varphi'(\hat\nu)+\log(\Gamma(\hat\nu)))\sqrt{\hat\nu\varphi'(\hat\nu)-1}.$$

Hence $\hat\mu$ has approximately a gamma density with scale parameter μ and shape parameter $n\nu$, and is approximately independent of $\hat\nu$. One can easily verify that up to a multiplicative constant the density approximation for $\hat\mu$ is exact. Fig. 16 demonstrates that the approximation for the density of $\hat\nu$ is very close.

The numeric approximation is calculated by sampling 10,000 sets of 10 $\Gamma(1,1)$ random variables and calculating T^1 and T^2 for each set. The density for $\hat\nu$ might then be approximated by noting from (150) that since $\mathcal{H}_T(\hat\nu)-\log(\hat\nu)=(T^2/n)^{‘}-\log(T^1/n)$, the density of $\hat\nu$ is the density of $(T^2/n)-\log(T^1/n)$ times the first derivative of $\varphi(\hat\nu)-\log(\hat\nu)$. The density of $(T^2/n)-\log(T^1/n)$ can be approximated by fitting a spline to the midpoints of the bars of the histogram of $(T^2/n)-\log(T^1/n)$. However, since $\varphi(\hat\nu)-\log(\hat\nu)$ from (150) has a pole at zero a better approximation is derived by noting that $h(\hat\nu)=\exp((T^2/n)-\log(T^1/n))$, where $h(\hat\nu)=\exp(\varphi(\hat\nu)-\log(\hat\nu))$. The density is then approximated by fitting a spline to the midpoints of the bars of the histogram of $\exp((T^2/n)-\log(T^1/n))$ and multiplying the result by h'. Also shown is the normal approximation to the density of $\hat\nu$, based on the large sample approximations to the mean, ν, and to the variance, $n^{-1}(-1/\nu+\varphi'(\nu))^{-1}=(n[\pi^2/6-1])^{-1}$.

Formula (149) holds in some non-exponential cases, as shown by Durbin (1980). Suppose $\boldsymbol{X}$ is a random vector of length n, from density $f_{\boldsymbol{X}}(\boldsymbol{x};\boldsymbol{\theta})$ for $\boldsymbol{\theta}\in\mathfrak{P}\subset\mathfrak{R}^{\mathfrak{d}}$, and suppose $\boldsymbol{T}=\boldsymbol{t}(\boldsymbol{X})$ is a sufficient estimator of $\boldsymbol{\theta}$. By the definition of sufficiency, the density for $\boldsymbol{X}$ can be factored as $f_{\boldsymbol{X}}(\boldsymbol{x};\boldsymbol{\theta})=g(\boldsymbol{t};\boldsymbol{\theta})h(\boldsymbol{x})$, and for any $\boldsymbol{\theta}$ and $\boldsymbol{\theta}_0\in\mathfrak{P}$, $g(\boldsymbol{t};\boldsymbol{\theta}_0)=g(\boldsymbol{t};\boldsymbol{\theta})f(\boldsymbol{x},\boldsymbol{\theta}_0)/f(\boldsymbol{x},\boldsymbol{\theta})$. In particular,

$$g(\boldsymbol{t};\boldsymbol{\theta}_0)=g(\boldsymbol{t};\boldsymbol{t})[f(\boldsymbol{x},\boldsymbol{\theta}_0)/f(\boldsymbol{x},\boldsymbol{t})].$$

This is a non-cumulant generating function analogy to exponential tilting to the mean. Here since $\boldsymbol{t}$ estimates $\boldsymbol{\theta}$ one might expect $\boldsymbol{t}$ to lie near the mean of $g(.;\boldsymbol{t})$ and hence for $g(\boldsymbol{t};\boldsymbol{t})$ to be easy to approximate. Durbin (1980) uses the normal approximation to the mean to show that

$$g(\boldsymbol{t};\boldsymbol{\theta}_0)=(n/2\pi)^{\mathfrak{d}/2}\det\left[\imath(\boldsymbol{t})\right]^{1/2}\mathcal{L}(\boldsymbol{\theta}_0)/\mathcal{L}(\boldsymbol{t})\{1+O(1/n)\}. \qquad (151)$$

Density of the MLE of the Scale Parameter in a Gamma(1,1) Distribution

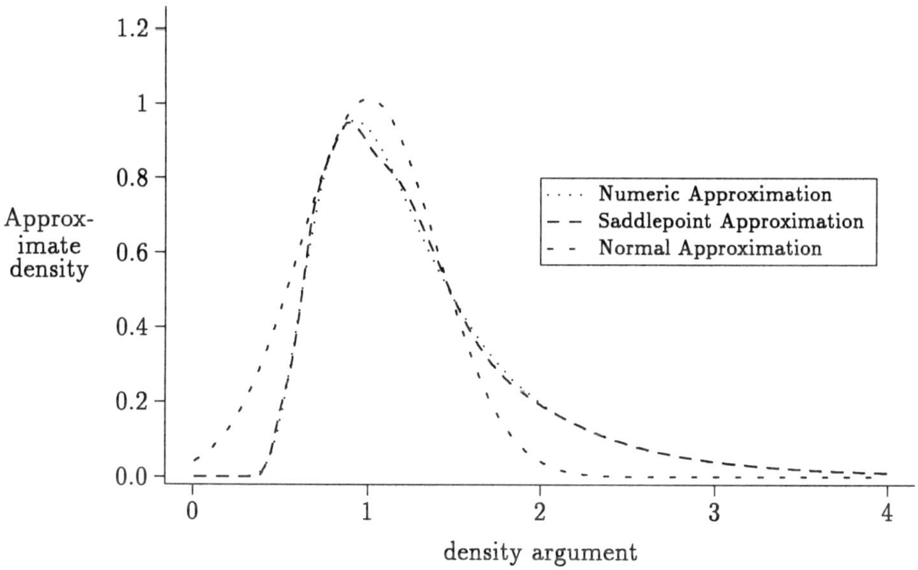

Fig. 16.

If $\boldsymbol{X}$ had a cumulant generating function $\mathcal{K}_{\boldsymbol{X}}$ then $i(\boldsymbol{t})$ could be replaced by $\mathcal{K}''_{\boldsymbol{X}}(\boldsymbol{t})$; if $\boldsymbol{X}$ came from an exponential family this would be $\mathcal{H}''_{\boldsymbol{X}}(\boldsymbol{t}+\boldsymbol{\theta})$. Durbin (1980) gives four cases in which (151) holds. The first and most important is:

Theorem 8.2.1: *Suppose that the sufficient statistic vector $\boldsymbol{T}$ with a distribution depending on n estimates $\boldsymbol{\theta}$ with a bias that is uniformly $O(n^{-1})$ in a neighborhood of $\boldsymbol{\theta}_0$, that $n\mathrm{Var}[\boldsymbol{T}]$ converge to a positive definite matrix uniformly in a neighborhood of $\boldsymbol{\theta}_0$, and that $\boldsymbol{T}$ have a valid Edgeworth expansion to order $o(1/n)$. Then (149) holds with $\hat{\boldsymbol{\theta}} = \boldsymbol{t}$.*

8.3. Ancillarity

More subtle expansions of use in conditional inference require a review of the background on sufficiency and ancillarity. A statistic $\boldsymbol{T} = \boldsymbol{t}(\boldsymbol{X})$ calculated from raw data $\boldsymbol{X}$ and used to make inference on $\boldsymbol{\theta}$ is called sufficient if the density generating $\boldsymbol{X}$ can be factored as $g(\boldsymbol{t};\boldsymbol{\theta})h(\boldsymbol{x})$. The implication is that inference on $\boldsymbol{\theta}$ should be based on $\boldsymbol{T}$ and not on information contained in $\boldsymbol{X}$ but not reflected in $\boldsymbol{T}$. Examples are:

a. If X_i are independent and identically distributed observations from a $N(\mu, \sigma^2)$ distribution, then their density is

$$f_{\boldsymbol{X}}(\boldsymbol{x};\mu,\sigma) = \exp\left(-n\frac{\mu^2}{2\sigma^2} + \frac{\mu}{\sigma^2}\sum x_i - \frac{1}{\sigma^2}\sum_i x_i^2 - n\log(\sigma) - \frac{n}{2}\log(2\pi)\right)$$

Hence $\boldsymbol{T} = (\sum_i X_i, \sum_i X_i^2)^\top$ is sufficient.

b. If $\boldsymbol{X}_i$ are independent and identically distributed random vectors from a natural multivariate exponential family, of the form $f_{\boldsymbol{X}}(\boldsymbol{x};\boldsymbol{\theta}) = \exp(\boldsymbol{\theta}^\top \boldsymbol{x} - \mathcal{H}(\boldsymbol{\theta}) - \mathcal{G}(\boldsymbol{x}))$, then $\boldsymbol{T} = \sum_i \boldsymbol{X}_i$ is sufficient.

c. If X_i are independent draws form a binomial distribution with success probability π and number of trials N_i, with N_i known, then $\sum_i X_i$ is sufficient.

For any statistical model the complete raw data set is always a sufficient statistic. The principal of simplifying analysis by examining a sufficient reduction of the data heuristically leads to choosing, among those sufficient statistics available, the smallest or minimal sufficient statistic, where the minimal sufficient statistic is defined to be any sufficient statistic that is a function of any other sufficient statistic.

Related to the concept of sufficiency is the concept of ancillarity. If $\boldsymbol{T}$ is a minimal sufficient statistic and can be written as

$$\boldsymbol{T} = \begin{pmatrix} \boldsymbol{Y} \\ \boldsymbol{A} \end{pmatrix}$$

where the distribution of $\boldsymbol{A}$ does not depend on $\boldsymbol{\theta}$, then $\boldsymbol{A}$ is called ancillary. Many authors as early as Fisher (1934) have argued for inference conditional on ancillary statistics. For example, if X has a binomial distribution with success probability π and number of trials N, with N generated from some probability model not depending on π, then $\boldsymbol{T} = (X, Y)^\top$ is sufficient and Y is ancillary. Then $\mathrm{P}_\pi[X = x, Y = y] = \exp(x \log(\pi/(1-\pi)) + y \log(1-\pi) + \log(\binom{y}{x}))$. See Fig. 17.

8.4. Approximate Ancillarity

For a multivariate $(\mathfrak{k}, \mathfrak{d})$ curved exponential family assigning probabilities to a random vector $\boldsymbol{T}$ taking values in a convex set $\mathfrak{T} \subset \Re^{\mathfrak{k}}$ according to the density

$$f(\boldsymbol{t};\theta) = \exp(\boldsymbol{t}^\top \boldsymbol{\eta}(\boldsymbol{\theta}) - \mathcal{H}(\boldsymbol{\theta}) - \mathcal{G}(\boldsymbol{t})) \text{ with } \boldsymbol{\eta} : \mathfrak{H} \subset \Re^{\mathfrak{d}} \to \Re^{\mathfrak{k}},\ \mathfrak{H} \text{ open}, \tag{152}$$

a sufficient statistic is $\boldsymbol{T}$. In this case no shorter sufficient statistic is available, except when $\boldsymbol{\eta}$ is very simple, even though when $\mathfrak{k} > \mathfrak{d}$ the sufficient statistic vector is longer than the parameter vector. Efron (1975) discusses the geometry of such families, and produces an asymptotic expansion for the variance of the maximum likelihood estimator. This will be discussed later in the context of efficiency.

Often times greater asymptotic precision can be obtained by using conditioning to reduce the dimensionality of the conditioned sampling distribution to that of the unknown parameter. Assume that the curved exponential family can be embedded in a full exponential family; that is, suppose that there exists an open set $\mathfrak{Q} \supset \boldsymbol{\eta}(\mathfrak{H})$ such that

$$\mathcal{H}^*(\boldsymbol{\eta}) = \int_{-\infty}^{\infty} \cdots \int_{-\infty}^{\infty} \exp(\boldsymbol{t}^\top \boldsymbol{\eta} - \mathcal{G}(\boldsymbol{t}))\, d\boldsymbol{t} < \infty\ \forall \boldsymbol{\eta} \in \mathfrak{Q}.$$

Then the cumulant generating function for the larger family exists. This assumption insures that the variance matrix for the sufficient statistic vector exists, and insures

Example of trivial curved exponential family.

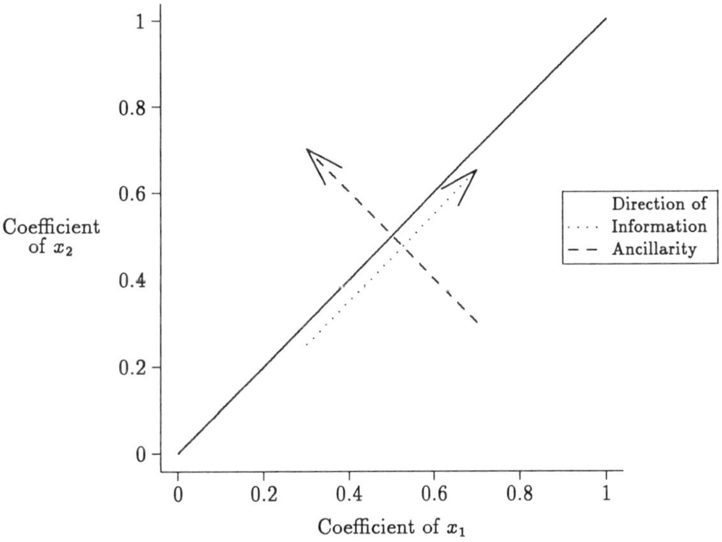

Fig. 17a.

Example of normal correlation curved exponential family.

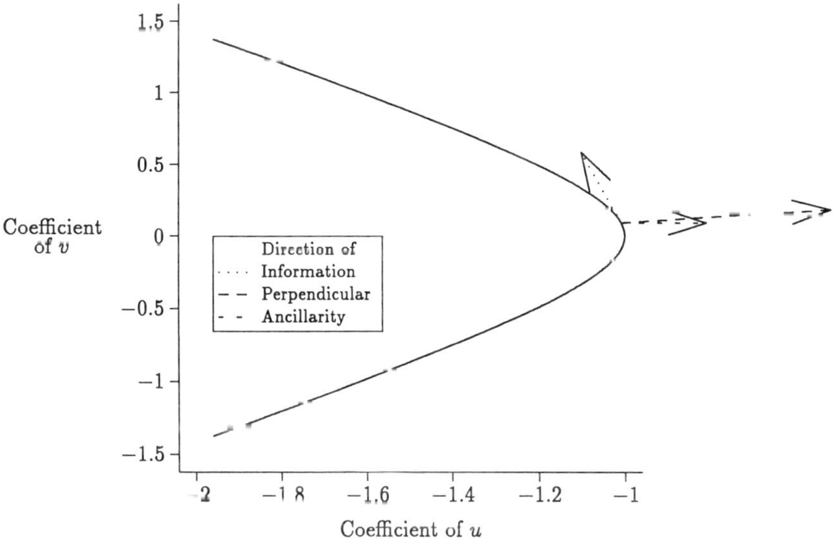

Fig. 17b.

that the chain rule can be used for differentiating. It also insures that $\mathcal{H}^*$ is twice-differentiable, and that the second derivative matrix is positive-definite at all $\boldsymbol{\eta} \in \mathfrak{Q}$. Assume further that the full-model likelihood equation $\mathcal{H}^{*\prime}(\boldsymbol{\eta}) = \boldsymbol{t}$ has a solution in $\mathfrak{Q}$ for all $\boldsymbol{t} \in \mathfrak{T}$. Set $\boldsymbol{\tau}(\boldsymbol{\theta}) = \mathrm{E}_{\boldsymbol{\theta}}[\boldsymbol{T}]$ and $\boldsymbol{\Sigma}(\boldsymbol{\theta}) = \mathrm{Var}_{\boldsymbol{\theta}}[\boldsymbol{T}]$. Since $\boldsymbol{\tau}(\boldsymbol{\theta}) = \mathcal{H}^{*\prime}(\boldsymbol{\eta}(\boldsymbol{\theta}))$, then

$$\partial\boldsymbol{\tau}/\partial\boldsymbol{\theta} = \mathcal{H}^{*\prime\prime}(\boldsymbol{\eta}(\boldsymbol{\theta}))\partial\boldsymbol{\eta}/\partial\boldsymbol{\theta} = \boldsymbol{\Sigma}\partial\boldsymbol{\eta}/\partial\boldsymbol{\theta}. \tag{153}$$

The range of $\boldsymbol{\tau}(\boldsymbol{\theta})$ then represents the part of the data space that can be explained by the data. The observed information $\jmath(\boldsymbol{\theta})$ and the Fisher information $\imath(\boldsymbol{\theta})$ in this family are given by

$$\jmath(\boldsymbol{\theta}) = -\sum_j (T^j - \tau^j)\partial^2\eta_j/(\partial\boldsymbol{\theta}^\top\partial\boldsymbol{\theta}) + (\partial\boldsymbol{\tau}/\partial\boldsymbol{\theta})^\top\partial\boldsymbol{\eta}/\partial\boldsymbol{\theta}$$

$$\imath(\boldsymbol{\theta}) = (\partial\boldsymbol{\tau}/\partial\boldsymbol{\theta})^\top\partial\boldsymbol{\eta}/\partial\boldsymbol{\theta} = (\partial\boldsymbol{\eta}/\partial\boldsymbol{\theta})^\top\boldsymbol{\Sigma}\partial\boldsymbol{\eta}/\partial\boldsymbol{\theta}. \tag{154}$$

Barndorff–Nielsen (1980) suggests taking as an ancillary $\boldsymbol{A}$ perpendicular to $\boldsymbol{\eta}(\hat{\theta})$ in the metric defined locally by the covariance matrix of $\boldsymbol{T}$. The desired ancillary should have a distribution such that as $\hat{\boldsymbol{\theta}}$ becomes more and more precise, then $\boldsymbol{A}$ converges to a non-degenerate distribution. In fact, it can be made to have approximately zero mean and unit variance, if it is given by

$$\boldsymbol{A} = \boldsymbol{B}(\hat{\boldsymbol{\theta}})(\boldsymbol{T} - \boldsymbol{\tau}(\hat{\boldsymbol{\theta}})), \tag{155}$$

where $\boldsymbol{B}(\boldsymbol{\theta}) = [(\partial\boldsymbol{\tau}/\partial\boldsymbol{\theta})^{\perp\top}\boldsymbol{\Sigma}(\partial\boldsymbol{\tau}/\partial\boldsymbol{\theta})^{\perp}]^{-\frac{1}{2}}(\partial\boldsymbol{\tau}/\partial\boldsymbol{\theta})^{\perp\top}$. This will be proven below. The matrix $(\partial\boldsymbol{\tau}/\partial\boldsymbol{\theta})^{\perp}$ is chosen such that

$$(\partial\boldsymbol{\tau}/\partial\boldsymbol{\theta})^{\perp}(\partial\boldsymbol{\tau}/\partial\boldsymbol{\theta})^\top = 0 \text{ and } \det\left[(\partial\boldsymbol{\tau}/\partial\boldsymbol{\theta})^{\perp}\boldsymbol{\Sigma}(\partial\boldsymbol{\tau}/\partial\boldsymbol{\theta})^{\perp\top}\right] = \det[\imath]/\det[\boldsymbol{\Sigma}]; \tag{156}$$

in other words $\partial\boldsymbol{\tau}/\partial\boldsymbol{\theta}$ is chosen to have rows orthogonal according to an inner product determined by $\boldsymbol{\Sigma}$. The second condition in (156) is equivalent to

$$\det\left[\begin{pmatrix}\partial\boldsymbol{\eta}/\partial\boldsymbol{\theta}\\(\partial\boldsymbol{\tau}/\partial\boldsymbol{\theta})^{\perp}\end{pmatrix}\right] = \det[\imath]/\det[\boldsymbol{\Sigma}],$$

as is seen by expanding the determinant for

$$\begin{pmatrix}\partial\boldsymbol{\eta}/\partial\boldsymbol{\theta}\\(\partial\boldsymbol{\tau}/\partial\boldsymbol{\theta})^{\perp}\end{pmatrix}^\top \boldsymbol{\Sigma}\begin{pmatrix}\partial\boldsymbol{\eta}/\partial\boldsymbol{\theta}\\(\partial\boldsymbol{\tau}/\partial\boldsymbol{\theta})^{\perp}\end{pmatrix},$$

and using (153) and the second expression for $\imath(\boldsymbol{\theta})$. That such a matrix can be constructed for each value of $\boldsymbol{\theta}$ is an elementary result from linear algebra; that $(\partial\boldsymbol{\tau}/\partial\boldsymbol{\theta})^{\perp}$ can be expressed as a differentiable function of $\boldsymbol{\theta}$ is a deeper question. Construction of such a matrix will be outlined in the exercises. The key idea here is that while the matrix $\boldsymbol{B}(\hat{\boldsymbol{\theta}})$ is the function of the random quantity $\hat{\boldsymbol{\theta}}$ and hence is random, it varies far less than $\boldsymbol{T} - \boldsymbol{\tau}(\hat{\boldsymbol{\theta}})$.

To make these notions precise, suppose that $\boldsymbol{T}_n$ arises as the sum of n independent and identically distributed observations from the curved exponential family

(152) nested within a full exponential family. Let $\boldsymbol{\tau}_1(\boldsymbol{\theta})$ and $\boldsymbol{\Sigma}_1(\boldsymbol{\theta})$ be the mean and variance functions for each summand. Then $\boldsymbol{A}_n = \boldsymbol{B}(\hat{\boldsymbol{\theta}}_n)(\boldsymbol{T}_n - n\boldsymbol{\tau}_1(\hat{\boldsymbol{\theta}}_n))/\sqrt{n}$.

One might consider how far $\boldsymbol{A}_n$ is from being exactly ancillary. Note that

$$\begin{aligned}\boldsymbol{A}_n = {} & \boldsymbol{B}(\boldsymbol{\theta})(\boldsymbol{T}_n - n\boldsymbol{\tau}_1(\boldsymbol{\theta}))/\sqrt{n} + (\boldsymbol{B}(\hat{\boldsymbol{\theta}}_n) - \boldsymbol{B}(\boldsymbol{\theta}))(\boldsymbol{T}_n - n\boldsymbol{\tau}_1(\boldsymbol{\theta}))/\sqrt{n} \\ & + \sqrt{n}(\boldsymbol{B}(\hat{\boldsymbol{\theta}}_n) - \boldsymbol{B}(\boldsymbol{\theta}))(\boldsymbol{\tau}(\boldsymbol{\theta}) - \boldsymbol{\tau}(\hat{\boldsymbol{\theta}}_n)) + \sqrt{n}\boldsymbol{B}(\boldsymbol{\theta})(\boldsymbol{\tau}_1(\boldsymbol{\theta}) - \boldsymbol{\tau}_1(\hat{\boldsymbol{\theta}}_n))].\end{aligned}$$

The first term on the right has a distribution independent of $\boldsymbol{\theta}$, with zero mean and unit variance, to order $O_p(1/\sqrt{n})$. The remaining terms are all of size $O_p(1/\sqrt{n})$. Furthermore, under weak regularity conditions the moments of $\boldsymbol{A}_n$ will be differentiable functions of $\boldsymbol{\theta}$, and hence for changes in $\boldsymbol{\theta}$ of order $1/\sqrt{n}$ the mean, variance, and third cumulant will be constant to order $O(n^{-3/2})$, $O(n^{-2})$, and $O(n^{-5/2})$. Cox (1980) refers to this property as second order local ancillarity.

As will be proved below, the general formula for expansions of statistics conditional on the ancillary vector $\boldsymbol{A}$ of (155) is

$$f(\hat{\boldsymbol{\theta}}|\boldsymbol{a};\boldsymbol{\theta}) = (2\pi)^{-\mathfrak{k}/2}\det\left[\hat{j}\right]^{-\frac{1}{2}}(\mathcal{L}(\boldsymbol{\theta};\boldsymbol{t})/\mathcal{L}(\hat{\boldsymbol{\theta}};\boldsymbol{t}))\{1+O(1/n)\}.$$

In the case of full exponential $(\mathfrak{k},\mathfrak{k})$ families, with the canonical parameterization (152) with $\mathfrak{d} = \mathfrak{k}$ and $\boldsymbol{\eta}(\boldsymbol{\theta}) = \boldsymbol{\theta}$, this result is proved above as (149). Here the ancillary $\boldsymbol{A}$ is of length 0, and no effective conditioning happens.

When $\mathfrak{k} > \mathfrak{d}$ ancillarity really has teeth; conditioning on the ancillary is a requirement, not just an option, since a one-to-one map is needed to transform to the maximum likelihood estimator, and an approximate ancillary is the logical choice.

As an example of a curved exponential family, consider inference on a normal correlation presented by Barndorff–Nielsen (1980). Suppose that $(X_i, U_i)^\top$ are independent and identically distributed with distribution

$$N\left(0, \begin{pmatrix} 1 & \theta \\ \theta & 1 \end{pmatrix}\right).$$

Then $\boldsymbol{T} = (\frac{1}{2}\sum_i(X_i^2 + U_i^2), \sum_i X_iU_i)$ is sufficient, and

$$f_{X,U}(\boldsymbol{x},\boldsymbol{u};\theta) = (2\pi)^{-n/2}(1-\theta^2)^{-n/2}\exp(\theta t_2 - t_1/(1-\theta)).$$

Straight forward calculations show that

$$\mathrm{E}\left[\boldsymbol{T}\right] = n\begin{pmatrix} 1 \\ \theta \end{pmatrix},\ \mathrm{Var}\left[\boldsymbol{T}\right] = n\begin{pmatrix} 1+\theta^2 & 3\theta \\ 3\theta & 1+2\theta^2 \end{pmatrix}.$$

Hence $\partial\boldsymbol{\tau}/\partial\theta = (0, n)$, and so we can take $(\partial\boldsymbol{\tau}/\partial\theta)^\perp = (1, 0)$, yielding an ancillary statistic that is an affine function of T^1. Specifically, use $\boldsymbol{A} = (1+\theta^2)^{-1/2}(T^1 - n)/\sqrt{n}$.

The ancillary $\boldsymbol{A}$ is not uniquely determined by (155), because of the lack of uniqueness in the definition of perpendicular compliments used above. This non-uniqueness is a general drawback of inference conditional on ancillary statistics. Furthermore, calculation of any particular choice of $\boldsymbol{A}$ might prove troublesome. These problems might be avoided by constructing a statistic $\boldsymbol{S}$ such that $(\boldsymbol{S}, \boldsymbol{A})$ are sufficient statistics and such that $\boldsymbol{S}$ and $\boldsymbol{A}$ are approximately independent at some parameter value of interest. McCullagh (1984) shows that such a $\boldsymbol{S}$ is uniquely defined in one-dimensional problems, and is easily calculated from the likelihood ratio statistic. This work results in a more general proof of Barndorff–Nielsen's formula. The construction and proof crucially uses multivariate tensor notation beyond the scope of this work. McCullagh (1987) provides details.

8.5. Barndorff–Nielsen's Formula

This section is devoted to proving that Barndorff–Nielsen's formula holds for certain approximately ancillary statistics. First a lemma:

Lemma 8.5.1: *If $\boldsymbol{A}$ is constructed as in (155), then the Jacobian* $\det\left[\partial \boldsymbol{T}/\partial(\hat{\boldsymbol{\theta}}, \boldsymbol{A})\right]$ *is given by*

$$\det\left[\jmath(\hat{\boldsymbol{\theta}})\right] \det\left[\imath(\hat{\boldsymbol{\theta}})\right]^{-1/2} \det\left[\boldsymbol{\Sigma}(\hat{\boldsymbol{\theta}})\right]^{1/2}.$$

Proof: The maximum likelihood estimator $\hat{\boldsymbol{\theta}}(\boldsymbol{t})$ satisfies $\boldsymbol{t}^{\top}(\partial\boldsymbol{\eta}/\partial\boldsymbol{\theta}) - \mathcal{H}'(\hat{\boldsymbol{\theta}}) = 0$. Since

$$\mathcal{H}'(\boldsymbol{\theta}) = \mathcal{H}^{*\prime}(\boldsymbol{\eta}(\boldsymbol{\theta}))(\partial\boldsymbol{\eta}/\partial\boldsymbol{\theta}) = \boldsymbol{\tau}(\boldsymbol{\theta})(\partial\boldsymbol{\eta}/\partial\boldsymbol{\theta}),$$

the expression relating $\boldsymbol{T}$ to $\hat{\boldsymbol{\theta}}$ and $\boldsymbol{A}$ is

$$(\boldsymbol{T}-\boldsymbol{\tau})^{\top}\begin{pmatrix}\frac{\partial\boldsymbol{\eta}}{\partial\boldsymbol{\theta}} & \frac{\partial\boldsymbol{\tau}}{\partial\boldsymbol{\theta}}^{\perp}\end{pmatrix}\begin{pmatrix}\boldsymbol{I} & 0 \\ 0 & [(\partial\boldsymbol{\tau}/\partial\boldsymbol{\theta})^{\perp\top}\boldsymbol{\Sigma}(\partial\boldsymbol{\tau}/\partial\boldsymbol{\theta})^{\perp}]^{-\frac{1}{2}}\end{pmatrix} = (0, \boldsymbol{A}^{\top}),$$

with all functions of $\boldsymbol{\theta}$ evaluated at $\hat{\boldsymbol{\theta}}$. The quantity $(\partial\boldsymbol{\tau}/\partial\boldsymbol{\theta})^{\perp}$ is the result of taking the derivative of the mean function $\boldsymbol{\tau}$, and constructing a matrix perpendicular to it. Letting $\boldsymbol{A}^{*} = \boldsymbol{A}[(\partial\boldsymbol{\tau}/\partial\boldsymbol{\theta})^{\perp\top}\boldsymbol{\Sigma}(\partial\boldsymbol{\tau}/\partial\boldsymbol{\theta})^{\perp}]^{\frac{1}{2}}$,

$$(\boldsymbol{T}-\boldsymbol{\tau})^{\top}\begin{pmatrix}\frac{\partial\boldsymbol{\eta}}{\partial\boldsymbol{\theta}} & \frac{\partial\boldsymbol{\tau}}{\partial\boldsymbol{\theta}}^{\perp}\end{pmatrix} = (\mathbf{o}, \boldsymbol{A}^{*}).$$

Differentiating these relations with respect to $\hat{\boldsymbol{\theta}}$ and $\boldsymbol{A}^{*}$, we find that

$$\left(\frac{\partial\boldsymbol{T}}{\partial(\hat{\boldsymbol{\theta}}, \boldsymbol{A}^{*})} - \begin{pmatrix}\partial\boldsymbol{\tau}/\partial\boldsymbol{\theta} & \mathbf{o}\end{pmatrix}\right)^{\top}\begin{pmatrix}\frac{\partial\boldsymbol{\eta}}{\partial\boldsymbol{\theta}} & \frac{\partial\boldsymbol{\tau}}{\partial\boldsymbol{\theta}}^{\perp}\end{pmatrix} + \sum_{j}(T^{j}-\tau^{j})\begin{pmatrix}\frac{\partial^{2}\eta_{j}}{\partial\boldsymbol{\theta}^{\top}\partial\boldsymbol{\theta}} & * \\ \mathbf{o} & \mathbf{o}\end{pmatrix} = \begin{pmatrix}\mathbf{o} & \mathbf{o} \\ \mathbf{o} & \boldsymbol{I}\end{pmatrix}.$$

The quantity $*$ above is the result of differentiating row j of $(\partial\boldsymbol{\tau}/\partial\boldsymbol{\theta})^{\perp}$ with respect to $\boldsymbol{\theta}$. Then

$$\frac{\partial \boldsymbol{T}}{\partial(\hat{\boldsymbol{\theta}},\boldsymbol{A}^{*})}^{\top}\begin{pmatrix}\frac{\partial\boldsymbol{\eta}}{\partial\boldsymbol{\theta}} & \frac{\partial\boldsymbol{\tau}}{\partial\boldsymbol{\theta}}^{\perp}\end{pmatrix} = \begin{pmatrix}\mathbf{o} & \mathbf{o}\\ \mathbf{o} & \boldsymbol{I}\end{pmatrix} + \begin{pmatrix}\frac{\partial\boldsymbol{\tau}}{\partial\boldsymbol{\theta}} & \mathbf{o}\end{pmatrix}^{\top}\begin{pmatrix}\frac{\partial\boldsymbol{\eta}}{\partial\boldsymbol{\theta}} & \frac{\partial\boldsymbol{\tau}}{\partial\boldsymbol{\theta}}^{\perp}\end{pmatrix} - \sum_j (T^j - \tau^j)\begin{pmatrix}\frac{\partial^2\eta_j}{\partial\boldsymbol{\theta}^{\top}\partial\boldsymbol{\theta}} & *\\ \mathbf{o} & \mathbf{o}\end{pmatrix}$$
$$= \begin{pmatrix}\mathbf{o} & \mathbf{o}\\ \mathbf{o} & \boldsymbol{I}\end{pmatrix} + \begin{pmatrix}(\partial\boldsymbol{\tau}/\partial\boldsymbol{\theta})^{\top}(\partial\boldsymbol{\eta}/\partial\boldsymbol{\theta}) & *\\ \mathbf{o} & \mathbf{o}\end{pmatrix}^{\top} + \begin{pmatrix}j(\hat{\boldsymbol{\theta}}) - i(\hat{\boldsymbol{\theta}}) & *\\ \mathbf{o} & \mathbf{o}\end{pmatrix}$$
$$= \begin{pmatrix}j(\hat{\boldsymbol{\theta}}) & *\\ \mathbf{o} & \boldsymbol{I}\end{pmatrix}, \tag{157}$$

by (154). Here $*$ marks parts of the matrix that can be ignored when taking its determinant since the corresponding lower block is $\mathbf{o}$. The determinant of (157) is $\det\left[j(\hat{\boldsymbol{\theta}})\right]$, and hence the determinant of $(\partial\boldsymbol{T}/\partial(\hat{\boldsymbol{\theta}},\boldsymbol{A}^{*}))$ is $\det\left[j(\hat{\boldsymbol{\theta}})\right]\det\left[\boldsymbol{\Sigma}(\hat{\boldsymbol{\theta}})\right] \times \det\left[i(\hat{\boldsymbol{\theta}})\right]^{-1}$. The Jacobian of the transformation from $\boldsymbol{A}^{*}$ to $\boldsymbol{A}$ is $\det\left[i(\hat{\boldsymbol{\theta}})\right]^{1/2} \times \det\left[\boldsymbol{\Sigma}(\hat{\boldsymbol{\theta}})\right]^{-1/2}$, and hence the Jacobian of the transformation from $\boldsymbol{t}$ to $(\hat{\boldsymbol{\theta}},\boldsymbol{A})$ is $\det\left[j(\hat{\boldsymbol{\theta}})\right]\det\left[i(\hat{\boldsymbol{\theta}})\right]^{-1/2}\det\left[\boldsymbol{\Sigma}(\hat{\boldsymbol{\theta}})\right]^{1/2}$.

Q.E.D

The next lemma expresses the exact joint distribution of the maximum likelihood estimator $\hat{\boldsymbol{\theta}}$ and an associated statistic $\boldsymbol{U}$ such that $(\hat{\boldsymbol{\theta}},\boldsymbol{U})$ are minimally sufficient, as the product of Barndorff–Nielsen's formula and an error term.

Lemma 8.5.2: *Suppose that $(\hat{\boldsymbol{\theta}},\boldsymbol{U})$ are minimally sufficient for the model (152), and $\boldsymbol{A}$ is a function of $\boldsymbol{U}$, possibly depending on $\hat{\boldsymbol{\theta}}$ as well. Let*

$$H(\hat{\boldsymbol{\theta}},\boldsymbol{U}) = \det\left[\partial\boldsymbol{T}/\partial(\hat{\boldsymbol{\theta}},\boldsymbol{U})\right]\sqrt{(2\pi)^{\partial}/\det\left[j(\hat{\boldsymbol{\theta}})\right]}\exp(\boldsymbol{t}^{\top}\boldsymbol{\eta}(\hat{\boldsymbol{\theta}}) - \mathcal{H}(\hat{\boldsymbol{\theta}}) - \mathcal{G}(\boldsymbol{t}))$$

Then $f_{\hat{\boldsymbol{\Theta}},\boldsymbol{A}}(\hat{\boldsymbol{\theta}},\boldsymbol{a};\boldsymbol{\theta}) = (2\pi)^{-\partial/2}[H(\hat{\boldsymbol{\theta}},\boldsymbol{U})/(\partial\boldsymbol{U}/\partial\boldsymbol{A})](\hat{\boldsymbol{\theta}},\boldsymbol{a})\det\left[j(\hat{\boldsymbol{\theta}})\right]^{1/2}\mathcal{L}(\boldsymbol{\theta};\boldsymbol{t})/\mathcal{L}(\hat{\boldsymbol{\theta}};\boldsymbol{t})$.

Proof: This holds from a straight-forward application of the usual change of variables theorem.

Q.E.D

Applying this to the affine ancillary calculated earlier:

Theorem 8.5.3: *If $\boldsymbol{T}$ arises as the mean of independent and identically distributed observations from the curved exponential family (152) nested within a full exponential family, and if $\boldsymbol{A}$ is constructed as in (155), then the expansion holds:*

$$f_{\hat{\boldsymbol{\Theta}}|\boldsymbol{A}}(\hat{\boldsymbol{\theta}}|\boldsymbol{a};\boldsymbol{\theta}) = (2\pi)^{-\partial/2}\det\left[j(\hat{\boldsymbol{\theta}})\right]^{1/2}\mathcal{L}(\boldsymbol{\theta};\boldsymbol{t})/\mathcal{L}(\hat{\boldsymbol{\theta}};\boldsymbol{t}) + O_p(n^{-1}). \tag{158}$$

Proof: By Lemma 8.5.2, with $\boldsymbol{A}$ constructed as before and $\boldsymbol{U} = \boldsymbol{A}$, it suffices to show that $H(\hat{\boldsymbol{\theta}},\boldsymbol{A}) = 1 + O_p(n^{-1})$. The quantity $\mathcal{G}(\boldsymbol{t})$ is approximated by reference to the saddlepoint expansion for the full exponential family model $f_T(\boldsymbol{t};\boldsymbol{\eta})$, which shows

$$\exp(\boldsymbol{t}^\top\boldsymbol{\eta} - \mathcal{H}^*(\boldsymbol{\eta}) - \mathcal{G}(\boldsymbol{t})) = c\det\left[\mathcal{H}^{*\prime\prime}(\tilde{\boldsymbol{\eta}})\right]^{-1/2}\exp(\boldsymbol{t}^\top(\boldsymbol{\eta}-\tilde{\boldsymbol{\eta}}) - \mathcal{H}^*(\boldsymbol{\eta}) + \mathcal{H}^*(\tilde{\boldsymbol{\eta}}))\times \left\{1 + b(\tilde{\boldsymbol{\eta}})/[2n] + O(n^{-2})\right\}$$

where $\tilde{\boldsymbol{\eta}}$ satisfies $\mathcal{H}^{*\prime}(\tilde{\boldsymbol{\eta}}) = \boldsymbol{t}$. Then

$$\exp(-\mathcal{G}(\boldsymbol{t})) = c\det\left[\mathcal{H}^{*\prime\prime}(\tilde{\boldsymbol{\eta}})\right]^{-\frac{1}{2}}\exp(-\boldsymbol{t}^\top\tilde{\boldsymbol{\eta}} + \mathcal{H}^*(\tilde{\boldsymbol{\eta}}))\{1 + b(\tilde{\boldsymbol{\eta}})/[2n] + O(n^{-2})\}.$$

Hence by Lemma 8.5.1,

$$H(\hat{\boldsymbol{\theta}},\boldsymbol{A}) = \left[\frac{\det\left[\boldsymbol{\Sigma}(\hat{\boldsymbol{\theta}})\right]\det\left[\jmath(\hat{\boldsymbol{\theta}})\right]}{\det\left[\mathcal{H}^{*\prime\prime}(\tilde{\boldsymbol{\eta}})\right]\det\left[\imath(\hat{\boldsymbol{\theta}})\right]}\right]^{\frac{1}{2}}\exp(\boldsymbol{t}^\top\boldsymbol{\eta}(\hat{\boldsymbol{\theta}}) - \mathcal{H}(\hat{\boldsymbol{\theta}}) - \boldsymbol{t}^\top\tilde{\boldsymbol{\eta}} + \mathcal{H}^*(\tilde{\boldsymbol{\eta}}))\times \left\{1 + b(\tilde{\boldsymbol{\eta}})/[2n] + O(n^{-2})\right\} \tag{159}$$

If $\boldsymbol{T}$ is the mean of independent and identically distributed random vectors from (152), then the matrix $\boldsymbol{\Sigma}$ in the definition of $\boldsymbol{B}$ has a factor of n^{-1}, and the functions $(\partial\boldsymbol{\tau}/\partial\boldsymbol{\theta})^\perp$ are unchanged. Then $\boldsymbol{T}$ can be expressed as an invertible function of $\hat{\boldsymbol{\theta}}$ and $\boldsymbol{A}/\sqrt{n}$ near $(\boldsymbol{\theta},\mathbf{o})$ with no other dependence on n. Since $\sqrt{n}(\boldsymbol{T}-\boldsymbol{\tau})$ converges in distribution, or in other words $\boldsymbol{T}-\boldsymbol{\tau} = O_p(1/\sqrt{n})$, then $\boldsymbol{A} = O_p(1)$. Expanding the quantity in large brackets in (159) about $(\hat{\boldsymbol{\theta}},\mathbf{o})$ we determine an expansion of the form $1 + n^{-1/2}\sum_l A^l c_l(\hat{\boldsymbol{\theta}},\mathbf{o}) + O(\|\boldsymbol{A}\|^2/n)$, where c_l are continuous in both arguments, and $\boldsymbol{A} = (A^1,\ldots,A^{\mathfrak{k}-\mathfrak{d}})$. Since $\hat{\boldsymbol{\theta}} = \boldsymbol{\theta} + O_p(1/\sqrt{n})$, as will be shown in the section on the Bartlett's correction, $\hat{\boldsymbol{\theta}}$ can be replaced by $\boldsymbol{\theta}$, incurring an error no larger than $O_p(1/n)$. A similar treatment shows that the final factor in large brackets is of the form $c(\boldsymbol{A}) + O_p(1/n)$.

Q.E.D

Formula (158) is known as Barndorff–Nielsen's formula. One might ask whether this formula might hold exactly. Consider the case of $(\mathfrak{k},\mathfrak{d})$ curved exponential families in which $\mathfrak{d} = \mathfrak{k} - 1$. Barndorff–Nielsen (1984) explores the applicability of this formula for other ancillary statistics, and notes that exactness holds for approximate ancillary statistics $\boldsymbol{V}$ satisfying $H(\hat{\boldsymbol{\theta}},\boldsymbol{V}) = \partial U/\partial\boldsymbol{V}$, and presents an example in which this differential equation can be solved to yield an ancillary for which Barndorff–Nielsen's formula holds exactly.

More generally when $\mathfrak{d} = \mathfrak{k} - 1$ Barndorff–Nielsen (1984) constructs an ancillary for which this formula holds to $O(n^{-3/2})$. This might be constructed in the above proof by multiplying the factors containing the errors, $\{1 + n^{-1/2}\sum_l A^l c_l(\hat{\boldsymbol{\theta}},\mathbf{o}) + O(\|A\|^2/n)\}\{1 + b(\tilde{\boldsymbol{\eta}})/[2n] + O(n^{-2})\}$, extracting the next order term in n, and constructing an ancillary that sets this term to zero. This may be achieved as a linear combination of components of components of $\boldsymbol{A}$, times a scalar function of $\boldsymbol{A}$, because of the presence of the term $\sum_l A^l c_l(\hat{\boldsymbol{\theta}},\mathbf{o})$ in the error. Barndorff–Nielsen

(1984) suggests using a directed log likelihood ratio statistic formed by multiplying the log likelihood ratio statistic by the unit vector in the direction of the affine ancillary. Specifically, let $\hat{\Omega} = \sqrt{2n[\boldsymbol{T}^{\top}(\tilde{\boldsymbol{\eta}} - \hat{\boldsymbol{\eta}}) - \mathcal{H}(\hat{\boldsymbol{\theta}}) + \mathcal{H}^{*}(\tilde{\boldsymbol{\eta}})]}$, with the sign given by the sign of $\hat{\eta}_1$. Define the directed log likelihood ratio statistic to be

$$\left|\hat{\Omega}\right| \boldsymbol{A} / \|\boldsymbol{A}\| . \tag{160}$$

The appropriateness in the case of a one-dimensional affine ancillary is demonstrated in the following theorem. In this case a careful analysis of the above-mentioned linear combination is avoided by transforming to the density of the maximum likelihood estimator and the directed log likelihood ratio statistic immediately, and exploiting the occurrence of the likelihood ratio statistic in the exponent of the term in brackets in (159).

Theorem 8.5.4: *If $\boldsymbol{T}$ arises as the sum of n independent and identically distributed observations from the $(\mathfrak{k}, \mathfrak{k}-1)$ curved exponential family (152) nested within a full exponential family, and $\hat{\Omega}$ is given as above, then there exists an additive correction to $\hat{\Omega}$ depending on $\hat{\boldsymbol{\theta}}$ yielding $\hat{\Omega}^{\dagger}$ such that $\hat{\Omega}^{\dagger}$ has a marginal normal density to $O(n^{-1})$, and such that*

$$f_{\hat{\Theta}|\hat{\Omega}^{\dagger}}(\boldsymbol{\theta}|\hat{\omega}^{\dagger};\boldsymbol{\theta}) = (2\pi)^{-0/2} \det\left[j(\hat{\boldsymbol{\theta}})\right]^{1/2} \mathcal{L}(\boldsymbol{\theta};\boldsymbol{t})/\mathcal{L}(\hat{\boldsymbol{\theta}};\boldsymbol{t}) + O_p(n^{-3/2}).$$

Proof: In this theorem A is scalar, since the parameter vector has one fewer component than has the sufficient statistic. Again this theorem will be verified by examining $H(\hat{\boldsymbol{\theta}}, \hat{\Omega}/\sqrt{n})$. Then $\partial(\hat{\Omega}/\sqrt{n})/\partial \boldsymbol{T} = [\tilde{\boldsymbol{\eta}} - \boldsymbol{\eta}(\hat{\boldsymbol{\theta}})]/(\hat{\Omega}/\sqrt{n})$ when $\hat{\Omega} \neq 0$. Since $\hat{\Omega} \neq 0$ and $A \neq 0$ if $\tilde{\boldsymbol{\eta}} - \boldsymbol{\eta}(\hat{\boldsymbol{\theta}}) \neq 0$, and since $\hat{\boldsymbol{\theta}}$ and $\tilde{\boldsymbol{\eta}} - \boldsymbol{\eta}(\hat{\boldsymbol{\theta}})$ are differentiable functions of $\boldsymbol{T}$, the derivative is defined even when $\hat{\Omega} = 0$. Furthermore, as a consequence of (157),

$$\partial(\hat{\boldsymbol{\theta}}, A)/\partial \boldsymbol{T} = \begin{pmatrix} j^{-1}(\hat{\boldsymbol{\theta}}) & \mathbf{0} \\ * & \boldsymbol{I} \end{pmatrix} \begin{pmatrix} \dfrac{\partial \boldsymbol{\eta}}{\partial \boldsymbol{\theta}} & \dfrac{\partial \boldsymbol{\tau}^{\perp}}{\partial \boldsymbol{\theta}} \end{pmatrix}^{\top},$$

and hence $\partial\hat{\boldsymbol{\theta}}/\partial \boldsymbol{T} = j^{-1}(\hat{\boldsymbol{\theta}})(\partial\boldsymbol{\eta}/\partial\boldsymbol{\theta})^{\top}$. Here i and j are expected and observed information for a sample of size 1, evaluated at $\boldsymbol{T}$. Therefore

$$\partial(\hat{\boldsymbol{\theta}}, \hat{\Omega}/\sqrt{n})/\partial \boldsymbol{T} = \begin{pmatrix} j^{-1}(\hat{\boldsymbol{\theta}})(\partial\boldsymbol{\eta}/\partial\boldsymbol{\theta})^{\top} \\ [\tilde{\boldsymbol{\eta}} - \hat{\boldsymbol{\eta}}]^{\top}/(\hat{\Omega}/\sqrt{n}) \end{pmatrix} = \begin{pmatrix} j(\hat{\boldsymbol{\theta}}) & 0 \\ 0 & (\hat{\Omega}/\sqrt{n}) \end{pmatrix}^{-1} \begin{pmatrix} (\partial\boldsymbol{\eta}/\partial\boldsymbol{\theta})^{\top} \\ [\tilde{\boldsymbol{\eta}} - \hat{\boldsymbol{\eta}}]^{\top} \end{pmatrix},$$

and

$$H(\hat{\boldsymbol{\theta}}, (\hat{\Omega}/\sqrt{n})) = \det\left[\begin{pmatrix} (\partial\boldsymbol{\eta}/\partial\boldsymbol{\theta})^{\top} \\ \sqrt{n}(\tilde{\boldsymbol{\eta}} - \hat{\boldsymbol{\eta}})^{\top}/\hat{\Omega} \end{pmatrix}\right]^{-1} \det\left[j(\hat{\boldsymbol{\theta}})\right]^{1/2} \phi(\hat{\Omega})\left\{1 + \frac{b(\tilde{\boldsymbol{\eta}})}{2n} + O(n^{-2})\right\}.$$

The ratio $(\tilde{\boldsymbol{\eta}} - \boldsymbol{\eta}(\hat{\boldsymbol{\theta}}))/(\hat{\Omega}/\sqrt{n})$ must be expanded as a function of $\hat{\Omega}/\sqrt{n}$ and $\hat{\boldsymbol{\theta}}$. Only the lead term will be used. Differentiability of the function will first be demonstrated, and need only be demonstrated at $\hat{\Omega} = 0$. Since $\boldsymbol{T} = \mathcal{H}^{*\prime}(\tilde{\boldsymbol{\eta}})$, $\boldsymbol{T}$ is a differentiable function of $\tilde{\boldsymbol{\eta}}$, having a differentiable inverse. Hence $\hat{\boldsymbol{\eta}}$ is a differentiable function of $\tilde{\boldsymbol{\eta}}$. Since

$$\hat{\Omega}^2/n = -\mathcal{H}^{*\prime}(\tilde{\boldsymbol{\eta}})^\top(\hat{\boldsymbol{\eta}} - \tilde{\boldsymbol{\eta}}) - \mathcal{H}^*(\tilde{\boldsymbol{\eta}}) - \mathcal{H}^*(\hat{\boldsymbol{\eta}}) = \tfrac{1}{2}(\tilde{\boldsymbol{\eta}} - \hat{\boldsymbol{\eta}})^\top \mathcal{H}^{*\prime\prime}(\hat{\boldsymbol{\eta}})(\tilde{\boldsymbol{\eta}} - \hat{\boldsymbol{\eta}}) + O(\|\tilde{\boldsymbol{\eta}} - \hat{\boldsymbol{\eta}}\|^3),$$

then $\hat{\Omega}/\sqrt{n}$ has a power series expansion in $\tilde{\boldsymbol{\eta}} - \hat{\boldsymbol{\eta}}$ near any $\hat{\boldsymbol{\eta}} \in \boldsymbol{\eta}(\mathfrak{H})$, and when $\hat{\boldsymbol{\eta}} = \tilde{\boldsymbol{\eta}}$ then $\hat{\Omega} = 0$. Hence $\det\left[\partial \boldsymbol{T}/\partial(\hat{\boldsymbol{\theta}}, \hat{\Omega}/\sqrt{n})\right]$ has a power series representation as $\hat{\Omega} \to 0$. The lead term in this series will now be calculated. Choose any $\hat{\boldsymbol{\theta}} \in \mathfrak{H}$. By the definition of $\hat{\boldsymbol{\eta}}$, $[\mathcal{H}^{*\prime}(\tilde{\boldsymbol{\eta}}) - \mathcal{H}^{*\prime}(\hat{\boldsymbol{\eta}})](d\boldsymbol{\eta}/d\boldsymbol{\theta}) = 0$, and hence $(\tilde{\boldsymbol{\eta}} - \hat{\boldsymbol{\eta}})^\top \mathcal{H}^{*\prime\prime}(\hat{\boldsymbol{\eta}})(d\boldsymbol{\eta}/d\boldsymbol{\theta}) = O(\|\tilde{\boldsymbol{\eta}} - \hat{\boldsymbol{\eta}}\|^2)$. Then

$$\det\left[\begin{pmatrix} \left(\dfrac{\partial \boldsymbol{\eta}}{\partial \boldsymbol{\theta}}\right)^\top \\ [\tilde{\boldsymbol{\eta}} - \hat{\boldsymbol{\eta}}]^\top \end{pmatrix} \boldsymbol{\Sigma} \begin{pmatrix} \dfrac{\partial \boldsymbol{\eta}}{\partial \boldsymbol{\theta}} & \tilde{\boldsymbol{\eta}} - \hat{\boldsymbol{\eta}} \end{pmatrix}\right] = \det\left[\begin{pmatrix} i(\hat{\boldsymbol{\theta}}) & O(\|\tilde{\boldsymbol{\eta}} - \hat{\boldsymbol{\eta}}\|^2) \\ O(\|\tilde{\boldsymbol{\eta}} - \hat{\boldsymbol{\eta}}\|^2) & \dfrac{\hat{\Omega}^2}{n}(1 + O(\|\tilde{\boldsymbol{\eta}} - \hat{\boldsymbol{\eta}}\|) \end{pmatrix}\right],$$

and at $\hat{\Omega} = 0$,

$$\det\left[\begin{pmatrix} (\partial \boldsymbol{\eta}/\partial \boldsymbol{\theta})^\top \\ (\tilde{\boldsymbol{\eta}} - \hat{\boldsymbol{\eta}})^\top/(\hat{\Omega}/\sqrt{n}) \end{pmatrix}\right]^{-1} = \det[\boldsymbol{\Sigma}]^{1/2} / \det\left[i(\hat{\boldsymbol{\theta}})\right]^{1/2},$$

so from (159), $H(\hat{\boldsymbol{\theta}}, \hat{\Omega}/\sqrt{n}) = \phi(\hat{\Omega})\left(1 + c_1(\hat{\boldsymbol{\theta}})\hat{\Omega}/\sqrt{n} + c_2(\hat{\boldsymbol{\theta}})\hat{\Omega}^2/n + O(1/n\sqrt{n})\right)$, for some functions $c_1(\hat{\boldsymbol{\theta}})$ and $c_2(\hat{\boldsymbol{\theta}})$. Letting $\hat{\Omega}^\dagger = \hat{\Omega} - c_1(\hat{\boldsymbol{\theta}})/\sqrt{n}$, and applying Lemma 8.5.2, there exists a function $c_3(\hat{\boldsymbol{\theta}})$ such that

$$f_{\hat{\Theta}, \hat{\Omega}^\dagger}(\hat{\boldsymbol{\theta}}, \hat{\omega}^\dagger; \boldsymbol{\theta}) = (2\pi)^{-\mathfrak{d}/2}[\mathcal{L}(\boldsymbol{\theta}; \boldsymbol{t})/\mathcal{L}(\hat{\boldsymbol{\theta}}; \boldsymbol{t})]\phi(\hat{\omega}^\dagger)\{1 + c_3(\hat{\boldsymbol{\theta}}, \hat{\omega}^\dagger/\sqrt{n})/n + O(n^{-3/2})\}.$$

The additional cross terms added into the exponent of $\phi(\hat{\omega}^\dagger)$ is exactly that necessary to remove the term $c_1(\hat{\boldsymbol{\theta}})\hat{\omega}/\sqrt{n}$. The theorem is completed by noting that $c_3(\hat{\boldsymbol{\theta}}, \hat{\omega}^\dagger/\sqrt{n}) = c_3(\boldsymbol{\theta}, \hat{\omega}^\dagger/\sqrt{n}) + O(1/\sqrt{n})$.

Q.E.D

When $\mathfrak{d} < \mathfrak{k} - 1$ this construction may be applied sequentially to produce a multivariate ancillary with the same properties.

Barndorff–Nielsen (1986) applies these techniques to general statistical models represented by submanifolds of larger models. Barndorff–Nielsen (1990b) derives conditional cumulative distribution function approximations for the maximum likelihood estimator. Approximate integration techniques like those of §5.5 are used to integrate (158). The ancillary to be conditioned on is the directed likelihood ratio (160) for testing the the adequacy of (152) within the ambient full exponential family. The result is an approximation of the form (89), where $\hat{\omega}$ and $\hat{z}$ are computed as in §5.1, except that $\hat{z} = j(\hat{\boldsymbol{\theta}})^{-1/2}(\hat{\ell}_{;1} - \ell_{;1})$ where $\ell_{;1}(\boldsymbol{\theta}; \hat{\boldsymbol{\theta}}, \boldsymbol{a})$ is the partial derivative of $\ell_n(\boldsymbol{\theta}; \hat{\boldsymbol{\theta}}, \boldsymbol{a})$ with respect to $\hat{\boldsymbol{\theta}}$.

See §5.4 for an alternative rescaled signed root of the likelihood ratio statistic, derived from the r^* approximation.

8.6. Transformation Families

Consider the example presented by Reid (1988), citing Fisher (1934), in which $X_1, \ldots, X_n$ are generated independently with density $f((x-\mu)/\sigma)/\sigma$; here f is presumed known but the location and scale parameters μ and σ are to be estimated. The family of distributions generated as the unknown parameters are allowed to vary is not necessarily an exponential family. Such families are in general known as transformation families (Barndorff–Nielsen, 1978, Barndorff–Nielsen, 1983). In the case of transformation families Barndorff–Nielsen's formula also holds, as this example will illustrate. Once $\hat{\mu}$ and $\hat{\sigma}$ are determined, $\boldsymbol{A} = \left((X_{(1)} - \hat{\mu})/\hat{\sigma}, \ldots, (X_{(n)} - \hat{\mu})/\hat{\sigma}\right)$ are treated as ancillary. The transformation $(\hat{\mu}, \hat{\sigma}, a^1, ..., a^{\mathfrak{k}-2}) \rightarrow \boldsymbol{x}$ has a Jacobian of the form $\hat{\sigma}^{n-2}$ times a function of $\boldsymbol{a}$. Hence the joint density of $(\hat{\mu}, \hat{\sigma}, A_1, ..., A_{\mathfrak{k}-2})$ is $g(\boldsymbol{a})\hat{\sigma}^{n-2}\sigma^{-n}\prod_i f((a^i\hat{\sigma} + \hat{\mu} - \mu)/\sigma)$. Integrating with respect to $\hat{\mu}$ removes dependence on μ and also removes one factor of σ; integrating with respect to $\hat{\sigma}$ removes the remaining factors of σ, resulting in a distribution of $\boldsymbol{A}$ free of unknown parameters and justifying treating $\boldsymbol{A}$ as ancillary. Since $\mathcal{L}(\hat{\mu}, \hat{\sigma}; \boldsymbol{x}) = \hat{\sigma}^{-n}\prod_i f(a^i)$, and $\mathcal{L}(\mu, \sigma; \boldsymbol{x}) = \prod_i f((a^i\hat{\sigma} + \hat{\mu} - \mu)/\sigma$, then the density of $(\hat{\mu}, \hat{\sigma})$ conditional on $\boldsymbol{A}$ can be written as $c(\boldsymbol{a}) \det\left[j(\hat{\mu}, \hat{\sigma})\right]^{-1} \mathcal{L}(\mu, \sigma; \boldsymbol{x})/\mathcal{L}(\hat{\mu}, \hat{\sigma}; \boldsymbol{x})$, thus verifying Barndorff–Nielsen's formula in this case.

The data X_i have the cumulative distribution function $F(g_{\mu,\sigma}(x))$, with F known, the function $g_{\mu,\sigma}(x) = (x-\mu)/\sigma$, and the parameters μ and σ unknown. Define an operation $\circ$ on $\mathfrak{G} = \{g_{\mu,\sigma} : \mu \in \mathfrak{R}, \sigma > 0\}$ to be composition. Then

$$(g_{\mu_2,\sigma_2} \circ g_{\mu_1,\sigma_1})(x) = g_{\mu_2,\sigma_2}((x-\mu_1)/\sigma_1) = (x - \mu_1 - \sigma_1\mu_2)/\sigma_1\sigma_2 = g_{\mu_1+\sigma_1\mu_2,\sigma_1\sigma_2}(x)$$

Hence $\mathfrak{G}$ has the following 4 properties.

a. $g_{\mu_2,\sigma_2} \circ g_{\mu_1,\sigma_1} \in \mathfrak{G}$.

b. $g_{0,1} \circ g_{\mu,\sigma} = g_{\mu,\sigma}$.

c. $g_{-\mu/\sigma,1/\sigma} \circ g_{\mu,\sigma} = g_{0,1}$.

d. $(g_{\mu_3,\sigma_3} \circ g_{\mu_2,\sigma_2}) \circ g_{\mu_1,\sigma_1} = g_{\mu_3,\sigma_3} \circ (g_{\mu_2,\sigma_2} \circ g_{\mu_1,\sigma_1})$.

A structure $(\mathfrak{G}, \circ)$ with these four properties is called a group; a group which is a set of functions on another space where the operation is composition is said to be a group acting on the other space. The general model, then, has X_i distributed independently with common cumulative distribution function $F(g^{-1}(x))$ for a known cumulative distribution function F on a sample space $\mathfrak{X}$ and an unknown $g \in \mathfrak{G}$ for the group $\mathfrak{G}$ acting on $\mathfrak{X}$. Equivalently, $X_i = g(Y_i)$ with Y_i drawn from F. The objective is to draw inference on which element $g \in \mathfrak{G}$ is involved. Generally, the group $\mathfrak{G}$ and cumulative distribution function F are said to comprise a transformation family. For a fuller exposition see Fraser (1968) or Fraser (1979). Barndorff–Nielsen's formula for all such models is exact. If $F(g(x))$ is different for each distinct $g \in \mathfrak{G}$, or equivalently, if $g(Y_i)$ has a different distribution for each $g \in \mathfrak{G}$, and one can find $\hat{g} \in \mathfrak{G}$ maximizing the likelihood, then the ancillary quantity is $\boldsymbol{A}$ with components $A^i = \hat{g}^{-1}(X_i)$. Barndorff–Nielsen (1980) treats the more general case when $g(y)$ is a distinct function of y for each $g \in \mathfrak{G}$, and Barndorff–Nielsen (1983) treats the still more general case when this assumption is dropped.

8.7. Bartlett's Correction

For a sufficient statistic $\boldsymbol{T}$ with density $f_T(\boldsymbol{t}, \boldsymbol{\theta})$ depending on a measure of sample size n, with the unknown parameter $\boldsymbol{\theta} \in \mathfrak{P} \subset \mathfrak{R}^{\mathfrak{d}}$, let $\ell_n(\boldsymbol{\theta}; \boldsymbol{t}) = \log(f_T(\boldsymbol{t}, \boldsymbol{\theta}))$, and define the likelihood ratio statistic

$$W_n(\boldsymbol{\theta}) = 2(\ell_n(\hat{\boldsymbol{\theta}}; \boldsymbol{T}_n) - \ell_n(\boldsymbol{\theta}; \boldsymbol{T}_n)), \tag{161}$$

where $\hat{\boldsymbol{\theta}}$ is the maximum likelihood estimator of $\boldsymbol{\theta}$. Then under wide regularity conditions, $\mathrm{P}\left[W_n(\boldsymbol{\theta}) \leq w\right] \approx F_{\mathfrak{d}}(w)$ where $F_{\mathfrak{d}}$ is the χ^2 cumulative distribution function on $\mathfrak{d}$ degrees of freedom. Serfling (1980) demonstrates this by expanding $\ell_n(\hat{\boldsymbol{\theta}}; \boldsymbol{T}_n) - \ell_n(\boldsymbol{\theta}; \boldsymbol{T}_n)$ in $\boldsymbol{\theta} - \hat{\boldsymbol{\theta}}$ about $\hat{\boldsymbol{\theta}}$. By the definition of $\hat{\boldsymbol{\theta}}$ the first order term is zero; then

$$\begin{aligned} 2(\ell_n(\hat{\boldsymbol{\theta}}; \boldsymbol{T}_n) - \ell_n(\boldsymbol{\theta}; \boldsymbol{T}_n)) &= (\boldsymbol{\theta} - \hat{\boldsymbol{\theta}}) \jmath_n(\hat{\boldsymbol{\theta}})(\boldsymbol{\theta} - \hat{\boldsymbol{\theta}}) + o_p(1) \\ &= (\boldsymbol{\theta} - \hat{\boldsymbol{\theta}}) \imath_n(\boldsymbol{\theta})(\boldsymbol{\theta} - \hat{\boldsymbol{\theta}}) + o_p(1) \sim \chi^2_{\mathfrak{d}}. \end{aligned}$$

Here $\jmath_n(\hat{\boldsymbol{\theta}})$ is the observed information matrix evaluated at the maximum likelihood estimator; that is, the negative of the matrix of second derivatives, and $\imath_n(\boldsymbol{\theta})$ is the expected information matrix evaluated at the true parameter vector, which is equal to the asymptotic inverse variance matrix. When data arise from the sum of independent and identically distributed random variables,

$$\jmath_n = O_p(n), \; \imath_n = O_p(n), \; \boldsymbol{\theta} - \hat{\boldsymbol{\theta}} = O_p(1/\sqrt{n}). \tag{162}$$

Relations (162) hold if:

a. The distribution of the maximum likelihood estimator is multivariate normal with the usual variance matrix to the correct order, which is true in turn if
   1. f has three derivatives for all data values,
   2. These are bounded by integrable functions near the true parameter values,
   3. For any $\boldsymbol{\theta}$, $0 < \mathrm{E}_{\boldsymbol{\theta}}\left[\partial f(X, \boldsymbol{\theta})/\partial \boldsymbol{\theta}\right]^2 < \infty$.

b. $\imath_n(\boldsymbol{\theta})$ and $\jmath_n(\boldsymbol{\theta})$ are sufficiently close; order $o_p(n)$ component-wise is sufficient. This is proved by noting that $\jmath_n(\boldsymbol{\theta})$ obeys the law of large numbers.

c. $\jmath_n(\boldsymbol{\theta})$ and $\jmath_n(\hat{\boldsymbol{\theta}})$ are $o_p(n)$ apart;

this result follows from expanding about $\boldsymbol{\theta}$ and using dominated convergence methods.

Bartlett (1953) notes that generally the $\chi^2_{\mathfrak{d}}$ approximation to the distribution of the log likelihood ratio statistic does not hold to $O(1/n)$. Since $W_n(\boldsymbol{\theta})$ has approximately the $\chi^2_{\mathfrak{d}}$ distribution, $\mathrm{E}_{\boldsymbol{\theta}}\left[W_n(\boldsymbol{\theta})\right] \approx \mathfrak{d}$. As a first refinement to the elementary χ^2 approximation to the distribution of $W_n(\boldsymbol{\theta})$, consider multiplicative rescalings, called Bartlett's corrections, moving the entire distribution closer to the asymptotic approximation. This approach is presented below.

Suppose that $\boldsymbol{X}_i$ are distributed independent and identically distributed, giving rise to the log likelihood

$$\ell_n(\boldsymbol{\theta}; \boldsymbol{X}_1, \ldots, \boldsymbol{X}_n) = \sum_{i=1}^{n} \ell_n(\boldsymbol{\theta}; \boldsymbol{X}_i),$$

where $\boldsymbol{\theta} \in \mathfrak{P} \subset \mathfrak{R}^{\mathfrak{d}}$. The next theorem investigates the existence and value of the multiplicative correction, and investigates the distribution of the transformed statistic. In general these questions are hard to answer, and in this section attention is restricted to the exponential family case, in which $\ell_n(\boldsymbol{\theta}; \boldsymbol{x}) = \boldsymbol{\theta}^{\top}\boldsymbol{T}(\boldsymbol{x}) - \mathcal{H}(\boldsymbol{\theta})$.

Theorem 8.7.1: *When the likelihood ratio test statistic $W_n(\boldsymbol{\theta})$ given by (161) arising from n independent and identically distributed random vectors $\boldsymbol{X}_j$ from the exponential family (120) is divided by the factor $1 - b(\boldsymbol{\theta})/n$, then the resulting statistic $W_n'(\boldsymbol{\theta}) = W_n(\boldsymbol{\theta})/(1 - b(\boldsymbol{\theta})/n)$ has a χ^2 density to order $O(n^{-3/2})$. Here $b(\boldsymbol{\theta}) = (3\rho_4(\boldsymbol{\theta}) - 3\rho_{13}^2(\boldsymbol{\theta}) - 2\rho_{23}^2(\boldsymbol{\theta}))/12$, exactly as in §6.3. where the invariants given are for the individual vectors $\boldsymbol{X}_j$, as in (103).*

Proof: Recall that

$$f_T(\boldsymbol{t}; \boldsymbol{\theta}) = (2\pi)^{-\mathfrak{d}/2} \det\left[j_n(\hat{\boldsymbol{\theta}})\right]^{-1/2} \exp(\ell_n(\boldsymbol{\theta}) - \ell_n(\hat{\boldsymbol{\theta}}))(1 + b(\hat{\boldsymbol{\theta}})/(2n)) + O(n^{-2})$$

by (148), where $\boldsymbol{T}_n = \sum_{j=1}^{n} \boldsymbol{X}_j$. As before,

$$f_{\hat{\boldsymbol{\theta}}}(\hat{\boldsymbol{\theta}}; \boldsymbol{\theta}) = (2\pi)^{-\mathfrak{d}/2} \det\left[j_n(\hat{\boldsymbol{\theta}})\right]^{1/2} \exp(-W(\boldsymbol{\theta})/2)(1 + b(\hat{\boldsymbol{\theta}})/(2n)) + O(n^{-2}) \quad (163)$$

here $j(\boldsymbol{\theta})$ is the observed information matrix, and $W(\boldsymbol{\theta})$ is calculated as in (161) with $\boldsymbol{T}_n$ the sufficient statistic vector necessary to yield a maximum likelihood estimator $\hat{\boldsymbol{\theta}}$. Since this is an exponential family, $j(\boldsymbol{\theta})$ is free of $\boldsymbol{T}$ and $j(\boldsymbol{\theta}) = \mathcal{H}_T''(\boldsymbol{\theta})$.

In the multivariate case this proof requires an integration over a $\mathfrak{d}-1$ dimensional surface of $\hat{\boldsymbol{\theta}}$ on which W is constant. The proof given here is for the simpler univariate case, and is found in Barndorff-Nielsen and Cox (1979). Since $W = 2((\hat{\theta} - \theta)T + n(\mathcal{H}(\theta) - \mathcal{H}(\hat{\theta}))$, and $T = n\mathcal{H}'(\hat{\theta})$, then $W = 2n((\hat{\theta} - \theta)\mathcal{H}'(\hat{\theta}) + \mathcal{H}(\theta) - \mathcal{H}(\hat{\theta}))$. Then $dW/d\hat{\theta} = 2n(\hat{\theta} - \theta)\mathcal{H}''(\hat{\theta}) + 2n\mathcal{H}'(\hat{\theta}) - 2n\mathcal{H}'(\hat{\theta}) = 2n(\hat{\theta} - \theta)\mathcal{H}''(\hat{\theta})$, and

$$f_W(w; \theta) = \frac{1}{2\sqrt{2\pi w}} \exp(-\tfrac{1}{2}w) \sum \left|\frac{w/n}{(\hat{\theta} - \theta)^2 \mathcal{H}''(\hat{\theta})}\right|^{1/2} \left\{1 + \frac{b(\hat{\theta})}{2n} + O(n^{-2})\right\}.$$

The sum above is taken over both values of $\hat{\theta}$ yielding the same value for w.

Since $\mathcal{H}(\hat{\theta}) = \sum_{j=0}^{\infty} \mathcal{H}^{(j)}(\theta)(\hat{\theta}-\theta)^j/j!$ then

$$W/n = 2\left[\mathcal{H}(\theta) + (\hat{\theta}-\theta)\sum_{j=1}^{\infty}\mathcal{H}^{(j)}(\theta)(\hat{\theta}-\theta)^{j-1}/(j-1)! - \sum_{j=0}^{\infty}\mathcal{H}^{(j)}(\theta)(\hat{\theta}-\theta)^j/j!\right]$$

$$= 2\left[\sum_{j=2}^{\infty}\mathcal{H}^{(j)}(\theta)(\hat{\theta}-\theta)^j/(j(j-2)!)\right]. \tag{164}$$

Then

$$(W/n)/(\theta-\hat{\theta})^2\mathcal{H}''(\hat{\theta}) = \frac{\sum_{j=2}^{\infty}\mathcal{H}^{(j)}(\theta)(\theta-\hat{\theta})^j/j!}{\sum_{j=2}^{\infty}\mathcal{H}^{(j)}(\theta)(\theta-\hat{\theta})^j/(j(j-2)!)},$$

and

$$\sqrt{(W/n)/(\theta-\hat{\theta})^2\mathcal{H}''(\hat{\theta})} \tag{165}$$

has a power series expansion for $\hat{\theta}$ near θ. The lead term is one. If the square root is taken to both sides of (164) and one solves for $\hat{\theta}$ then powers of $\sqrt{W/n}$ can replace those of $\hat{\theta}$ in (165) yielding a power series for $\sqrt{(W/n)/(\theta-\hat{\theta})^2\mathcal{H}''(\hat{\theta})}$ in powers of $\sqrt{W/n}$. The solution will depend on whether θ is greater or less than $\hat{\theta}$. The coefficient of the first order term of $\sqrt{W/n}$ in (165) will vary in sign according to which side of the true parameter $\hat{\theta}$ is on, since to first order (164) is quadratic. Higher order terms will also vary in magnitude. Hence $\sqrt{(W/n)/(\theta-\hat{\theta})^2\mathcal{H}''(\hat{\theta})} = 1 \pm c_1\sqrt{W/n} + (c_2 \pm c_3)W/n + O(n^{-3/2})$, and $[(\theta-\hat{\theta})^2\mathcal{H}''(\hat{\theta})]^{-1/2} = \sqrt{n/W}(1 \pm c_1\sqrt{W/n} + (c_2 \pm c_3)W/n + O(n^{-3/2}))$, and

$$f_W(w;\theta) = \frac{q_1(w)}{2}\left[\sum\left(1 \pm c_1\sqrt{\frac{w}{n}} + (c_2 \pm c_3)\frac{w}{n}\right)\left(1 + \frac{b(\theta) + e_n\sqrt{\frac{w}{n}}}{2n}\right) + O(n^{-3/2})\right]$$

$$= \frac{q_1(w)}{2}\left[\sum\left(1 + c_2\frac{w}{n}\right)\left(1 + \frac{b(\theta)}{2n}\right) + O(n^{-3/2})\right]$$

$$= q_1(w)\left(1 + (2c_2 w + b(\theta))/2n + O(n^{-3/2})\right),$$

where $q_1(w) = \phi(+\sqrt{w})/\sqrt{w}$ is the density of a χ^2 variable with one degree of freedom, and e_n is the derivative of $b(\hat{\theta})$ with respect to w/n, evaluated somewhere between θ and $\hat{\theta}$, and under the conditions of Theorem 4.6.1, this is bounded. As the summation is taken over the two values corresponding to the positive and negative values for $\pm$, these terms cancel. Since f_W must integrate to 1, and since the norming constants for the χ^2 distributions with 1 and 3 degrees of freedom are the same, $2c_2 + b(\theta) = 0$, and

$$f_W(w;\theta) = q_1(w)(1 + b(\theta)(1-w)/2n + O(n^{-3/2})).$$

The density of $W' = W/(1-b(\theta)/n)$ is $f_{W'}(w') = f_W(w'(1-b(\theta)/n))\,(1-b(\theta)/n)$. Hence

$$f_{W'}(w') = q_1(w')\exp(w'b(\theta)/2n)\,(1-b(\theta)/n)^{1/2}\left(1+\frac{b(\theta)-w'b(\theta)}{2n}+O(n^{-3/2})\right).$$

Then

$$\begin{aligned} f_{W'}(w') &= q_1(w')\left(1+w'b(\theta)/[2n]+O(n^{-2})\right)\left(1-b(\theta)/[2n]+O(n^{-2})\right)\times \\ &\quad \left(1+(1-w')b(\theta)/[2n]+O(n^{-3/2})\right) \\ &= q_1(w')\left(1+O(n^{-3/2})\right). \end{aligned}$$

Q.E.D

In the general multivariate case, recall (163). If A is a positive definite matrix then for any $c>0$, $\boldsymbol{x}^\top A^\top A\boldsymbol{x} = c$ is an ellipse in $\Re^{\mathfrak{d}}$, with surface area $2\left(\pi^{\mathfrak{d}/2}\right)/\Gamma(\mathfrak{d}/2)\det[A]^{-1}c^{\mathfrak{d}/2-1/2}$. Hence, if $d\sigma$ is the surface area measure,

$$\begin{aligned} \int_{\hat{\theta}\text{ yielding }w}\det\left[j(\hat{\boldsymbol{\theta}})\right]^{1/2}d\sigma &= \int_{\hat{\boldsymbol{\theta}}|(\hat{\boldsymbol{\theta}}-\boldsymbol{\theta})^\top j(\hat{\boldsymbol{\theta}})(\hat{\boldsymbol{\theta}}-\boldsymbol{\theta})=w}\det\left[j(\hat{\boldsymbol{\theta}})\right]^{1/2}d\sigma \\ &\approx w^{\mathfrak{d}/2-1/2}/\Gamma(\mathfrak{d}/2). \end{aligned}$$

As in the univariate case it is sufficient to show that an integrable term of order $O(n^{-1})$ exists. From the order $O(n^{-1})$ term in the multivariate saddlepoint expansion one can show that to order $O(n^{-3/2})$ the resulting density for W is $(1-\mathfrak{d}b(\boldsymbol{\theta})/n)q_{\mathfrak{d}}(w)+(\mathfrak{d}b(\boldsymbol{\theta})/n)q_{\mathfrak{d}+2}(w)$, and that the density of $W' = W/(1-b(\boldsymbol{\theta})/n)$ is $q_{\mathfrak{d}}(w')$ to order $O(n^{-3/2})$, where $q_{\mathfrak{d}}$ is the χ^2 density on $\mathfrak{d}$ degrees of freedom. Again $b(\boldsymbol{\theta}) = (3\rho_4(\boldsymbol{\theta})-3\rho_{13}^2(\boldsymbol{\theta})-2\rho_{23}^2(\boldsymbol{\theta}))/12$. References for this material are Reid (1988), Barndorff–Nielsen and Cox (1979), and Barndorff–Nielsen and Cox (1984).

Another approach is presented by Lawley (1956), McCullagh and Cox (1986), and McCullagh (1987). This involves constructing approximately normal random variables $\boldsymbol{Y}$ such that $W = \boldsymbol{Y}^\top j(\boldsymbol{\theta})\boldsymbol{Y}$, and such that $\boldsymbol{Y}$ has a non-zero mean. The resulting W has a non-central χ^2 distribution with cumulants $\kappa_l = (1-c/n)^l 2^{l-1}(l-1)!\mathfrak{d}+O(n^{-2})$, implying that W' has the same cumulants as the χ^2 distribution, to order $O(n^{-2})$. See also Cordeiro and Paula (1989).

The extension to cases of compound null hypotheses is more difficult. In many such cases, the parameter of interest $\boldsymbol{\theta}$ is written as $(\boldsymbol{\alpha},\boldsymbol{\beta})$, where the null hypothesis is $\boldsymbol{\alpha}=\boldsymbol{\alpha}_0$. Often the test statistic is $W(\alpha) = 2(\ell_n(\hat{\alpha},\hat{\boldsymbol{\beta}})-\ell_n(\boldsymbol{\alpha},\tilde{\boldsymbol{\beta}}(\boldsymbol{\alpha})))$ where $\tilde{\boldsymbol{\beta}}(\boldsymbol{\alpha})$ maximizes ℓ_n for a fixed value of $\boldsymbol{\alpha}$. Barndorff–Nielsen and Cox (1984) do this on a case-by-case basis, by determining if necessary the existence of an appropriate ancillary statistic allowing use of the Barndorff–Nielsen's formula to estimate the density of the constrained maximum likelihood estimator. The Bartlett factor is determined from the difference in norming function for these two approximations; if $b_1(\hat{\alpha},\hat{\boldsymbol{\beta}})$ is the norming function for the density of the full maximum likelihood estimator, and $b_2(\boldsymbol{\alpha}_0, \tilde{\boldsymbol{\beta}}(\boldsymbol{\alpha}_0))$ is the norming constant for the density of the constrained maximum likelihood estimator, as in (163), then the Bartlett factor is $1 - (b_1(\hat{\boldsymbol{\alpha}}, \hat{\boldsymbol{\beta}}) - b_2(\boldsymbol{\alpha}_0, \tilde{\boldsymbol{\beta}}(\boldsymbol{\alpha}_0))/(n\mathfrak{d})$ where $\mathfrak{d}$ is the length of $\boldsymbol{\alpha}$, the change in degrees of freedom as extra parameters are estimated. The functions b may often be calculated from saddlepoint expansions to the densities of the sufficient statistics.

Proofs that the Bartlett's correction is as above follow a general pattern:

1. Transform the saddlepoint density approximation,

$$f_W(w;\boldsymbol{\theta}) = \frac{1}{2}\frac{\exp(-w/2)}{(\sqrt{2\pi w})^{\mathfrak{d}}} \int \det\left[\jmath(\hat{\boldsymbol{\theta}})\right]^{-1/2} \mathfrak{d}s(\hat{\boldsymbol{\theta}}) \left(1 + b(\boldsymbol{\theta})/2n + O(n^{-3/2})\right).$$

2. Expand the integrand in $\sqrt{W/n}$. The constant term is 1, the first order term is 0, and the second term is of form $b_2 w/n$. Find b_2 either using a careful expansion, or by using next correction term in the density of the sufficient statistic, and the fact that the corrected density integrates to $1 + O(n^{-3/2})$.
3. Show that $(1 - c/2n)f_W(w(1 - c/2n))$ is a χ^2 density to order $O(n^{-3/2})$.

As an example consider testing $H_0 : \mu = \mu_0$ in the normal model with known variance $N(\mu, 1)$. Since $\ell_n(\mu; x) = (\mu x - \mu^2/2)$, then $W(\mu_0) = (x - \mu_0)^2$. Hence the distribution of $W(\mu_0)$ is exactly χ^2. The basic saddlepoint approximation to the density of $\bar{x}$ is exact; that is, $b(\mu) = 0$.

As another example consider testing the scale parameter in a gamma distribution with known shape α. The density here is $\gamma^\alpha x^{\alpha-1}\exp(-\gamma x)/\Gamma(\alpha)$; we assume that α is known. The log likelihood for a sample of size n is $-(\gamma\sum_i X_i) + n\alpha\log(\gamma) + (\alpha - 1)\sum_i \log(X_i) - n\log(\Gamma(\alpha))$, its derivative is $n(-\bar{X} + \alpha/\gamma)$, and the maximum likelihood estimator for γ is $\hat{\gamma} = \alpha/\bar{X}$. Then $W(\gamma) = n[\alpha(1 - \log(\bar{X}) + \log(\alpha)) + \gamma\bar{X} - \log(\gamma))]$. The cumulant generating function and its derivatives are

$$\mathcal{K}(\beta) = -\alpha\log(1 - \beta/\gamma), \quad \mathcal{K}^{(j)}(\beta) = (j-1)!\alpha(\gamma - \beta)^{-j}.$$

Hence $b(\gamma)$ is

$$(3\rho_4 - 5\rho_3^2)/12 = [3 \times 6(\alpha/\alpha^2) - 5 \times 4(\alpha^2/\alpha^3)]/12 = -1/[6\alpha],$$

which is independent of γ. Recall from §4.8 that the Gamma distribution, with the normal and Inverse Gaussian, are the only distributions with constant third and fourth standardized cumulants. Bartlett's correction is

$$1 + 1/(6n\alpha). \tag{166}$$

The expectation of the likelihood ratio statistic may also be calculated exactly. Recall that the density of the mean of n independent $\Gamma(\gamma, \alpha)$ random variables has a $\Gamma(n\gamma, n\alpha)$ distribution. The likelihood for γ is now $-\left(n\gamma\bar{X}\right) + \alpha n(\log(n) + \log(\gamma)) + (\alpha n - 1)\log(\bar{X}) - \log(\Gamma(\alpha n))$, and the likelihood ratio statistic is

$$2n\left(-\gamma\bar{X} + \alpha(1 - \log(\alpha)) + \alpha\log(\gamma) + \alpha\log(\bar{X})\right).$$

This expression is linear in $\bar{X}$ and $\log(\bar{X})$. The expectation of these quantities can be calculated analytically: $\mathrm{E}\left[\bar{X}\right] = \alpha/\gamma$ and $\mathrm{E}\left[\log(\bar{X})\right] = \psi(\alpha n) - \log(n\gamma)$, where $\psi(x) = \frac{d}{dx}\log(\Gamma(x))$ is the di-gamma function. Hence the expectation of $W(\gamma)$ is

$$2\alpha n\left(\log(\alpha) + \log(n) - \psi(\alpha n)\right).$$

The approximation (166) agrees with this exact value very closely (Fig. 18).

Evaluation of Bartlett's Correction for Gamma(1,1) Mean Parameter

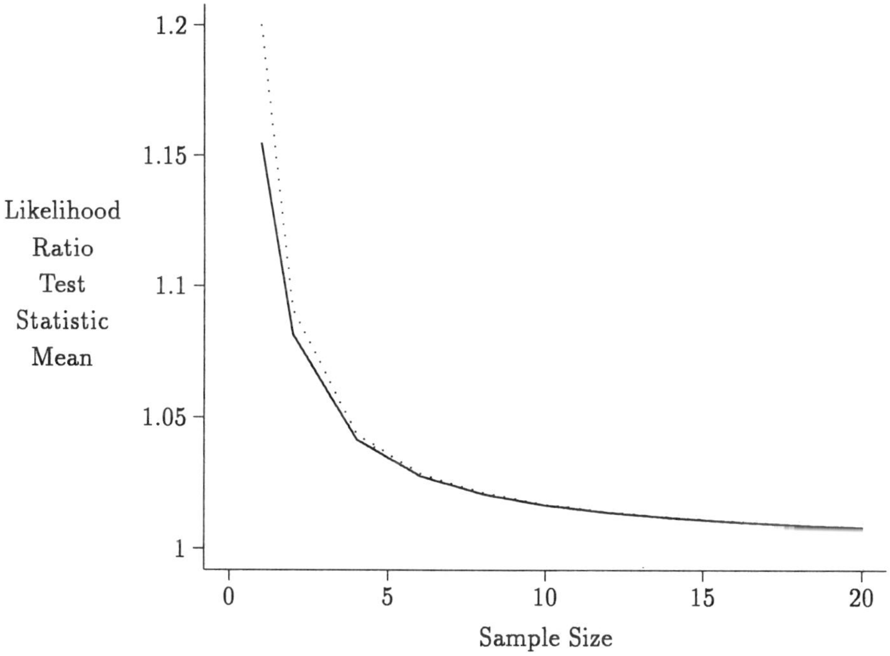

Fig. 18.

These derivations are difficult in for lattice distributions. Examples of lattice distributions for $\boldsymbol{T}_n$ might arise when the underlying random variables $\boldsymbol{X}_1, \ldots, \boldsymbol{X}_n$ for which $\boldsymbol{T}_n$ is a summary are lattice variables. Examples are:

1. When $\boldsymbol{X}_i$ is a d-long vector of zeros except for a one in one place; multinomial probabilities are to be estimated, and the sufficient statistics are $\boldsymbol{T}_n$ are the sum of the $\boldsymbol{X}_i$ giving cell totals.
2. When X_i are response variables in a logistic regression corresponding to covariates $\boldsymbol{z}_i$; the sufficient statistic is $\boldsymbol{T}_n = \boldsymbol{Z}^\top \boldsymbol{X}$ where $\boldsymbol{Z}$ is the matrix with the $\boldsymbol{z}_i$ as rows, and $\boldsymbol{X}$ is the vector $(X_1, \ldots, X_n)$. When the ratios of entries of $\boldsymbol{Z}$ are all rational, $\boldsymbol{T}_n$ is a lattice variable; otherwise it is not a lattice variable but has a singular distribution.

In the present circumstance the variable of real interest is $W(\boldsymbol{\theta})$ rather than $\boldsymbol{T}_n$; in general $W(\boldsymbol{\theta})$ will not have a lattice distribution, but a singular non-lattice distribution in which $\limsup_{|\boldsymbol{t}|\to\infty} \phi_{W(\boldsymbol{\theta})}(\boldsymbol{t}) = 1$ and the Edgeworth series to order $O(n^{-1})$ and smaller is not valid, and for which no continuity correction exists to give the series better asymptotic properties. Since $W(\boldsymbol{\theta})$ does not have a density, transformation techniques like those used in the continuous case do not apply, and one is forced to proceed by using asymptotic techniques in $\boldsymbol{T}_n$ space rather than in $W(\boldsymbol{\theta})$-space. Of interest are calculations of quantities like $\mathrm{P}\left[\boldsymbol{T}_n \in W(\boldsymbol{\theta})^{-1}[0, w]\right]$. Since these regions are approximately elliptical with the form $\mathfrak{C}' = \{\boldsymbol{t}|\boldsymbol{t}^\top \Sigma^{-1}\boldsymbol{t} \leq c\}$, the ability to approximate probabilities for such sets should be an upper limit on the ability to approximate the cumulative distribution function of $W(\boldsymbol{\theta})$. Frydenberg and Jensen (1989) provide calculations indicating that while Bartlett's correction does not improve the asymptotic error rate, in many cases it does improve accuracy. Comments on the difficulties in approximating the distribution of the Wald statistic carry over to the likelihood ratio statistic as well.

As an example of Wald and likelihood ratio testing in multivariate discrete cases, consider the problem of simultaneously testing values of means in two independent Poisson distributions. Note the similarity in the Wald and likelihood ratio rejection regions (Fig. 19).

Wald and Likelihood Ratio Acceptance Regions for the the Poisson Distribution

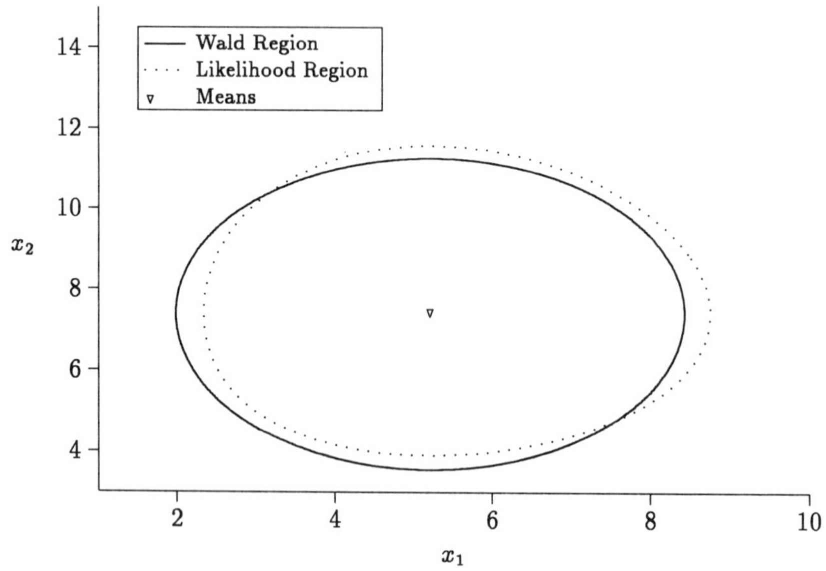

Fig. 19.

Wald and Likelihood Ratio CDFs for the Poisson Distribution

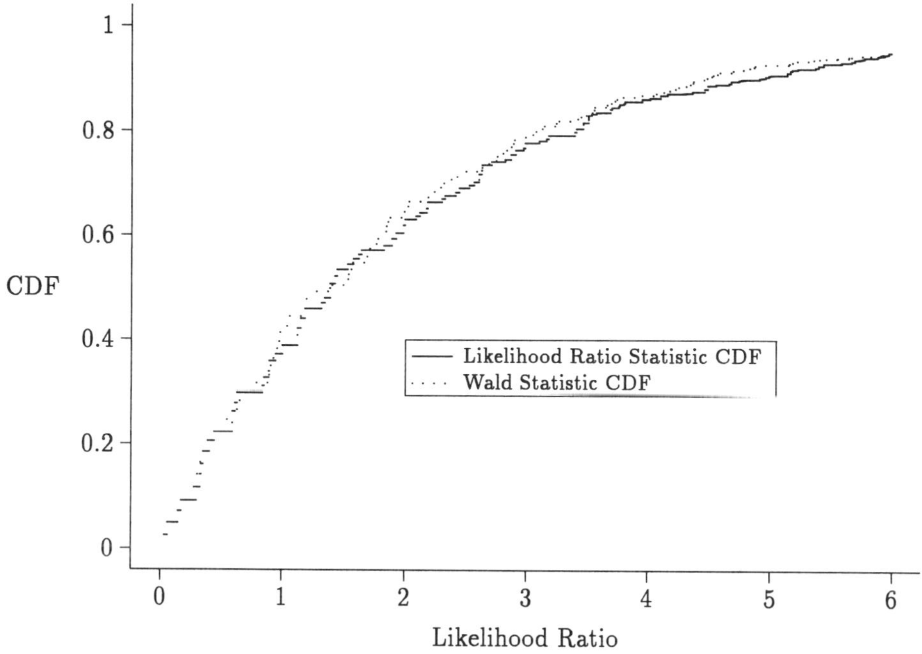

Fig. 20

Note also the irregular pattern of jumps in both cumulative distribution functions arising because points enter the expanding acceptance regions at irregular intervals (Fig. 20).

Bartlett's corrections may also be applied in less standard contexts; DiCiccio, Hall, and Romano (1991) apply it to empirical likelihoods.

8.8. Exercises

1. Consider the logistic regression model of question 2 of §7.8. Sketch contours for the density of the maximum likelihood estimator assuming the true regression parameters are the observed value of the maximum likelihood estimator.
2. Give an expression for the distribution of the maximum likelihood estimator in a Poisson regression model, similar to the above but with y a Poisson variable whose mean is the exponential of linear covariates.

9
Other Topics

This chapter contains miscellaneous material applying saddlepoint and other distribution function approximations to statistical inference. An approximation for the root of an estimating equation is presented. Series approximation methods in Bayesian inference and resampling will also be discussed.

9.1. Applications to Estimating Equations

Many statistical estimates are constructed by finding the root of a function depending on the value of a parameter and the unknown data. For instance, maximum likelihood estimators are found by equating the sum of the derivatives of the log likelihood contributions for the various items sampled to zero. Daniels (1983) provides saddlepoint approximations to the distributions of roots of estimating equations.

Suppose $\boldsymbol{X}_j$ are independent and identically distributed according to a distribution indexed by a scalar parameter θ, for $j = 1, ..., n$. Let $M(\boldsymbol{x}, \theta)$ be a function such that $\mathrm{E}[M(\boldsymbol{X}, \theta)] = 0$ for all θ, and such that M is decreasing in θ. Then estimate θ as the root $\hat{\theta}$ of $\sum_{j=1}^{n} M(\boldsymbol{X}_j, \hat{\theta}) = 0$. The estimator $\hat{\theta}$ is known as an M-estimate. Serfling (1980) discusses the asymptotic properties of these estimators. Under general conditions, one can, for instance, demonstrate asymptotic normality.

Define $\bar{M}_n(a) = \sum_{j=1}^{n} M(\boldsymbol{X}_j, a)/n$; then by the monotonicity of M,

$$\mathrm{P}\left[\hat{\theta} > a\right] = \mathrm{P}\left[\bar{M}(a) > 0\right],$$

and the problem of tail areas for $\hat{\theta}$ reduces to finding probabilities that $\bar{M}_n(a) > 0$ as a varies. The question of uniformity of error bounds is more difficult in this case than in the case of cumulative distribution function approximation, since as a varies the distribution whose cumulative distribution function at 0 is approximated changes. Uniformity properties in this case are not as widely studied as they are in the case of tail probabilities from one distribution.

Let $\mathcal{K}(\beta; a)$ be the cumulant generating function of $\bar{M}_1$. This may be difficult to calculate. Then

$$\mathrm{P}\left[\hat{\theta} > a\right] - \mathrm{P}\left[\bar{M} > 0\right] = \frac{1}{2\pi i}\int_C \exp(n\mathcal{K}(\beta; a))\beta^{-1}\, d\beta, \qquad (167)$$

where the path C is chosen to have a positive real part at each point. Taking the derivative with respect to a, the density of $\hat{\theta}$ is $-(n/2\pi i)\int_C \exp(n\mathcal{K}(\beta; a))\mathcal{K}_2(\beta; a) \times \beta^{-1}\, d\beta$. Here $\mathcal{K}_2(\beta; a)$ represents the derivative of $\mathcal{K}$ with respect to a. Since $\mathcal{K}(0; a) = 0$ for all a, the singularity at zero is removable. Approximating this integral using steepest descent methods gives the density approximation

$$-\sqrt{\frac{n}{2\pi\mathcal{K}_{11}(\hat{\beta}; a)}}\mathcal{K}_2(\hat{\beta}; a)\hat{\beta}^{-1}\exp(n\mathcal{K}(\hat{\beta}; a)).$$

Here $\mathcal{K}_{11}(\beta; a)$ represents the second derivative of $\mathcal{K}$ with respect to a. Alternatively, the Robinson and Lugannani and Rice approximations may be used directly on (167) to calculate tail probabilities.

These methods may be applied to maximum likelihood estimators, giving approximations to the unconditional distribution of the estimator; recall that the Barndorff–Nielsen's formula in general gives an approximation to distribution of the maximum likelihood estimator conditional on an approximate ancillary statistic.

As a second example, consider the problem of drawing inference on the ratio ρ of two means from the ratio of the observations. If X and Y are exponential random variables with means μ_X and μ_Y, then the joint likelihood is $-X/\mu_X - Y/\mu_Y - \log(\mu_X) - \log(\mu_Y)$. The maximum likelihood estimator $\hat{\rho}$ of μ_Y/μ_X is Y/X, and hence satisfies $M(X, Y, \hat{\rho}) = Y - \hat{\rho}X = 0$; note that $\mathrm{E}\left[M(X, Y, \rho)\right] = 0$. Then $\mathcal{K}(\beta, a) = -\log(1 - \mu_X\beta) - \log(1 - \rho\mu_X\beta)$. Partial derivatives with respect to β are $\mathcal{K}_1(\beta, a) = \mu_X/(1 - \mu_X\beta) - \mu_X\hat{\rho}/(1 + \mu_X\hat{\rho}\beta)$, and

$$\mathcal{K}_{11}(\beta, \hat{\rho}) = \mu_X^2(\mu_X\beta - 1)^{-2} + \mu_X^2\hat{\rho}^2(1 + \mu_X\hat{\rho}\beta)^{-2}.$$

Then $\hat{\beta} - (-1 + \hat{\rho})/(2\mu_X\hat{\rho})$, and so $\mathcal{K}_{11}(\hat{\beta}, \hat{\rho}) = 8\mu_X^2\hat{\rho}^2/\left(1 + \hat{\rho}\right)^2$. Also, $\mathcal{K}_2(\hat{\beta}, \hat{\rho}) = -2(1 + \hat{\rho})^{-1}\mu_X$. The approximation to the density of $\hat{\rho}$ has the form

$$2^{2n-1}\sqrt{n/\pi}\,\hat{\rho}^{n-1}\left(1 + \hat{\rho}\right)^{-2n}.$$

Field and Ronchetti (1990) discuss saddlepoint approximations for estimating equations in detail and provide many additional references.

9.2. Applications to Bayesian Methods

Consider the Bayesian paradigm in which π is a prior distribution on a parameter space $\mathfrak{P} \subset \mathfrak{R}^k$. Data represented by the random vector $\boldsymbol{T}$ is generated according to a density $f_T(\boldsymbol{t}, \boldsymbol{\theta})$ depending on $\boldsymbol{\theta} \in \mathfrak{P}$. Let $\ell(\boldsymbol{\theta}; \boldsymbol{t}) = \log(f_T(\boldsymbol{t}, \boldsymbol{\theta}))$ be the corresponding likelihood. The posterior density for $\boldsymbol{\theta}$ conditional on $\boldsymbol{T}$ is then

$$\frac{\exp(\ell(\boldsymbol{\theta}; \boldsymbol{t}))\pi(\boldsymbol{\theta})}{\int_{\mathfrak{P}} \exp(\ell(\boldsymbol{\theta}; \boldsymbol{t}))\pi(\boldsymbol{\theta})\, d\boldsymbol{\theta}}.$$

An approximation to posterior moments

$$\int_{\mathfrak{P}} g(\boldsymbol{\theta})\exp(\ell(\boldsymbol{\theta}; \boldsymbol{t}))\pi(\boldsymbol{\theta})\, d\boldsymbol{\theta} \Big/ \int_{\mathfrak{P}} \exp(\ell(\boldsymbol{\theta}; \boldsymbol{t}))\pi(\boldsymbol{\theta})\, d\boldsymbol{\theta}$$

for a function $g(\boldsymbol{\theta})$ is desired. For example, g might represent the value of one component of $\boldsymbol{\theta}$ when a marginal posterior distribution is desired. The method to be employed here is known as Laplace's method, and is analogous to saddlepoint methods (Kass, 1988).

The idea in one dimension is as follows: Recall that saddlepoint methods involve an approximation to an integral of $\exp(t\beta - \mathcal{K}_T(\beta))$ over one of a variety of possible complex paths. The path chosen ran through $\hat{\beta}$, the point along the real axis where $\exp(t\beta - \mathcal{K}_T(\beta))$ was at a minimum, and ran along the path of steepest descent so that the value of the integrand would be as small as possible outside of a region where $t\beta - \mathcal{K}_T(\beta)$ is roughly quadratic and the integrand could be well-approximated by a normal characteristic function. The density approximation is achieved by doing the approximating integral analytically.

The Laplace method applies to integrals of the form

$$\int_{\mathfrak{R}} \exp(-\mathcal{B}(\theta))d\theta. \tag{168}$$

The path of integration here is fixed, but the point at which to center the approximation is not. The mode $\hat{\theta}$ of the integrand is chosen, and the integrand is replaced by $c_1 \exp(-c_2^2(\theta - \hat{\theta})^2/2$. Since

$$\int_{\mathfrak{R}} (2\pi)^{-1/2} c_2 \exp(-c_2^2(\theta - \hat{\theta})^2/2)\, d\theta = 1,$$

the integral of the approximating integrand is $c_1\sqrt{2\pi}/c_2$. Here $c_1 = \exp(-\mathcal{B}(\hat{\theta}))$ and $c_2^2 = \mathcal{B}''(\hat{\theta})$, yielding the approximation

$$\sqrt{2\pi/\mathcal{B}''(\hat{\theta})}\exp(-\mathcal{B}(\hat{\theta})). \tag{169}$$

In the $\mathfrak{k}$-dimensional case the approximation is $(2\pi)^{\mathfrak{k}/2} \det \left[\mathcal{B}''(\hat{\theta})\right]^{-1/2} \exp(-\mathcal{B}(\hat{\theta}))$.

Tierney and Kadane (1986) apply these methods to the problem of approximating moments of a function g of the parameter. This involves approximating two integrals of the form (168), with the functions $\mathcal{B}(\theta) = -\log(\pi(\theta)) - \ell(\theta, \boldsymbol{T})$ and $\mathcal{B}^*(\theta) = -\log(g(\theta)) - \log(\pi(\theta)) - \ell(\theta, \boldsymbol{T})$. This moment is approximated by taking the ratio of the two integral approximations (169). If $\hat{\theta}$ minimizes $\mathcal{B}$ and $\hat{\theta}^*$ minimizes $\mathcal{B}^*$, the approximation is $\exp(\mathcal{B}(\hat{\theta}) - \mathcal{B}^*(\hat{\theta}^*))\sqrt{\det\left[\mathcal{B}''(\hat{\theta})\right] / \det\left[\mathcal{B}^{*\prime\prime}(\hat{\theta}^*)\right]}$.

Asymptotics come into play when the distribution of $\boldsymbol{T}$ depends on a parameter n indexing sample size, and when n increases; fortunately the Laplace approximation works well for moderate values of n. As n increases, ℓ_n becomes the dominant term in both $\mathcal{B}$ and $\mathcal{B}^*$. Formal asymptotic treatments of these integrals rest on the fact that $\ell_n(\theta, \boldsymbol{T}) = O_p(n)$ and can be expressed as $n\ell(\theta) + h_n(\theta) - \log(\pi(\theta))$. By the weak law of large numbers, this holds if the likelihood arises from independent and identically distributed random variables.

Using an analogue to the method of steepest descent, these methods can be demonstrated accurate to $O(n^{-2})$; see Kass, Tierney, and Kadane (1990).

When these methods are applied to deriving the marginal posterior of θ_1, the first component of $\boldsymbol{\theta}$, one proceeds as follows: First, fix a value for θ_1 at which to evaluate the posterior. Let $\tilde{\boldsymbol{\theta}}(\theta_1)$ solve

$$\ell_n'(\tilde{\boldsymbol{\theta}}) + \pi'(\tilde{\boldsymbol{\theta}})/\pi(\tilde{\boldsymbol{\theta}}) = 0 \tag{170}$$

where differentiation is with respect to all but the first components of $\boldsymbol{\theta}$, and the first component is fixed. Let $\hat{\boldsymbol{\theta}}$ solve (170) for all first order derivatives without θ_1 fixed. The posterior approximation is the ratio of the appropriate integral approximation, and can be expressed as

$$\frac{(2\pi)^{-1/2} \det\left[\ell_n^{(-)\prime\prime}(\tilde{\boldsymbol{\theta}}(\theta_1))\right]^{-1/2} \exp(\ell_n(\tilde{\boldsymbol{\theta}}(\theta_1)) + \log(\pi(\tilde{\boldsymbol{\theta}}(\theta_1))))}{\det\left[\ell_n''(\hat{\boldsymbol{\theta}})\right]^{-1/2} \exp(\ell_n(\hat{\boldsymbol{\theta}}) + \log(\pi(\hat{\boldsymbol{\theta}}))},$$

where $\ell_n^{(-)\prime\prime}$ denotes the Hessian matrix with entries corresponding to the first component removed. When simplified this yields

$$(2\pi)^{-1/2}\left(\det\left[\ell_n''(\hat{\boldsymbol{\theta}})\right] / \det\left[\ell_n^{(-)\prime\prime}(\tilde{\boldsymbol{\theta}}(\theta_1))\right]\right)^{1/2} \exp(\ell_n(\tilde{\boldsymbol{\theta}}(\theta_1)) - \ell_n(\hat{\boldsymbol{\theta}}))\frac{\pi(\tilde{\boldsymbol{\theta}}(\theta_1))}{\pi(\hat{\boldsymbol{\theta}})}. \tag{171}$$

Note the parallels between this and the double saddlepoint conditional density approximation (124).

Davison (1986) uses similar methods to approximate the unmarginalized posterior

$$\pi(\boldsymbol{\theta}|\boldsymbol{Y}) = \frac{f_{\boldsymbol{Y}}(\boldsymbol{Y};\boldsymbol{\theta})\pi(\boldsymbol{\theta})}{\int f_{\boldsymbol{Y}}(\boldsymbol{Y};\boldsymbol{\theta})\pi(\boldsymbol{\theta})\, d\boldsymbol{\theta}}.$$

Laplace's methods are again used to approximate the denominator as:

$$(2\pi)^{k/2} \det\left[\ell_n''(\hat{\boldsymbol{\theta}})\right]^{-1/2} \exp(\ell_n(\hat{\boldsymbol{\theta}}) + \log(\pi(\hat{\boldsymbol{\theta}})),$$

where $\hat{\boldsymbol{\theta}}$ solves (170), and ratio is approximated as

$$(2\pi)^{-k/2} \det\left[\ell_n''(\hat{\boldsymbol{\theta}})\right]^{1/2} \exp(\ell_n(\boldsymbol{\theta}) - \ell_n(\hat{\boldsymbol{\theta}}))\frac{\pi(\boldsymbol{\theta})}{\pi(\hat{\boldsymbol{\theta}})}(1 + O(n^{-1})).$$

Note the formal equivalence to Barndorff–Nielsen's formula (149); in this case, however, it is $\boldsymbol{\theta}$ and not $\hat{\boldsymbol{\theta}}$ that is random. Wong and Li (1992) apply asymptotic methods to the problem of generating marginal distributions for marginal distributions of a non–linear function of the parameter vector.

As an example, Tierney and Kadane (1986) consider the analysis of a survival model for the Stanford heart transplant data; following Turnbull, Brown, and Hu (1974), they divide patients into groups $i = 1$ if they receive a transplant and $i = 2$ if they receive no transplant and model the survival time without transplant T_{ij} for individual j in group i as exponential with parameter ϕ_{ij}, and for those in group 1, the survival times after replacement Z_j as exponential $\theta\phi_{ij}$. Of particular interest is the posterior distribution for θ. Tierney and Kadane (1986) apply a uniform prior to this data, and compare the results of the approximation (171) to the values given by 60 point Gaussian quadrature and the normal approximation. They find close agreement between (171) and quadrature, and substantial disagreement between these and the normal approximation.

9.3. Applications to Resampling Methods

Suppose that $\{X_i\}$ are independent and identically distributed with unknown cumulative distribution function F. One desires to estimate a characteristic θ of F; for example, $\theta(F)$ might be the mean $\int x\, dF$. Often times one uses as an estimate $\theta(\hat{F})$, the corresponding characteristic of $\hat{F}$, the empirical cumulative distribution function; $\hat{F}(x) = \frac{1}{n}\#\{X_i \leq x\}$. The cumulative distribution function of the difference of the resulting estimate from the true value is given by $G(y) = \mathrm{P}_F\left[\theta(\hat{F}) - \theta(F) \leq y\right]$ and inference on $\theta(F)$ would ideally be done using $G(y)$. Usually $G(y)$ is unavailable, and the estimated cumulative distribution function $\hat{G}(y) = \mathrm{P}_{\hat{F}}\left[\theta(\hat{F}) - \theta(F) \leq y\right]$ is used in its place. Calculating quantities depending on $\hat{G}(y)$ exactly is often impractical; integrals with respect to $d\,\hat{G}(y)$ are often approximated using sampling techniques. Davison and Hinkley (1988) suggest instead using saddlepoint methods. The cumulant generating function associated with $\hat{F}$ is $\hat{\mathcal{K}}(\beta) = \log(\sum_{i=1}^{n} \exp(\beta X_i)/n)$.

Confidence intervals for $\theta(F)$ might be constructed as follows: Ideally one desires statistics $U(\boldsymbol{X})$ and $L(\boldsymbol{X})$ such that

$$\mathrm{P}_F\left[\theta(\hat{F}) - \theta(F) \leq L(\boldsymbol{X})\right] = \alpha_1, \quad \mathrm{P}_F\left[\theta(\hat{F}) - \theta(F) \geq U(\boldsymbol{X})\right] = \alpha_2$$

for all F under consideration; this implies that $(\theta(\hat{F}) + L(\boldsymbol{X}), \theta(\hat{F}) + U(\boldsymbol{X}))$ forms a $1 - \alpha_1 - \alpha_2$ confidence interval. As an approximation to this method, one might use L and U such that

$$\mathrm{P}_{\hat{F}}\left[\theta(\hat{F}) - \theta(F) \leq L\right] = \alpha_1, \quad \mathrm{P}_{\hat{F}}\left[\theta(\hat{F}) - \theta(F) \geq U\right] = \alpha_2. \tag{172}$$

The construction (172) may exhibit poor coverage properties, since L and U are chosen for one possible cumulative distribution function, the empirical one, rather than for all distribution functions under consideration. One solution, proposed by

Efron (1982) for estimating the mean, is to embed $\hat{F}$ in the exponential family $\hat{F}(x,\theta)$ putting mass $\exp(X_i\eta)/\sum_j \exp(X_j\eta)$ on X_i. Here $\eta(\theta)$ is chosen so that

$$\sum_j X_j \exp(X_j\eta(\theta))/\sum_j \exp(X_j\eta(\theta)) = \theta.$$

The resulting likelihood is called the exponential empirical likelihood; Jing and Wood (1996) discuss its properties with respect to Bartlett's correction. Here then, $\theta(\hat{F}) = \bar{X}$. Within this exponential family, the sample mean is sufficient for θ, and $1-\alpha_1-\alpha_2$ confidence intervals for θ can be taken as those θ such that

$$\mathrm{P}_{F(.,\theta)}\left[\theta(\hat{F}) \geq t\right] = \alpha_1, \quad \mathrm{P}_{F(.,\theta)}\left[\theta(\hat{F}) \leq t\right] = \alpha_2. \tag{173}$$

Here t is the observed characteristic value. One might calculate these probabilities under the distribution $\hat{F}(.;\theta)$ rather than F; Davison and Hinkley (1988) suggest approximating these quantities using saddlepoint methods. The cumulant generating function associated with $\hat{F}(.;\theta)$ is

$$\hat{\mathcal{K}}(\beta;\theta) = \log(\sum \exp[(\beta+\eta(\theta))X_i]) - \log(\sum \exp[\eta(\theta)X_i])$$

and

$$\hat{\mathcal{K}}'(\beta;\theta) = \sum X_i \exp[(\beta+\eta(\theta))X_i]/\sum \exp[(\beta+\eta(\theta))X_i].$$

The saddlepoint $\hat{\beta}$ is defined by $\hat{\mathcal{K}}'(\hat{\beta};\theta) = \bar{X}$. Hence $\hat{\beta}+\eta(\theta) = 0$. The Lugannani and Rice approximation (89) can then be used to approximate tail areas and hence approximate L and U in (173). Tail areas then have the form $1-\Phi(\hat{\omega})+\phi(\hat{\omega})[1/\hat{z} - 1/\hat{\omega}]$; here $\hat{z} = -\eta(\theta)\sqrt{\Sigma(X_i-\bar{X})^2}$ and $\hat{\omega} = \sqrt{2n[\log(\Sigma\exp(X_i\eta(\theta)/n)) - \eta(\theta)\bar{X}]}$, with $\hat{\omega}$ given the same sign as $\hat{z}$.

A quantity more nearly pivotal may be created by removing the effect of properties of the underlying distribution of higher order than the mean is approximate studentization. Suppose H is the cumulative distribution function of $\theta(\hat{F}) - \theta(F)$. Then $H^{-1}(\theta(\hat{F}) - \theta(F))$ is distributed uniformly on the unit interval. Confidence intervals might then be generated by comparing $\hat{H}^{-1}(\theta(\hat{F})-\theta(F))$ to a uniform distribution, where $\hat{H}$ denotes the cumulative distribution function of $\theta(\hat{F})-\theta(\hat{F})$. Although in general H would be analytically intractable even if $\hat{F}$ were not, in practice both are intractable, and $\hat{H}$ might be approximated through a two-step sampling scheme known as double-bootstrapping. Davison and Hinkley (1988) suggest using saddlepoint methodology here too to avoid sampling. Theoretical justifications for much of this is found in Wang (1990a).

Similarly, Davison and Hinkley (1988) suggest using the empirical cumulant generating function $\hat{\mathcal{K}}$ to do non-parametric inference on M-estimates.

Related are non-parametric permutation tests. Suppose that $X_1, ..., X_n \sim F(x)$, and one wishes to test whether F is symmetric about θ_0. The test is usually performed conditional on the size of deviations from the proposed symmetry point.

Let

$$a_i = (X_i - \theta_0)/\sqrt{\sum_j (X_i - \theta_0)^2} \text{ and } T = \sum a_i. \tag{174}$$

The distribution of T conditional on the absolute values of the a_i is now that of $\sum V_i |a_i|$, where the V_i take the values of positive and negative 1 with probability half each. Asymptotic techniques can then be used to construct tests and generate p-values. Construction of the cumulant generating function is straight forward.

Suppose that

$$X_1, \ldots, X_{n_B} \sim F(x - \theta_1)$$
$$X_{n_B+1}, \ldots, X_{n_B+n_A} \sim F(x - \theta_2)$$

and we wish to test whether $\theta_1 - \theta_2 = \delta_0$. Here set $a_k = (Y_k - \bar{Y})/\sqrt{\sum_{i=1}^{n_B+n_A}(Y_i - \bar{Y})^2}$ where $Y_k = X_k - \delta_0$ if $k \leq n_B$ and X_k otherwise. Inference is conditional on the a_k; here the test statistic is $\sum_{k=1}^{n_B} a_k$ whose conditional distribution is that of $\sum_{k=1}^{n_B} a_{S_k}$ where S_j are n_B random selections without replacement from $1, \ldots, n_B + n_A$. Robinson (1982) gives an approximation to the cumulant generating function and shows that the tail approximation (86) works well here.

Here the cumulative distribution function has jumps of regular size, since the underlying distribution has atoms all equally likely. Unfortunately the spacing between atoms is non-uniform, and in general the limit distribution will be a singular distribution but not a lattice distribution. The most straight-forward continuity correction is half the distance to the neighboring atom. Because finding the next larger atom might be quite difficult, Robinson suggests an approximate continuity correction using the average atom separation and the constant size of cumulative distribution function jumps.

9.4. Applications to Efficiency

In this section I present expansions of the form (1) for efficiencies of a method of inference. Suppose that two statistical procedures are considered, and that k_n observations are required to give the second procedure the same precision as is realized with n observations from the first procedure; then the relative efficiency of these two procedures defined to be k_n/n, and the asymptotic relative efficiency and the asymptotic deficiency are defined to be the first–order and second–order terms respectively in the expansion of k_n/n:

$$k_n/n = \text{ARE} + \text{ADEF}/n + o(n^{-1}). \tag{175}$$

Here equality of precision may refer to equality of mean square errors of two estimates, or powers of two tests for a similar alternative. The first procedure is preferable if ARE > 1, and the second procedure is preferable if ARE < 1. Fisher (1925) considered discrimination between estimation procedures when the asymptotic relative efficiency is unity. In this case the first procedure is preferable if ADEF > 0, and the second procedure is preferable if ADEF < 0. Relation (175) implies that ADEF $= \lim_{n\to\infty} k_n - n$. Hodges and Lehmann (1970) define the deficiency of the two procedures to be $k_n - n$; the asymptotic deficiency of the second procedure relative to the first is then ADEF, when this limit exists. For further details and references, see Kolassa (1996c). Consider assessing an asymptotically normal and asymptotically unbiased estimator of a parameter. Take as a definition of the efficiency of this estimator to be the ratio of the Cramér–Rao lower bound for its variance to the actual achieved variance. Kolassa (1996c) provides alternate definitions and examples. Estimates with efficiency closer to one are preferable to those with lower efficiency. Estimators whose asymptotic efficiency is unity are called first–order efficient.

One might also define the efficiency of one estimator relative to another to be the inverse of the ratio of their variances, and the asymptotic relative efficiency to be the limit of this inverse ratio. When the two estimators have variances approximately proportional to n in large samples, this definition of asymptotic relative efficiency coincides with the definition in terms of relative sample sizes needed to give equivalent precision.

As a simple example, consider the problem of estimating a mean of a population with finite variance, using a sample of n independent observations. If one procedure estimates the mean as the sample mean, and the second procedure estimates the mean as the sample mean with the first observation ignored, then $k_n = n + 1$, the relative efficiency is $(n + 1)/n$, and the asymptotic relative efficiency is 1. The deficiency is then 1 for all values of n, and so the asymptotic deficiency is also 1.

Fisher (1925) argues heuristically that maximum likelihood estimators are first–order efficient, with variances, to first order, given by the Fisher information in the whole sample, and that loss in efficiency incurred by other estimators might be measured by the correlation of these other estimators to the maximum likelihood estimator, or alternately by the differences between the whole sample information and the information in the sampling distribution of the estimator. Other authors have made these claims rigorous, and some of these results will be reviewed below. Wong (1992) presents a more thorough rigorous review. Rao (1962, 1963) uses the correlation between estimators to build a definition for second order efficiency which is equivalent to the Fisher information difference, under certain regularity conditions.

Higher–order asymptotic expansions for the mean squared error for the maximum likelihood estimator can be generated. Since expansions for the mean squared error are related to expansions for the information content of the maximum likelihood estimator, and the information expansion is simpler, the information expansion will be considered first. Efron (1975) uses methods from differential geometry to define the statistical curvature γ_θ of an inference problem, at a potential parameter value θ_0, to be rate of change in $b(\theta) = \left(1 - \mathrm{Cor}\left[\dot{l}(\theta, \boldsymbol{X}), \dot{l}(\theta_0, \boldsymbol{X})\right]^2\right)^{1/2}$ per unit change in the arclength, evaluated at $\theta = \theta_0$. The arclength between θ_0 and θ is defined by $(\mathrm{E}_\theta\left[U(\theta_0, \boldsymbol{X})\right] - \mathrm{E}_{\theta_0}\left[U(\theta_0, \boldsymbol{X})\right])/\mathrm{Var}_{\theta_0}\left[U(\theta_0, \boldsymbol{X})\right]^{1/2}$. The quantity $b(\theta)$ has the interpretation as the unexplained fraction of standard deviation of $U(\theta, \boldsymbol{X})$

given $U(\theta_0, \boldsymbol{X})$. This curvature is related to the loss of efficiency when inference procedures designed for local alternatives are applied globally. Consider a family of distributions on a sample space $\mathfrak{X}$ parameterized by θ taking values in $\mathfrak{P} \subset \mathfrak{R}$, and suppose that $\boldsymbol{X} \in \mathfrak{X}^n$ is a vector of n independent and identically distributed variables X_j. Let $\ell_n(\theta; \boldsymbol{X})$ be the log likelihood for $\boldsymbol{X}$. Let $\ddot{\ell}_n(\theta, \boldsymbol{X})$ be the second derivative of the log likelihood with respect to θ. Let $\imath_n(\theta) = -\mathrm{E}_\theta\left[\ddot{\ell}_n(\theta, \boldsymbol{X})\right]$ be the Fisher information in the sample $\boldsymbol{X}$, let $\imath_n^{\hat{\theta}}(\theta)$ be Fisher's information for the sampling distribution of $\hat{\theta}$, the maximum likelihood estimator for θ, and let $\imath_1(\theta)$ be the Fisher information for X_1. If γ_θ^1 is the curvature defined for the distribution of a random variable X_1, and γ_θ^n is the curvature calculated for the distribution of $\boldsymbol{X}$, then $\gamma_\theta^1 = \gamma_\theta^n/\sqrt{n}$. One may show that $\lim_{n\to\infty}(\imath_n(\theta) - \imath_n^{\hat{\theta}}(\theta)) = \imath_1(\theta)(\gamma_\theta^1)^2$, and hence $\imath_n^{\hat{\theta}}(\theta)/n = \imath_1(\theta) - \imath_1(\theta)\gamma_\theta^1/n + o(n^{-1})$, giving an asymptotic expansion for the average information is contained in a maximum likelihood estimator. Efron (1975) also produced an asymptotic expansion for the variance of the maximum likelihood estimator at a parameter value θ_0, which contains the statistical curvature and additional terms involving the curvature of the bias of the result of one scoring iteration, and the bias of the maximum likelihood estimator at θ_0. These terms are all of size $O(n^{-2})$, and the error is of size $o(n^{-2})$. Asymptotic comparisons of powers of families of tests having exactly or approximately the same significance level have been examined by many authors. Generally these investigations have considered the problem of testing a null hypothesis that a statistical parameter θ takes a null value, of the from $H_0 : \theta = \theta_0$, using two competing tests T_n^1 and T_n^2, indexed by a parameter n generally indicating sample size. Their critical values t_n^1 and t_n^2 satisfy

$$\mathrm{P}\left[T_n^i \geq t_n^i; H_0\right] = \alpha \text{ for } i = 1, 2. \tag{176}$$

Single–number measures of efficiency often times compare powers of tests whose sizes, exactly or approximately, are fixed and identical. For consistent tests, and a fixed alternative hypothesis, distribution functions for the test statistics, or asymptotic approximations to these distribution functions, indicate an identical first–order asymptotic power of unity. Distinguishing between such tests, then, requires a local measure of relative efficiency such as Pitman efficiency, which is the ratio of sample sizes necessary to give the same power against a local alternative. That is, alternatives of the from $H_A : \theta = \theta_0 + \epsilon/c_n$, where $c_n \to \infty$, are considered, and the limit $\lim_n(k_n/n)$ is desired, where

$$\mathrm{P}\left[T_n^1 \geq t_n^1; H_A\right] = \mathrm{P}\left[T_{k_n}^2 \geq t_n^2; H_A\right]. \tag{177}$$

Hodges and Lehmann (1970) apply their concept of deficiency to the problem of comparing tests in cases in which sizes can be calculated exactly and in which the asymptotic relative efficiency is unity.

Often times exact expressions for the probabilities in (176) and (177) are unavailable. In such cases the critical value, as well as the power, usually must be approximated. Pfanzagl (1980) notes that asymptotic comparisons of power are only interesting when significance levels of the tests agree to the same asymptotic order, and achieves this equality of size through a process of studentization, in the presence of nuisance parameters. Such equality of size might be obtained using a Cornish–Fisher expansion to calculate the critical value for the test. Albers, Bickel, and van Zwet (1976) apply Edgeworth series to calculate the powers of nonparametric tests. The primary difficulty in such cases arises from the discrete nature of the distributions to be approximated. Pfaff and Pfanzagl (1985) present applications of Edgeworth and saddlepoint expansions to the problem of approximating power functions for test statistics with continuous distributions; they find that the Edgeworth series is more useful for analytic comparisons of power, while the saddlepoint methods give more accurate numerical results. Kolassa (1996c) gives further references.

Asymptotic expansions of cumulative distribution functions can also be used to calculate asymptotic relative efficiencies and deficiencies. Those applications discussed here are under local alternatives. First–order approximations are generally sufficient to calculate asymptotic relative efficiencies; deficiency calculations generally require that second order terms be included as well. Taniguchi (1991) calculates these powers to examine cases when the deficiency is zero and a third–order counterpart of (175) is required; comparisons may then be based on the final coefficient.

9.5. Exercises

1. Calculate the tail probability approximations for the root of an estimating equation by approximating the inversion integral (167) using
   a. The method of Robinson (§5.2)
   b. The method of Lugannani and Rice (§5.3).
2. Consider the following changes in systolic blood pressure after administration of a drug: -9, -4, -21, -3, -20, -31, -17, -26, -26, -10, -23, -33, -19, -19, -23 (Cox and Snell, 1981, p. 72). Assume that these observations come from a symmetric distribution with point of symmetry θ.
   a. Approximate the 5% and 95% points t_L and t_U of the null distribution of the statistic T in (174), assuming $\theta = \theta_0$, using perhaps (89).
   b. Invert the statement $\mathrm{P}\left[t_L \leq T(\theta) \leq t_U\right] = 90\%$ to generate a 90% confidence interval for θ.

10
Computational Aids

This chapter contains code for doing some of the calculations presented here using the computer algebra package *Mathematica*. Lines surrounded by parenthesis star are comments. Material here is generally in the same order as it appears in the text, except that generally lattice material in the text was at the end of chapters; here it follows more naturally immediately after the continuous analogues. Code presented here is the minimal code necessary to perform many of the calculations in the text. Andrews and Stafford (1993) and Stafford and Andrews (1993) present more sophisticated *Mathematica* code.

```
(* Stirling's Expansion of §1.2.*)
gn[n_]=Normal[Series[Gamma[n],{n,Infinity,30}]]
uu=Series[Simplify[(gn[n] n^-n Exp[n] n^(1/2))],{n,Infinity,30}]
cl=CoefficientList[Normal[uu/Sqrt[2 Pi]]/.n->1/oon,oon]
n=1
approx=cl
pn=1
Do[pn=pn*n;approx[[i]]=cl[[i]]/pn+approx[[i-1]],{i,2,Length[approx]}]
error=approx-Gamma[n] n^-n Exp[n] n^(1/2)/Sqrt[2 Pi]
(* Define and exponentiate the MGF and CGF expansion of §2.1 for a
standardized sum.  Note rn is 1/sqrt(n).  Give conversions between
moments and cumulants. Give sample Fortran output.*)
K[x_]=Series[Sum[k[i]*x^i*rn^(i-2)/i!,{i,2,6}],{x,0,6}]
expK[x_]=Exp[K[x]]
M[x_]=Series[1+Sum[m[i]*x^i*rn^(i-2)/i!,{i,2,6}],{x,0,6}]
momentrule=Simplify[Solve[Simplify[LogicalExpand[M[x]==expK[x]]],
  {m[2],m[3],m[4],m[5],m[6]}]]
cumulantrule=Simplify[Solve[Simplify[LogicalExpand[M[x]==expK[x]]],
  {k[2],k[3],k[4],k[5],k[6]}]]
FortranForm[cumulantrule]
```

```
( *Write Edgeworth Series without expanding phi, from §3.1.*)
Remove[phi]
edgeworthseries=phi[x]*Series[Sum[PMlist[[i]]
   *hermitepolys[[i]],{i,1,Length[hermitepolys]}],{rn,0,6}]
Edgeworthseries=Phi[x]-phi[x]*Series[Sum[PMlist[[i]]
   *hermitepolys[[i-1]],{i,2,10}],{rn,0,3}]
TeXForm[Edgeworthseries]
(* Calculate Hermite Polynomials of §3.4 by constructing an empty list,
assigning phi to the first element, then recursively differentiating.*)
phipr={,,,,,,,,,,,,,,,,,}
phi[x_]=Exp[-x^2/2]/Sqrt[2*Pi]
phipr[[1]]=phi[x]
Do[phipr[[i+1]]=D[phipr[[i]],x],{i,1,Length[phipr]-1}]
hermitepolys=Simplify[phipr/phi[x]]* ((-1)^(1+Range[Length[phipr]]))
(* Define pseudocumulant g.f. of §3.4 and exponentiate to get
pseudomoments.  We need series to order 18 in x since we need powers of x
to order three times power in rn. *)
PK[x_]=Sum[k[i]*x^i*rn^(i-2)/i!,{i,3,18}]
PM=Expand[Normal[Series[Exp[PK[x]],{x,0,18}]]]
PMlist=CoefficientList[PM,{x}]
(* Do the same thing to get the pseudomoments for the Edgeworth series
with Sheppard-corrected cumulants of §3.16.*)
PE[x_]=Sum[e[2*i]*x^(2*i)*rn^(2*i)/(2*i)!,{i,1,5}]
PM=Expand[Normal[Series[Exp[PK[x]+PE[x]],{x,0,12}]]]
PMSlist=CoefficientList[PM,{x}]
(* Construct the Cornish-Fisher expansion of §3.13 by expanding the
Edgeworth Series about the normal approximation quantile and inverting.
Leave Hermite polynomials unexpanded (except for the first); express the
difference between the Edgeworth inverse and the Normal inverse as a
power series in rn.*)
Clear[phi]; Clear[z]; Clear[Phi]
ES[z_]=(Normal[Phi[z]-Series[Sum[PMlist[[i+1]]*(-1)^(i-1)
   *Derivative[i][Phi][z],{i,2,10}],{rn,0,2}]])
ESS=Simplify[(Expand[Normal[Series[ES[z+a]-Phi[a],{z,0,2}]]//.
   {Derivative[k_][Phi][z_]->Derivative[k-1][phi][z],
    Derivative[k_][phi][z_]->phi[z]*h[k]*(-1)^k,h[0]->1}])/phi[a]]
z=Sum[b[i]*rn^i,{i,1,3}]
ESSS=Expand[Series[ESS,{rn,0,2}]]
cfeqn=Simplify[ESSS==0]
cfrule=Solve[Simplify[LogicalExpand[cfeqn]],{b[1],b[2]}]
Do[cfrule=Simplify[cfrule/.h[i]->hermitepolys[[i+1]]],{i,0,7}]
(* Construct the saddlepoint expansion relative error in terms of invariants
by, as in §4.1, evaluating the Edgeworth series at 0 and adjusting the
coefficients for non unit variance. The coefficients beta are invariants.*)
spre=Simplify[edgeworthseries/.{x->0,k[r_]->beta[that,r] }]
```

```
(* Calculations for the proof that no distributions other than only
normal, inverse Gaussian, and gamma distributions have saddlepoint
approximations exact up to a constant of proportionality, from §4.8.*)
f[x]=(a x+b)^(-3/2)
3*D[D[f[x],x],x]-5 D[f[x],x]^2/f[x]
eq={,,,}
alphas=Expand[CoefficientList[spre,rn]]
dbetarule=Derivative[1,0][beta][that,r_]->Sqrt[K''[that]]*(
  beta[that,r+1]- (r/2)*beta[that,3]*beta[that,r])
eq[[1]]=Expand[24*(D[alphas[[3]],that]/.dbetarule)/Sqrt[K''[that]]]
eq[[2]]=Expand[(D[eq1,that]/.dbetarule)/Sqrt[K''[that]]]
eq[[3]]=Expand[(24*alphas[[3]])^2]
eq[[4]]=Expand[3*384*alphas[[5]]]
TeXForm[eq/.beta[that,r_]->beta[r]]
coefs={20*beta[that,3]/3,8,23/9,1}
TeXForm[coefs/.beta[that,r_]->beta[r]]
TeXForm[Simplify[Sum[coefs[[i]]*eq[[i]],{i,1,4}]]/.beta[that,r_]->beta[r]]
(* Calculations for the Bahadur and Ranga Rao and the Lugannani and Rice
expansions of §5.1 and §5.3.  As discussed in the text, an expansion
for K in terms of t is defined, the change t=exp(u) is performed, the
square root of the rescaled K is taken, the series is reversed, and a
derivative is taken.*)
nt=4
Ks[t_]=Sum[k[i]*(t-that)^i/i!,{i,2,3*nt}]
zrule={that->1/(invzhat*Sqrt[k[2]]),Sqrt[zhat^2]->zhat,
   Sqrt[invzhat^-2]->invzhat^(-1),invzhat*zhat->1}
k[i_]:=b[i]*k[2]^(i/2)/;i>2
dwsq[t_]=Series[Simplify[2*Ks[t]],{t,that,nt}]
dw[t_]=Simplify[Sqrt[dwsq[t]]]//.zrule
tt[dw_]=InverseSeries[dw[t],dw]
aa[dw_]=Simplify[D[tt[dw],dw]/tt[dw]]/.zrule
aa[dw_]=zhat^-1*Simplify[ Normal[Expand[zhat*(aa[dw]+aa[-dw])/2]]/.zrule]
cl=CoefficientList[aa[dw],dw]
Do[cl[[2*j+1]]=Apart[Simplify[cl[[2*j+1]]*((2*j)!)/j!]],
   {j,0,(Length[cl]-1)/2}]
brr=Exp[n*(K[that]-that*x)]/zhat*
   Sum[Simplify[zhat*cl[[2*j+1]]*(-1)^j/(2*n)^j],{j,0,(Length[cl]-1)/2}]
bb[dw_]=Normal[Series[(what+dw)^-1,{dw,0,nt-2}]]
bb[dw_]=Apart[(bb[dw]+bb[-dw])/2]
cc[dw_]=Series[aa[dw]-bb[dw],{dw,0,nt}]
cc[dw_]=Normal[Series[cc[dw],{dw,0,nt},{zhatinv,0,nt}]]
cl=CoefficientList[cc[dw],dw]
Do[cl[[2*j+1]]=Apart[Simplify[cl[[2*j+1]]*(-1)^j*(2*j)!/(2^j*j!)]],
   {j,1,(Length[cl]-1)/2}]
landr=1-Ncdf[what]+Exp[n*(K[that]-that*x)]*
   Sum[cl[[2*j+1]]/n^j,{j,0,(Length[cl]-1)/2}]
```

```
(* Commands to calculate terms in Robinson's expansion of §5.2.
I calculate integrals that are the upper half of the Laplace transform
of the normal density times Hermite polynomials.*)
II[j_]=If[j>0,(hermitepolys[[j]]/.x->0)/Sqrt[2 Pi]-a II[j-1],
   Exp[a^2/2](1-Ncdf[a])]
approx=Exp[-hatw^2/2](II[0]+kappa[3]/6 II[3])
(* Commands to calculate {Skates (1993)}expansion of §5.5.*)
P[x_]=Series[Sum[p[i]*x^i/i!,{i,2,7}],{x,0,7}]
Pp[x_]=D[P[x],x]
p[2]=1
Q[x_]=Series[Sum[q[i]*x^i/i!,{i,0,5}],{x,0,5}]
q[0]=1
X[y_]=Simplify[InverseSeries[Simplify[Sqrt[2*P[y]]]]]/.x->y
g[y_]=Simplify[Q[X[y]] y/Pp[X[y]]]
```

```
(* Now calculate the discrete analogues of §5.6.  Some care must be taken in
the treatment of Sinh, since otherwise the expansion won't simplify usefully.
Introduce new symbols for hyperbolic sine and cosine that Mathematica won't try
to expand except according to the derivative rules below.  These calculations
require definition of nt and tt[x] from above.*)
Derivative[i_][Ss][x_]:=Cc[x]/;OddQ[i]
Derivative[i_][Cc][x_]:=Ss[x]/;OddQ[i]
Derivative[i_][Ss][x_]:=Ss[x]/;EvenQ[i]
Derivative[i_][Cc][x_]:=Cc[x]/;EvenQ[i]
ztr={Ss[that/2]->ztil/(2*Sqrt[k[2]]),Sqrt[ztil^2]->ztil,
   Sqrt[invztil^-2]->invztil^(-1),invztil*ztil->1}
aa[dw_]=Simplify[ReplaceAll[D[tt[dw],dw]/(2*Ss[tt[dw]/2]),
   {Ss[1/(2*invzhat*Sqrt[k[2]])]->Ss[that/2],
   Cc[1/(2*invzhat*Sqrt[k[2]])]->Cc[that/2]}]]
aa[dw_]=zhat^-1*Simplify[Normal[Expand[zhat*(aa[dw]+aa[-dw])/2]]]
cl=CoefficientList[aa[dw],dw]
Do[cl[[2*j+1]]=Apart[Simplify[cl[[2*j+1]]*((2*j)!)/j!]],
   {j,0,(Length[cl]-1)/2}]
brr=Exp[n*(K[that]-that*x)]*
   Sum[Simplify[cl[[2*j+1]]*(-1)^j/(2*n)^j],{j,0,(Length[cl]-1)/2}]
bb[dw_]=Normal[Series[(what+dw)^-1,{dw,0,nt-2}]]
bb[dw_]=Apart[(bb[dw]+bb[-dw])/2]
cc[dw_]=Series[aa[dw]-bb[dw],{dw,0,nt}]
cc[dw_]=Normal[Series[cc[dw],{dw,0,nt},{zhatinv,0,nt}]]
cl=CoefficientList[cc[dw],dw]
Do[cl[[2*j+1]]=Apart[Simplify[cl[[2*j+1]]*(-1)^j*(2*j)!/(2^j*j!)/.ztr]],
   {j,0,(Length[cl]-1)/2}]
landr=1-Ncdf[what]+Exp[n*(K[that]-that*x)]*
   Sum[cl[[2*j+1]]/n^(j+1/2),{j,0,(Length[cl]-1)/2}]
(* Commands for the example of §7.2.*)
exppsi[theta1_,theta2_]=1-theta1-2 theta2+theta2(theta1+theta2)
Psi[theta1_,theta2_]=-Log[exppsi[theta1,theta2]]
tr=Simplify[Solve[D[Psi[theta1,theta2],theta2]==t2,theta2]][[1]]
lr={Log[a_*b_]->Log[a]+Log[b], Log[a_/c_]->Log[a]+-Log[c],
   Log[c_.*Power[a_,b_]]->b Log[a]+Log[c]}
var=Simplify[D[D[Psi[theta1,theta2], theta2],theta2]]
lvar=Log[var] //.lr
ep=Simplify[exppsi[theta1,theta2]/.tr]
K[theta1_]=Simplify[Simplify[(-theta2 t2-Log[ep]-(1/2) Log[var])/.tr]//.lr]
trans=( K[theta1]/.Times[Power[theta1, 2], Power[t2, 2]]->w^2-4)/.Sqrt[w^2]->w
Simplify[Simplify[trans]/.lr]
(* Commands for Reid's Gamma distribution ex. of §8.2.*)
L=(nu/mu)^(n*nu)*Exp[slx*(nu-1)]*Exp[-sx*nu/mu]/(Gamma[nu]^n)
l[mu_,nu_]=n(nu*Log[nu]-nu*Log[mu]+(nu-1)*T2-nu*T1/mu-Log[Gamma[nu]])
Simplify[{D[l[mu,nu],nu],D[l[mu,nu],mu]}]
Simplify[{ {D[D[l[mu,nu],nu],nu],D[D[l[mu,nu],mu],nu]},
  {D[D[l[mu,nu],nu],mu],D[D[l[mu,nu],mu],mu]}} ]
```

```
(* Calculate expected value of log(x) *)
eT2=Integrate[Simplify[Log[x]*L/.{n->1,sx->x,slx->Log[x]}],
   {x,0,Infinity}]
MLEdef={D[l[muhat,nuhat],nuhat]==0,D[l[muhat,nuhat],muhat]==0}
MLErules=Simplify[Solve[MLEdef, {T1,T2}]]
obsinfmat={{D[D[l[mu,nu],mu],mu],D[D[l[mu,nu],nu],mu]},
   {D[D[l[mu,nu],nu],mu],D[D[l[mu,nu],nu],nu]}}
det=Simplify[obsinfmat[[1,1]]*obsinfmat[[2,2]]-obsinfmat[[1,2]]^2]/.
   {mu->muhat,nu->nuhat}
sqrtdet=Sqrt[Simplify[det/.MLErules]][[1]]
exponent=Simplify[((l[mu,nu]-l[muhat,nuhat])/.MLErules)/n][[1]]
TeXForm[Simplify[Exp[n*exponent]*sqrtdet]]
(* Commands to calculate the exact correction factor for the mean of
the Gamma likelihood ratio statistic with fixed shape, from §8.7.*)
Llk[x_,n_]=(tau*n)^(n*alpha)*x^(n*alpha-1)*Exp[-tau x*n]/Gamma[alpha*n]
lrules={
   Log[a_/b_]->Log[a]-Log[b], Log[a_*c_]->Log[a]+Log[c],
   Log[a_*c_/b_]->Log[a]+Log[c]-Log[b], Log[a_^b_]->b*Log[a]}
grules={Gamma[1+x_]->x*Gamma[x],PolyGamma[0,x_]->psi[x]}
llk[x_,n_]=Simplify[Log[Llk[x,n]]]//.lrules
tauhat=Solve[D[llk[x,n],tau]==0,tau]
lrs=Simplify[-2*(llk[x,n]-(llk[x,n]/.tauhat[[1]]))/.lrules]
ex=Integrate[x*Llk[x,n],{x,0,Infinity}]/.grules
elx=Integrate[Log[x]*Llk[x,n],{x,0,Infinity}]
Elrs=Simplify[(lrs/.Log[x]->elx)/.x->ex/.lrules]/.grules
(* Perform calculations for the distribution of the ratio of means
from Daniels (1983) of §9.1.  The first three lines calculate
maximum likelihood estimates.  Moment and Cumulant generating
functions and derivatives are defined, the saddlepoint is calculated,
and the three factors for the approximation are defined. *)
l=-x/mux-y/muy-Log[mux]-Log[muy]
newl:=l/.muy->mux*rho
Solve[{D[newl,mux]==0,D[newl,rho]==0},{mux,rho}]
Kr[t_,rho_]=-Log[1-mux*t]-Log[1+mux*rho*t]
M[t_,rho_]=1/((1-mux*t)*(1+mux*rho*t))
Kr1[t_,rho_]=D[Kr[t,rho],t]
Kr11[t_,rho_]=D[Kr1[t,rho],t]
Kr2[t_,rho_]=D[Kr[t,rho],rho]
rl=Simplify[Solve[D[Kr[t,rho],t]==0,t]][[1]]
FirstSq=Simplify[n/(2*pi*Kr11[t,rho])/.rl]
Secnd=Simplify[-((1/t)*Kr2[t,rho])/.rl]
Third=Simplify[(M[t,rho]^n)/.rl]
that=t/.rl
approx=Expand[Sqrt[FirstSq]*Secnd*Third]//.
   {Sqrt[a_^2*b_.]->a*Sqrt[b],Sqrt[b_./(c_.*a_^2)]->Sqrt[b/c]/a,
   Sqrt[a_^(-2)+b_]->Sqrt[b]/a}
TeXForm[approx]
Remove[that]
```

Bibliography

Albers, W., Bickel, P.J., and van Zwet, W.R. (1976), "Asymptotic Expansions for the Power of Distribution Free Tests in the One-Sample Problem," *Annals of Statistics*, **4**, 108–156.

Albert, A., and Anderson, J.A. (1984), "Maximum Likelihood Estimates in Logistic Regression," *Biometrika*, **71**, 1–10.

Andrews, D.F., and Stafford, J.E. (1993), "Tools for the symbolic computation of asymptotic expansions," *Journal of the Royal Statistical Society Series B*, **55**, 613–627.

Bahadur, R.R., and Ranga Rao, R. (1960), "On deviations of the sample mean," *Annals of Mathematical Statistics*, **31**, 1015–1027.

Bak, J., and Newman, D.J. (1982), *Complex Analysis*, New York: Springer–Verlag.

Barndorff–Nielsen, O. E. (1990a), "A note on the standardized signed log likelihood ratio," *Scandinavian Journal of Statistics*, **17**, 157–160.

Barndorff–Nielsen, O.E., and Cox, D.R. (1979), "Edgeworth and Saddlepoint Approximations with Statistical Applications," *Journal of the Royal Statistical Society Series B*, **41**, 279–312.

Barndorff–Nielsen, O.E., and Cox, D.R. (1984), "Bartlett Adjustments to the Likelihood Ratio Statistic and the Distribution of the Maximum Likelihood Estimator," *Journal of the Royal Statistical Society Series B*, **46**, 483–495.

Barndorff–Nielsen, O.E., and Cox, D.R. (1989), *Asymptotic Techniques for Use in Statistics*, London: Chapman and Hall.

Barndorff–Nielsen, O.E. (1978), *Information and Exponential Families in Statistical Theory*, Chichester: John Wiley and Sons.

Barndorff–Nielsen, O.E. (1980), "Conditionality Resolutions," *Biometrika*, **67**, 293–310.

Barndorff–Nielsen, O.E. (1983), "On a Formula for the Distribution of the Maximum Likelihood Estimator," *Biometrika*, **70**, 343–365.

Barndorff–Nielsen, O.E. (1984), "On Conditionality Resolution and the Likelihood Ratio for Curved Exponential Models," *Scandinavian Journal of Statistics*, **11**, 157–170. Correction (1985).

Barndorff–Nielsen, O.E. (1986), "Inference on Full or Partial Parameters Based on the Standardized Signed Log Likelihood Ratio," *Biometrika*, **73**, 307–322.

Barndorff–Nielsen, O.E. (1990b), "Approximate Interval Probabilities," *Journal of the Royal Statistical Society Series B*, **52**, 485–496.

Bartlett, M.S. (1953), "Approximate Confidence Intervals: II. More than One Unknown Parameter," *Biometrika*, **40**, 306–317.

Bartlett, M.S. (1955), "Approximate Confidence Intervals: III. A Bias Correction," *Biometrika*, **42**, 201–204.

Beek, P. van (1972), "An Application of Fourier Methods to the Problem of Sharpening the Berry-Esseen Inequality," *Zeitshrift fur Wahrscheinlichkeitstheorie und Verwandte Gebiete*, **23**, 187–196.

Berman, E., Chacon, L., House, D., Koch, B.A., Koch, W.E., Leal, J., Løvtrup, S., Mantiply, E., Martin, A.H., Martucci, G.I., Mild, K.H., Monahan, J.C., Sandström, M., Shamsaifar, K., Tell, R., Trillo, M.A., Ubeda, A., and Wagner, P. (1990), "Development of Chicken Embryos in a Pulsed Magnetic Field," *Bioelectromagnetics*, **11**, 169–187.

Bhattacharya, R., and Denker, M. (1990), *Asymptotic Statistics,* Boston: Birkhäuser.

Bhattacharya, R.N., and Rao, R.R. (1976), *Normal Approximation and Asymptotic Expansions,* New York: Wiley.

Billingsley, P. (1986), *Probability and Measure,* New York: Wiley.

Blæsild, P., and Jensen, J.L. (1985), "Saddlepoint Formulas for Reproductive Exponential Models," *Scandinavian Journal of Statistics*, **12**, 193–202.

Bleistein, N. (1966), "Uniform Asymptotic Expansions of Integrals with Stationary Point near Algebraic Singularity," *Communications on Pure and Applied Mathematics*, **19**, 353–370.

Bochner, S., and Martin, W.T. (1948), *Several complex variables,* Princeton: Princeton University Press.

Chambers, J.M. (1967), "On methods of asymptotic approximation for multivariate distributions," *Biometrika*, **54**, 367–383.

Clarkson, D.B., and Jennrich, R.I. (1991), "Computing Extended Maximum Likelihood Estimates for Linear Parameter Models," *Journal of the Royal Statistical Society Series B*, **53**, 417–426.

Cordeiro, G.M., and Paula, G.A. (1989), "Improved likelihood ratio statistics for exponential family nonlinear models (Corr: V78 p935)," *Biometrika*, **76**, 93–100.

Cornish, E.A., and Fisher, R.A. (1937), "Moments and Cumulants in the Specification of Distributions," *International Statistical Review*, **5**, 307–322.

Cox, D.R., and Reid, N. (1987), "Parameter Orthogonality and Approximate Conditional Inference," *Journal of the Royal Statistical Society Series B*, **49**, 1–39.

Cox, D.R. (1980), "Local Ancillarity," *Biometrika*, **67**, 279–86.

Cramér, H. (1925), "On Some Classes of Series Used in Mathematical Statistics," *Proceedings of the Sixth Scandinavian Congress of Mathematicians*, 399–425.

Cramér, H. (1946), *Mathematical Methods of Statistics,* Princeton: Princeton University Press.

Daniels, H.E. (1954), "Saddlepoint Approximations in Statistics," *Annals of Mathematical Statistics*, **25**, 614–649.

Daniels, H.E. (1958), "Discussion of The Regression Analysis of Binary Sequences," *Journal of the Royal Statistical Society Series B*, **20**, 236–238.

Daniels, H.E. (1983), "Saddlepoint approximations for estimating equations," *Biometrika*, **70**, 89–96.

Daniels, H.E. (1987), "Tail Probability Approximations," *Review of the International Statistical Institute*, **55**, 37–46.

Davis, A.W. (1976), "Statistical distributions and multivariate Edgeworth populations," *Biometrika*, **63**, 661–670.

Davison, A. C. (1986), "Approximate predictive likelihood (Corr: V77 p667)," *Biometrika*, **73**, 323–332. Davison, A.C., and Hinkley, D.V. (1988), "Saddlepoint Approximations in Resampling Methods," *Biometrika*, **75**, 417–431.

Davison, A.C. (1988), "Approximate Conditional Inference in Generalized Linear Models," *Journal of the Royal Statistical Society Series B*, **50**, 445–461.

DiCiccio, T., Hall, P., and Romano, J. (1991), "Empirical likelihood is Bartlett-correctable," *Annals of Statistics*, **19**, 1053–1061.

DiCiccio, T.J., Martin, M.A., and Young, G.A. (1993), "Analytical approximations to conditional distribution functions," *Biometrika*, **80**, 781–790.

DiCiccio, T.J., and Martin, M.A. (1991), "Approximations of marginal tail probabilities for a class of smooth functions with applications to Bayesian and conditional inference," *Biometrika*, **78**, 891–902.

Durbin, J. (1980), "Approximations for Densities of Sufficient Estimators," *Biometrika*, **67**, 311–333.

Efron, B. (1975), "Defining the Curvature of a Statistical Problem (with Applications to Second Order Efficiency)," *Annals of Statistics*, **3**, 1189–1242.

Efron, B. (1982), *The Jackknife, the Bootstrap, and Other Resampling Plans,* Philadelphia: Society for Industrial and Applied Mathematics.

Esseen, C.G. (1945), "Fourier Analysis of Distribution Functions," *Acta Mathematica*, **77**, 1–125.

Esseen, C.G. (1956), "A Moment Inequality with an Application to the Central Limit Theorem," *Skandinavisk Aktuarietidskrift*, **39**, 160–170.

Feller, W. (1971), *An Introduction to Probability Theory and its Applications,* New York: Wiley, **II**.

Field, C., and Ronchetti, E. (1990), *Small Sample Asymptotics,* Hayward, CA: Institute of Mathematical Statistics.

Fisher, R.A., and Cornish,E.A. (1960), "The Percentile Points of Distributions Having Known Cumulants," *Technometrics*, **2**, 209–225.

Fisher, R.A. (1925), "Theory of Statistical Estimation," *Proceedings of the Cambridge Philosophical Society,* **22**, 700–725.

Fisher, R.A. (1934), "Two New Properties of Mathematical Likelihood," *Proceedings of the Royal Society of London Series A*, **144**, 285–307.

Fraser, D.A.S., Reid, N., and Wong, A. (1991), "Exponential Linear Models: A Two–Pass Procedure for Saddlepoint Approximation," *Journal of the Royal Statistical Society Series B*, **53**, 483–492.

Fraser, D.A.S. (1968), *The Structure of Inference,* New York: Wiley.

Fraser, D.A.S. (1979), *Inference and Linear Models,* New York: McGraw-Hill.

Frydenberg, M., and Jensen, J.L. (1989), "Is the 'Improved Likelihood Ratio Statistic' Really Improved in the Discrete Case?," *Biometrika*, **76**, 655–661.

Geman, S., and Geman, D. (1984), "Stochastic Relaxation, Gibbs Distributions and the Bayesian Restoration of Images," *IEEE Transactions on Pattern Analysis and Machine Intelligence*, **6**, 721–741.

Hall, P. (1983), "Inverting an Edgeworth Expansion," *Annals of Statistics*, **11**, 569–576.

Hall, P. (1992), *The Bootstrap and Edgeworth Expansion,* New York: Springer.

Haynsworth, E.V., and Goldberg, K. (1965), "Bernoulli and Euler Polynomials, Riemann Zeta Function," *Handbook of Mathematical Functions,* New York: Dover.

Hettmansperger, T.P. (1984), *Statistical Inference Based on Ranks,* Melbourne, FL: Krieger.

Hocking, R.R. (1985), *The Analysis of Linear Models,* Monterey, CA: Brooks/Cole.

Hodges, J.L.Jr., and Lehmann, E.L. (1970), "Deficiency," *Annals of Mathematical Statistics*, **41**, 783–801.

Jacobsen, M. (1989), "Existence and Unicity of MLEs in Discrete Exponential Family Distributions," *Scandinavian Journal of Statistics*, **16**, 335–349.

Jeffreys, H. (1962), *Asymptotic Approximations,* London: Oxford University Press.

Jensen, J. L. (1992), "The modified signed likelihood statistic and saddlepoint approximations," *Biometrika*, **79**, 693–703.

Jensen, J.L. (1988), "Uniform Saddlepoint Approximations," *Advances in Applied Probability*, **20**, 622–634.

Jensen, J.L. (1995), *Saddlepoint Approximations,* Oxford: Oxford Science Publications.

Jing, B.Y., and Wood, A.T.A (1996), "Exponential Empirical Likelihood is not Bartlett Correctable," *Annals of Statistics*, **24**, 365–369.

Kass, R.E., Tierney, L., and Kadane, J.B (1990), "The validity of posterior expansions based on Laplace's method," *Bayesian and Likelihood Methods in Statistics and Econometrics: Essays in Honor of George A. Barnard*, 473–488.

Kass, R.E. (1988), "Discussion of Saddlepoint Methods and Statistical Inference," *Statistical Science*, **3**, 234–236.

Kendall, M.G., Stuart, A., and Ord, J.K. (1987), *Kendall's Advanced Theory of Statistics,* New York: Oxford, **I**.

Kolassa, J.E., and McCullagh, P. (1990), "Edgeworth Series for Lattice Distributions," *Annals of Statistics*, **18**, 981–985.

Kolassa, J.E., and Tanner, M.A. (1994), "Approximate Conditional Inference in Exponential Families Via the Gibbs Sampler," *Journal of the American Statistical Association*, **89**, 697–702.

Kolassa, J.E. (1989), *Topics in Series Approximations to Distribution Functions,* Chicago: Unpublished doctoral dissertation.

Kolassa, J.E. (1991), "Saddlepoint Approximations in the Case of Intractable Cumulant Generating Functions," *Selected Proceedings of the Sheffield Symposium on Applied Probability,* Hayward, CA: Institute of Mathematical Statistics, 236–255.

Kolassa, J.E. (1992), "Confidence Intervals for Thermodynamic Constants," *Geochimica et Cosmochimica Acta*, **55**, 3543–3552.

Kolassa, J.E. (1994), "Small Sample Conditional Inference in Biostatistics," *Proceedings of the Symposium on the Interface,* Fairfax Station, VA: Interface Foundation of North America, **26**, 333–339.

Kolassa, J.E. (1995), "Monte Carlo Sampling in Multiway Contingency Tables," *Fiftieth Session of the ISI, Beijing, China. Invited paper.*

Kolassa, J.E. (1996a), "Infinite Parameter Estimates in Logistic Regression," *Scandinavian Journal of Statistics*, .

Kolassa, J.E. (1996b), "Higher-Order Approximations to Conditional Distribution Functions," *Annals of Statistics*, **24**, 353–365.

Kolassa, J.E. (1996c), "Asymptotics, Higher–Order," *Encyclopedia of Statistical Sciences,* : , .

Kolassa, J.E. (1996d), "Uniformity of Double Saddlepoint Conditional Probability Approximations," .

Kong, F., and Levin, B. (1996), "Edgeworth Expansions for the Conditional Distributions in Logistic Regression Models," *Journal of Statistical Planning and Inference*, **52**, 109–129.

Koning, A.J., and Does, R.J.M.M. (1988), "Approximating the Percentage Points of Simple Linear Rank Statistics with Cornish-Fisher Expansions," *Applied Statistics*, **37**, 278–284.

Konishi, S., Niki, N., and Gupta, A.K. (1988), "Asymptotic Expansions for the Distribution of Quadratic Forms in Normal Variables," *Annals of the Institute of Statistical Mathematics*, **40**, 279–296.

Lawley, D.N. (1956), "A General Method for Approximating to the Distribution of the Likelihood Ratio Criteria," *Biometrika*, **43**, 295–303.

Levin, B. (1990), "The Saddlepoint Correction in Conditional Logistic Likelihood Analysis," *Biometrika*, **77**, 275–285.

Levin, B., and Kong, F. (1990), "Bartlett's Bias Correction to the Profile Score Function is a Saddlepoint Corrrection," *Biometrika*, **77**, 219–221.

Le Cam, L.M. (1969), *Théorie Asymptique de la Décision Statistisque,* Montreal: Université de Montréal.

Lugannani, R., and Rice, S. (1980), "Saddle Point Approximation for the Distribution of the Sum of Independent Random Variables," *Advances in Applied Probability*, **12**, 475–490.

Mardia, K.V. (1970), "Measures of Multivariate Skewness and Kurtosis with Applications," *Biometrika*, **57**, 519–530.

Mathews, J., and Walker, R.L. (1964), *Mathematical Methods of Physics,* New York: W.A. Benjamin, Inc..

McCullagh, P., and Cox, D.R. (1986), "Invariants and Likelihood Ratio Statistics," *Annals of Statistics*, **14**, 1419–1430.

McCullagh, P., and Nelder, J.A. (1989), *Generalized Linear Models,* London: Chapman and Hall.

McCullagh, P., and Tibshirani, R. (1990), "A Simple Method for the Adjustment of Profile Likelihoods," *Journal of the Royal Statistical Society Series B*, **52**, 325–344.

McCullagh, P. (1984), "Local Sufficiency," *Biometrika*, **71**, 233–244.

McCullagh, P. (1986), "The Conditional Distribution of Goodness-of-Fit Statistics for Discrete Data," *Journal of the American Statistical Association*, **81**, 104–107.

McCullagh, P. (1987), *Tensor Methods in Statistics,* London: Chapman and Hall.

Pfaff, T., and Pfanzagl, J. (1985), "On the Accuracy of Asymptotic Expansions for Power Functions," *Journal of Statistical Computation and Simulation*, **22**, 1–25.

Pfanzagl, J. (1980), "Asymptotic Expansions in Parametric Statistical Theory," *Developments in Statistics,* New York: Academic Press, **3**, 1–98.

Prášková–Vizková, Z. (1976), "Asymptotic Expansion and a Local Limit Theorem for a Function of the Kendal Rank Correlation Coeficient," *Annals of Statistics*, **4**, 597–606.

Rao, C.R. (1962), "Efficient Estimates and Optimum Inference Procedures," *Journal of the Royal Statistical Society Series B*, **24**, 47–72.

Rao, C.R. (1963), "Criteria of Estimation in Large Samples," *Sankhyā*, **25**, 189–206.

Rao, R.R. (1960), *Some Problems in Probability Theory,* Calcutta: Unpublished doctoral thesis.

Rao, R.R. (1961), "On the Central Limit Theorem in R_k," *Bulletin of the American Mathematical Society*, **67**, 359–361.

Reid, N. (1988), "Saddlepoint Methods and Statistical Inference," *Statistical Science*, **3**, 213–238.

Reid, N. (1996), "Higher order results on the likelihood ratio statistic under model misspecification," *Canadian Journal of Statistics*, **?**, ?–?.

Robinson, J. (1982), "Saddlepoint approximations for permutation tests and confidence intervals," *Journal of the Royal Statistical Society Series B*, **44**, 91–101.

Rudin, W. (1973), *Functional Analysis,* New York: McGraw-Hill.

Rudin, W. (1976), *Principals of Mathematical Analysis,* New York: McGraw-Hill.

Serfling, R.J. (1980), *Approximation Theorems of Mathematical Statistics,* New York: Wiley.

Skates, S.J. (1993), "On Secant Approximations to Cumulative Distribution Functions," *Biometrika*, **80**, 223–235.

Skovgaard, I.M. (1987), "Saddlepoint Expansions for Conditional Distributions," *Journal of Applied Probability*, **24**, 875–887.

Skovgaard, I.M. (1989), "A Review of Higher Order Likelihood Methods," *Bulletin of the International Statistical Institute: Proceedings of the Forty-seventh Session,* Paris: International Statistical Institute, **III**, 331–351.

Skovgaard, I.M. (1990), *Analytic Statistical Models,* Hayward, CA: Institute of Mathematical Statistics.

Stafford, J.E., and Andrews, D.F. (1993), "A symbolic algorithm for studying adjustments to the profile likelihood," *Biometrika*, **80**, 715–730.

Strawderman, R.L., Cassella, G., and Wells, M.T. (1996), "Practical Small–Sample Asymptotics for Regression Problems," *Journal of the American Statistical Association*, **91**, 643–654.

Taniguchi, M. (1991), "Third-order asymptotic properties of a class of test statistics under a local alternative," *Journal of Multivariate Analysis*, **37**, 223–238.

Tanner, M.A. (1996), *Tools for Statistical Inference,* Heidelberg: Springer–Verlag.

Temme, N.M. (1982), "The Uniform Asymptotic Expansion of a Class of Integrals Related to Cumulative Distribution Functions," *SIAM Journal on Mathematical Analysis*, **13**, 239–252.

Thisted, R.A. (1988), *Elements of Statistical Computing,* New York: Chapman and Hall.

Tierney, L., and Kadane, J.B. (1986), "Accurate Approximations for Posterior Moments and Marginal Densities," *Journal of the American Statistical Association*, **81**, 82–87.

Turnbull, B.W., Brown, B.W., and Hu, M. (1974), "Survivorship Analysis of Heart Transplant Data," *Journal of the American Statistical Association*, **69**, 74–80.

Wallace, D.L. (1958), "Asymptotic Approximations to Distributions," *Annals of Mathematical Statistics*, **29**, 635–654.

Wang, S. (1990a), "Saddlepoint Approximations in Resampling Analysis," *Annals of the Institute of Statistical Mathematics*, **42**, 115–131.

Wang, S. (1990b), "Saddlepoint approximations for bivariate distributions," *Journal of Applied Probability*, **27**, 586–597.

Waterman, R.P., and Lindsay, B.G. (1996), "A Simple and Accurate Method for Approximate Conditional Inference in Linear Exponential Family Models," *Journal of the Royal Statistical Society Series B*, ?, .

Withers, C.S. (1988), "Nonparametric Confidence Intervals for Functions of Several Distributions," *Annals of the Institute of Statistical Mathematics*, **40**, 727–746.

Wong, W.H., and Li, B. (1992), "Laplace expansion for posterior densities of nonlinear functions of parameters," *Biometrika*, **79**, 393–398.

Wong, W.H. (1992), "On asymptotic efficiency in estimation theory," *Statistica Sinica*, **2**, 47–68.

Yarnold, J. (1972), "Asymptotic Approximations for the Probability that a Sum of Lattice Random Vectors Lies in a Convex Set," *Annals of Mathematical Statistics*, **43**, 1566–1580.

Author Index

Index

Lecture Notes in Statistics

For information about Volumes 1 to 51
please contact Springer-Verlag

Vol. 52: P.K. Goel, T. Ramalingam, The Matching Methodology: Some Statistical Properties. viii, 152 pages, 1989.

Vol. 53: B.C. Arnold, N. Balakrishnan, Relations, Bounds and Approximations for Order Statistics. ix, 173 pages, 1989.

Vol. 54: K.R. Shah, B.K. Sinha, Theory of Optimal Designs. viii, 171 pages, 1989.

Vol. 55: L. McDonald, B. Manly, J. Lockwood, J. Logan (Editors), Estimation and Analysis of Insect Populations. Proceedings, 1988. xiv, 492 pages, 1989.

Vol. 56: J.K. Lindsey, The Analysis of Categorical Data Using GLIM. v, 168 pages, 1989.

Vol. 57: A. Decarli, B.J. Francis, R. Gilchrist, G.U.H. Seeber (Editors), Statistical Modelling. Proceedings, 1989. ix, 343 pages, 1989.

Vol. 58: O.E. Barndorff-Nielsen, P. Bl¾sild, P.S. Eriksen, Decomposition and Invariance of Measures, and Statistical Transformation Models. v, 147 pages, 1989.

Vol. 59: S. Gupta, R. Mukerjee, A Calculus for Factorial Arrangements. vi, 126 pages, 1989.

Vol. 60: L. Gyorfi, W. Härdle, P. Sarda, Ph. Vieu, Nonparametric Curve Estimation from Time Series. viii, 153 pages, 1989.

Vol. 61: J. Breckling, The Analysis of Directional Time Series: Applications to Wind Speed and Direction. viii, 238 pages, 1989.

Vol. 62: J.C. Akkerboom, Testing Problems with Linear or Angular Inequality Constraints. xii, 291 pages, 1990.

Vol. 63: J. Pfanzagl, Estimation in Semiparametric Models: Some Recent Developments. iii, 112 pages, 1990.

Vol. 64: S. Gabler, Minimax Solutions in Sampling from Finite Populations. v, 132 pages, 1990.

Vol. 65: A. Janssen, D.M. Mason, Non-Standard Rank Tests. vi, 252 pages, 1990.

Vol 66: T. Wright, Exact Confidence Bounds when Sampling from Small Finite Universes. xvi, 431 pages, 1991.

Vol. 67: M.A. Tanner, Tools for Statistical Inference: Observed Data and Data Augmentation Methods. vi, 110 pages, 1991.

Vol. 68: M. Taniguchi, Higher Order Asymptotic Theory for Time Series Analysis. viii, 160 pages, 1991.

Vol. 69: N.J.D. Nagelkerke, Maximum Likelihood Estimation of Functional Relationships. V, 110 pages, 1992.

Vol. 70: K. Iida, Studies on the Optimal Search Plan. viii, 130 pages, 1992.

Vol. 71: E.M.R.A. Engel, A Road to Randomness in Physical Systems. ix, 155 pages, 1992.

Vol. 72: J.K. Lindsey, The Analysis of Stochastic Processes using GLIM. vi, 294 pages, 1992.

Vol. 73: B.C. Arnold, E. Castillo, J.-M. Sarabia, Conditionally Specified Distributions. xiii, 151 pages, 1992.

Vol. 74: P. Barone, A. Frigessi, M. Piccioni, Stochastic Models, Statistical Methods, and Algorithms in Image Analysis. vi, 258 pages, 1992.

Vol. 75: P.K. Goel, N.S. Iyengar (Eds.), Bayesian Analysis in Statistics and Econometrics. xi, 410 pages, 1992.

Vol. 76: L. Bondesson, Generalized Gamma Convolutions and Related Classes of Distributions and Densities. viii, 173 pages, 1992.

Vol. 77: E. Mammen, When Does Bootstrap Work? Asymptotic Results and Simulations. vi, 196 pages, 1992.

Vol. 78: L. Fahrmeir, B. Francis, R. Gilchrist, G. Tutz (Eds.), Advances in GLIM and Statistical Modelling: Proceedings of the GLIM92 Conference and the 7th International Workshop on Statistical Modelling, Munich, 13-17 July 1992. ix, 225 pages, 1992.

Vol. 79: N. Schmitz, Optimal Sequentially Planned Decision Procedures. xii, 209 pages, 1992.

Vol. 80: M. Fligner, J. Verducci (Eds.), Probability Models and Statistical Analyses for Ranking Data. xxii, 306 pages, 1992.

Vol. 81: P. Spirtes, C. Glymour, R. Scheines, Causation, Prediction, and Search. xxiii, 526 pages, 1993.

Vol. 82: A. Korostelev and A. Tsybakov, Minimax Theory of Image Reconstruction. xii, 268 pages, 1993.

Vol. 83: C. Gatsonis, J. Hodges, R. Kass, N. Singpurwalla (Editors), Case Studies in Bayesian Statistics. xii, 437 pages, 1993.

Vol. 84: S. Yamada, Pivotal Measures in Statistical Experiments and Sufficiency. vii, 129 pages, 1994.

Vol. 85: P. Doukhan, Mixing: Properties and Examples. xi, 142 pages, 1994.

Vol. 86: W. Vach, Logistic Regression with Missing Values in the Covariates. xi, 139 pages, 1994.

Vol. 87: J. Müller, Lectures on Random Voronoi Tessellations.vii, 134 pages, 1994.

Vol. 88: J. E. Kolassa, Series Approximation Methods in Statistics. Second Edition, ix, 183 pages, 1997.

Vol. 89: P. Cheeseman, R.W. Oldford (Editors), Selecting Models From Data: AI and Statistics IV. xii, 487 pages, 1994.

Vol. 90: A. Csenki, Dependability for Systems with a Partitioned State Space: Markov and Semi-Markov Theory and Computational Implementation. x, 241 pages, 1994.

Vol. 91: J.D. Malley, Statistical Applications of Jordan Algebras. viii, 101 pages, 1994.

Vol. 92: M. Eerola, Probabilistic Causality in Longitudinal Studies. vii, 133 pages, 1994.

Vol. 93: Bernard Van Cutsem (Editor), Classification and Dissimilarity Analysis. xiv, 238 pages, 1994.

Vol. 94: Jane F. Gentleman and G.A. Whitmore (Editors), Case Studies in Data Analysis. viii, 262 pages, 1994.

Vol. 95: Shelemyahu Zacks, Stochastic Visibility in Random Fields. x, 175 pages, 1994.

Vol. 96: Ibrahim Rahimov, Random Sums and Branching Stochastic Processes. viii, 195 pages, 1995.

Vol. 97: R. Szekli, Stochastic Ordering and Dependence in Applied Probability. viii, 194 pages, 1995.

Vol. 98: Philippe Barbe and Patrice Bertail, The Weighted Bootstrap. viii, 230 pages, 1995.

Vol. 99: C.C. Heyde (Editor), Branching Processes: Proceedings of the First World Congress. viii, 185 pages, 1995.

Vol. 100: Wlodzimierz Bryc, The Normal Distribution: Characterizations with Applications. viii, 139 pages, 1995.

Vol. 101: H.H. Andersen, M.Højbjerre, D. Sørensen, P.S.Eriksen, Linear and Graphical Models: for the Multivariate Complex Normal Distribution. x, 184 pages, 1995.

Vol. 102: A.M. Mathai, Serge B. Provost, Takesi Hayakawa, Bilinear Forms and Zonal Polynomials. x, 378 pages, 1995.

Vol. 103: Anestis Antoniadis and Georges Oppenheim (Editors), Wavelets and Statistics. vi, 411 pages, 1995.

Vol. 104: Gilg U.H. Seeber, Brian J. Francis, Reinhold Hatzinger, Gabriele Steckel-Berger (Editors), Statistical Modelling: 10th International Workshop, Innsbruck, July 10-14th 1995. x, 327 pages, 1995.

Vol. 105: Constantine Gatsonis, James S. Hodges, Robert E. Kass, Nozer D. Singpurwalla(Editors), Case Studies in Bayesian Statistics, Volume II. x, 354 pages, 1995.

Vol. 106: Harald Niederreiter, Peter Jau-Shyong Shiue (Editors), Monte Carlo and Quasi-Monte Carlo Methods in Scientific Computing. xiv, 372 pages, 1995.

Vol. 107: Masafumi Akahira, Kei Takeuchi, Non-Regular Statistical Estimation. vii, 183 pages, 1995.

Vol. 108: Wesley L. Schaible (Editor), Indirect Estimators in U.S. Federal Programs. viii, 195 pages, 1995.

Vol. 109: Helmut Rieder (Editor), Robust Statistics, Data Analysis, and Computer Intensive Methods. xiv, 427 pages, 1996.

Vol. 110: D. Bosq, Nonparametric Statistics for Stochastic Processes. xii, 169 pages, 1996.

Vol. 111: Leon Willenborg, Ton de Waal, Statistical Disclosure Control in Practice. xiv, 152 pages, 1996.

Vol. 112: Doug Fischer, Hans-J. Lenz (Editors), Learning from Data. xii, 450 pages, 1996.

Vol. 113: Rainer Schwabe, Optimum Designs for Multi-Factor Models. viii, 124 pages, 1996.

Vol. 114: C.C. Heyde, Yu. V. Prohorov, R. Pyke, and S. T. Rachev (Editors), Athens Conference on Applied Probability and Time Series Analysis Volume I: Applied Probability In Honor of J.M. Gani. viii, 424 pages, 1996.

Vol. 115: P.M. Robinson, M. Rosenblatt (Editors), Athens Conference on Applied Probability and Time Series Analysis Volume II: Time Series Analysis In Memory of E.J. Hannan. viii, 448 pages, 1996.

Vol. 116: Genshiro Kitagawa and Will Gersch, Smoothness Priors Analysis of Time Series. x, 261 pages, 1996.

Vol. 117: Paul Glasserman, Karl Sigman, David D. Yao (Editors), Stochastic Networks. xii, 298, 1996.

Vol. 118: Radford M. Neal, Bayesian Learning for Neural Networks. xv, 183, 1996.

Vol. 119: Masanao Aoki, Arthur M. Havenner, Applications of Computer Aided Time Series Modeling. ix, 329 pages, 1997.

Vol. 120: Maia Berkane, Latent Variable Modeling and Applications to Causality. vi, 288 pages, 1997.

Vol. 121: Constantine Gatsonis, James S. Hodges, Robert E. Kass, Robert McCulloch, Peter Rossi, Nozer D. Singpurwalla (Editors), Case Studies in Bayesian Statistics, Volume III. xvi, 487 pages, 1997.

Vol. 122: Timothy G. Gregoire, David R. Brillinger, Peter J. Diggle, Estelle Russek-Cohen, William G. Warren, Russell D. Wolfinger (Editors), Modeling Longitudinal and Spatially Correlated Data. x, 402 pages, 1997.

Vol. 123: D. Y. Lin and T. R. Fleming (Editors), Proceedings of the First Seattle Symposium in Biostatistics: Survival Analysis. xiii, 308 pages, 1997.

Vol. 124: Christine H. Müller, Robust Planning and Analysis of Experiments. x, 234 pages, 1997.

Vol. 125: Valerii V. Fedorov and Peter Hackl, Model-oriented Design of Experiments. viii, 117 pages, 1997.

Vol. 126: Geert Verbeke and Geert Molenberghs, Linear Mixed Models in Practice: A SAS-Oriented Approach. xiii, 306 pages, 1997.